Controlling Climate Cha

Controlling Climate Change is an unbiased and comprehensive overview that is free of jargon and solidly based on the findings of the Intergovernmental Panel on Climate Change (IPCC). It looks at what we can do to solve the problem of man-made climate change, working through the often confusing potential solutions. Readers will find answers to the vital questions:

- What will happen if climate change is not controlled?
- What is the magnitude of the challenge to avoid climate change?
- What is the role of prevention and adaptation?
- What measures can we take to control climate change and what do they cost?
- What policies are available to make it happen?

Bert Metz is a leading expert on climate change science and policy. As a former co-chair of the IPCC Working Group on Mitigation of Climate Change and an international climate change negotiator, his insider expertise provides a cutting edge, completely up-to-date assessment that gives the reader an insight to issues at the top of the political agenda. He leads the reader through the scenarios of ambitious actions to reduce greenhouse gas emissions and protect our forests, in the context of the challenges of countries to provide for economic growth and development and the need for societies to adapt to different climate conditions. Technical solutions, behavioural issues, costs, and policies and measures are discussed for each of the main economic sectors. The complexities of international climate negotiations are explained in a succinct manner. Damage to ecosystems, impact on food production, health and the very existence of a large proportion of the world's population are all tackled with an emphasis on the potential solutions.

Illustrations, tables of data, and extensive boxed examples motivate students to engage with this essential global debate. Questions for each chapter are available online for course instructors, so students can test their knowledge.

This textbook is ideal for any course on the consequences of climate change and its mitigation and adaptation. Written in accessible language with a minimum of technical jargon, it will also be valuable to anyone with an interest in what action should be taken to combat climate change, from the layman to scientists, professionals and policy makers.

Bert Metz has vast experience in the field of climate change policy. He served as the coordinator of climate policy at the Netherlands Ministry of Housing, Spatial Planning and Environment and chief negotiator for the Netherlands and the European Union in the international climate change negotiations from 1992 to 1998. He was elected co-chairman of the UN Intergovernmental Panel on Climate Change (IPCC) working group on climate change mitigation for the IPCC Third Assessment Report (1997–2002) and was re-elected for the Fourth Assessment report (2002–2008; in which period the IPCC received the 2007 Nobel Peace Prize). At the Netherlands Environmental Assessment Agency from 1998 to 2005, he led the group on climate change and global sustainability, publishing a large series of national and international policy analyses on climate change and sustainability. Since retiring, he is serving as advisor to the European Climate Foundation and other organizations. In 2009 he received the European Practitioner Achievement Award in applying environmental economics from the European Association for Environmental and Resource Economists.

Controlling Climate Change

BERT METZ

CAMBRIDGE UNIVERSITY PRESS
Cambridge, New York, Melbourne, Madrid, Cape Town, Singapore,
São Paulo, Delhi, Dubai, Tokyo

Cambridge University Press
The Edinburgh Building, Cambridge CB2 8RU, UK

Published in the United States of America by Cambridge University Press, New York

www.cambridge.org
Information on this title: www.cambridge.org/9780521764032

First published 2010

Printed in the United Kingdom at the University Press, Cambridge

A catalogue record for this publication is available from the British Library

ISBN 978-0-521-76403-2 Hardback
ISBN 978-0-521-74784-4 Paperback

Additional resources for this publication at www.cambridge.org/metz

This book is dedicated to my grandchildren Mare, Nica, Phine, Thijmen, Lotte, and Kiek

Contents

The colour plates will be found between pages 208 and 209.

Preface

This book is written to help people make sense of the discussion on climate change. In particular on the question of whether we can solve this problem. It is now generally accepted that our climate has changed and that it will further change due to our fossil fuel based economy, our transformation of the planet's surface, and the increasing number of people and their increasing wealth. But the confusion about the solutions is increasing.

Some people believe the only way is to change our way of life drastically. Give up our cars, give up our central heating, no more air travel. 'Back to the middle ages' so to speak. Some people believe that technology will give us abundant CO_2 free energy at low cost in the near future. Others think nuclear power is the only solution, because renewable energy and energy efficiency will never reduce CO_2 emissions strongly enough. For almost every possible solution to keep climate change under control there are problems to overcome. Biofuels can threaten food production and precious nature. Preserving forests may compete with land needed for food production. Wind turbines spoil the landscape. Nuclear reactors produce radioactive waste and increase the risk of proliferation of nuclear weapons. Capturing and storing CO_2 from coal fired power plants would make continued coal use compatible with stringent climate policy, but that would imply continuation of coal mining with its accidents and its air pollution. Energy efficient lamps seduce people to light up the garden. In a fuel efficient car you can drive further on the same amount of gasoline. If your refrigerator is energy efficient you can buy a bigger one. Furthermore, there are stories that these low carbon solutions will cost us a fortune and ruin our economy, that it will take a long time before they are commercially available and that for every reduction in CO_2 emissions that people in developed countries make there will be a much bigger increase in India and China. Many people have no clear picture of what it takes to solve the problem and even if it is possible to do so.

It is time to look at the facts. For that very reason this book leans heavily on the reports of the Intergovernmental Panel on Climate Change of the United Nations (IPCC). The IPCC was established in 1988 to assess and summarize our knowledge on climate change, its impacts, and ways to avoid it. Since then it has produced four big comprehensive reports and a number of smaller, more focused ones. It does its work by bringing together the best scientists, engineers, and economists of the world to critically look at all available publications. The reports of these authors are reviewed by hundreds of independent experts from around the world. At the end of the process the summary for policy makers is approved by the member governments of the IPCC by unanimity. The findings are formulated in a factual manner. No recommendations are given. The implications of certain policy decisions are outlined, but choices are always left to decision makers.

'Policy relevant, but not policy prescriptive' is the IPCC mantra. IPCC reports constitute therefore an authoritative and balanced picture of our knowledge.

Having been the co-chairman of the IPCC Working Group on Mitigation during the preparation of the Third (in 2001) and Fourth (in 2007) Assessment Report, I have seen the comprehensiveness, thoroughness and objectivity of the IPCC materials. I have therefore referred to IPCC findings extensively throughout this book. Where relevant new information was available that was not covered by the IPCC assessments, I used that. The book is my personal interpretation of the scientific facts and in no way constitutes or could be seen as being an IPCC product. In using the IPCC findings I relied on the painstaking work of the hundreds of authors that put the IPCC reports together. My task was to use their material and to tell the story of controlling climate change in a simple way. This means however that relevant details and considerations, carefully crafted statements about the uncertainty of IPCC findings, and precise references to the original literature sources are not found in this book. What I did was to point to specific sections of IPCC reports and other publications for further reading.

The book starts with a summary of our understanding of the climate system, the changes that are occurring, the prospects of further climate change, and the impact that will have on human and natural systems. It shows the huge risks of our current behaviour for the planet. It provides in a nutshell the rationale for the rest of the book that is devoted to the question of how to control climate change and limit it to manageable proportions. In chapter 2 I discuss the emissions of greenhouse gases, the main culprit of the man-made climate change we are facing. Chapter 3 looks at the question how much climate change the earth can handle and where we draw the line in terms of the amounts of greenhouse gases in the atmosphere. This chapter also shows the implications of keeping climate change under control: the need to drastically reduce our emissions of greenhouse gases, already in the short term. Before going into the major economic sectors and how they each can contribute to controlling climate change, Chapter 4 puts the problem firmly in a development context. It argues that climate change is in fact a development problem and that development in a more sustainable way also has to provide the solution. Chapters 5 to 9 then discuss specific contributions of energy supply, transportation, residential and commercial buildings, industry and waste management, and agriculture and forestry to the problem and to the solutions. Opportunities for emission reduction, the timeframe in which they are available, and the costs of achieving those reductions are discussed. Chapter 10 then puts all the bits together to present an overall economy-wide picture of strategies to keep climate change under control. It also deals with some of the cross cutting issues that are not covered by the sector based chapters. Finally, Chapters 11 and 12 address the critical question of how the opportunities to control climate change can be turned into reality, and it is made clear it will not happen automatically. Strong policy incentives are needed. Governments have a critical role to play domestically as well as internationally. The role of international agreements and the process of achieving them are therefore discussed extensively.

The book is written on the basis of the professional expertise that I gained in my capacity as IPCC Mitigation Working Group co-chair, my work for the Netherlands Environmental Assessment Agency, and before that as climate change negotiator for the government of the Netherlands. But my motivation goes beyond the wish to share this experience. What our current knowledge tells us is that we can control climate change. We cannot completely avoid further change and further negative impacts, but we can avoid the most serious impacts of climate change, so that things remain manageable. I would like people to understand that and to see that this is possible only if strong and decisive action is taken now.

What has driven me is the strong wish to leave a liveable planet to future generations. Being blessed with having two daughters and six grandchildren, they are the personification of this liveable future. My grandchildren will likely experience the climate of the 2080s and 2090s. They will personally face the turmoil in the world when climate change gets out of control. I want to make my small contribution to save them and their generation from that.

I could not have written this book without the painstaking work of hundreds of IPCC authors who put such excellent reference material together. Nor would it have been possible without the strong support of the management and my colleagues at the Netherlands Environmental Assessment Agency where I worked for the last 10 years and where I was given the time to produce this book. I also thank the staff of the Woods Institute for the Environment at Stanford University, where I was able to make a good start with this book and of the European Climate Foundation for facilitating the competition. And last but not least I would like to thank my long-time friend and companion Mieke Woerdeman for her never-ending support and understanding while putting this book together.

1 Climate change and its impacts: a short summary

What is covered in this chapter?

The climate has changed. Human beings are responsible. And the climate will change further as energy use, agriculture, deforestation, and industrial production continue to increase. In the course of this century it could get up to 6°C warmer, with more heat waves, droughts, floods, and storms. As a result a wide range of impacts can be expected. Food production and water availability will diminish. Nature will suffer, with a large percentage of species threatened with extinction. New health problems will arise. Coastal areas and river deltas will face more floods. The overall effect of this will be devastating for poor countries, undermining their efforts to eradicate poverty. But even rich countries will see the costs of these impacts rise to significant levels.

The climate has changed

Climate can be defined as 'average weather', so it covers averaged temperatures, rainfall and wind direction and speed. Usually this is averaged over a period of 30 years. Let's have a look at how temperatures, rainfall, and wind have changed since 1850[1].

From the temperature measurements across the world (see Box 1.1) it is clear that global average surface temperatures have gone up about 0.8°C since the the pre-industrial era (or since about 1850). This happened in two stages, between 1910 and 1940 (about 0.35°C) and from the 1970s till the present (more than 0.55°C), with a period of slight cooling (0.1°C) in between. The change is getting faster over time (see Figure 1.1). Eleven of the twelve years in the period 1995–2006 belong to the warmest since the beginning of instrumental temperature measurements in 1850. It is likely that temperatures are now higher than in the last 1300 years.

Over the last 50 years there has been a significant decrease in cold days and cold nights and a significant increase in warm days and nights and heat waves. In Europe the summer of 2003 was exceptionally warm, with record temperatures. The summer was 3.8°C warmer than the 1961–1990 average and 1.4°C warmer than any summer since 1780[2]. This was well beyond what can be expected of extreme events in an unchanged climate.

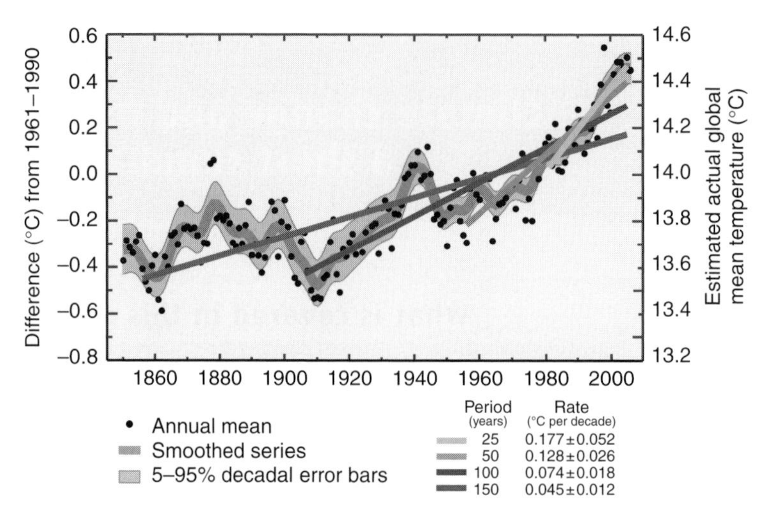

Figure 1.1 **Annual mean global temperatures 1850–2005 (dots and smoothed curve) and linear trends for the last 25, 50, 100, and 150 years (different lines).**
Source: IPCC Fourth Assessment report, Working Group I, figure TS.6.

Box 1.1 **Temperature measurements**

Average global surface temperatures are constructed from thousands of land based and ship based temperature measurements across the globe every day. They are corrected for additional urban warming (the so-called 'urban heat island effect'). Temperatures at higher altitudes are different. In the troposphere (up to 10 km) they are higher than at the surface. In the stratosphere (10–30 km) they are lower. This is exactly what the physical theory predicts. Satellites can measure the average over the whole atmosphere. Although there are uncertainties because of the integrated measurement and the fact that calibration of satellite instruments is complex, they are now fully consistent with the surface temperatures.
(Source: IPCC Fourth Assessment report, Working Group I, chapter 3, Frequently Asked Questions box 3.1)

Rainfall and snowfall (together: precipitation) patterns have also changed. On average precipitation has increased in Eastern North and South America, Northern Europe and Northern and Central Asia. In the Mediterranean, the Sahel, Southern Africa, and Southern Asia it has become drier. In addition heavy precipitation occurrences have increased in many areas, even in places where total amounts have decreased. This is caused by the higher amounts of water vapour in the atmosphere (the warmer the air, the more water vapour it can contain). Drought occurrences have increased as well in many areas as shown in Figure 1.2.

As far as wind is concerned, there is evidence that intensities of hurricanes in the North Atlantic have increased. The numbers of hurricanes have not increased. Wind patterns have also changed in many areas as a result of changes in storm tracks.

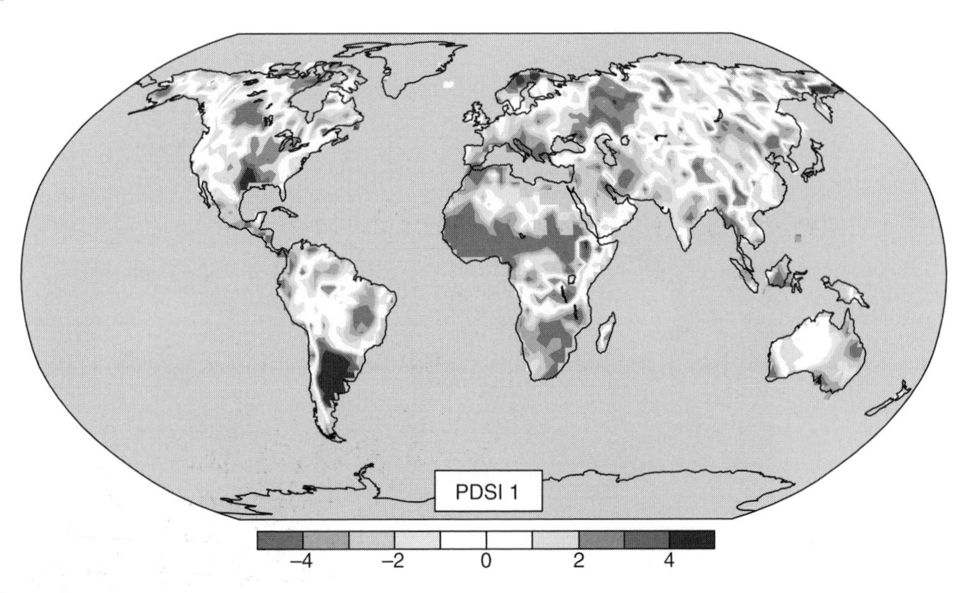

Figure 1.2

Change in drought index between 1900 and 2002.
Source: IPCC Fourth Assessment report, Working Group I, figure 1 from box FAQ3.2. See Plate 1 for colour version.

Are ice and snow cover and sea level consistent with the temperature trends?

Trends in snow and ice cover are consistent with global average temperature increase. Most mountain glaciers are getting smaller. Northern Hemisphere snow cover in winter and Arctic sea ice cover (see Box 1.2) and area of frozen ground in summer are declining. Glaciers, as well as the Greenland ice sheet, are getting smaller, even while snowfall on top is higher than before. The Antarctic sea ice cover and the Antarctic ice sheet do not yet show clear trends (see Figure 1.3 for some of these trends).

Box 1.2 Sea ice and land ice: the difference

Melting sea ice does not increase sea level, because the ice floats and displaces the same amount of water (check it with an ice cube in a glass of water!). Melting land ice (the Greenland ice sheet for example) does increase sea level. In Antarctica large chunks of sea ice have broken off over the years. These large sea ice plates do however provide some support for the land ice. It is uncertain if land ice would move faster towards the sea in such places.

A reduction in sea ice also reduces the reflection of sunlight. So the more sea ice is disappearing, the more sunlight is absorbed by the oceans, which speeds up warming. This is one of the so-called feedback mechanisms in the climate system.

The figure below shows how much lower the Arctic sea ice cover in 2005 and 2007 was compared to the average over the 1979–2000 period.

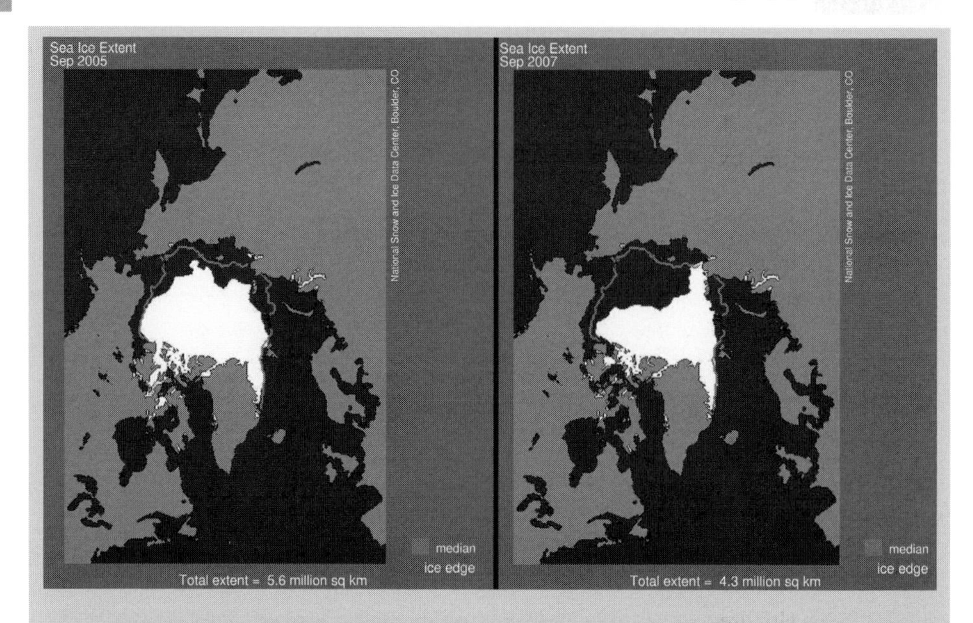

Minimum summer sea ice cover in 2005 (left) and 2007 (right); dotted line is average extent of sea ice between 1979 and 2000.

Source: National Snow and Ice Data Center, Boulder, Colorado, USA, http://nsidc.org/arcticseaicenews/.

(Source: IPCC Fourth Assessment report, Working Group I, box 4.1)

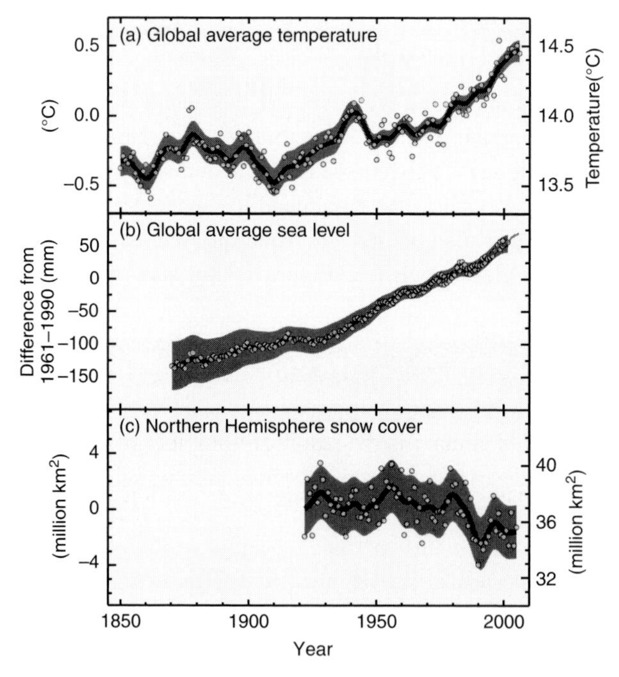

Figure 1.3 **Observed changes in (a) global mean surface temperature, (b) global average sea level, and (c) northern hemisphere November to March snow cover. All values are expressed as differences with the corresponding averages for the period 1961–1990. Shaded areas represent the uncertainty.**

Source: IPCC Fourth Assessment Report, Working Group I, figure SPM.3.

Sea level has been stable over the first 1900 years AD, but since 1900 it has been rising. Until 1990 this rise was about 1.7 mm per year, but since 1993 it has increased to 3 mm per year. Half this increase comes from melting of land ice (see also Box 1.2), the other half from expansion of sea water due to temperature increase (warmer water has a larger volume than cold water). This is fully consistent with the increase in global average temperatures. Annual fluctuations happen as a result of local weather conditions and human interventions in groundwater extraction and water storage reservoirs. Sea level rise is not the same everywhere, because of changes in ocean currents and local differences in ocean temperature and salinity. Rising or falling land can make a difference in specific locations.

Are observations of biological systems also consistent with the measurements of a changed climate?

Hundreds of studies were done on changes in fish, plankton, and algal populations, plants and trees, insects, and animals. Observations from these studies show a very strong correlation with the changes in climate that were discussed above[3]. Populations shift their ranges to areas where the climate has become favourable and disappear from areas where the climate is no longer appropriate. Often this means poleward movement of the ranges. Blooming occurs earlier. But it also means that mismatches are occurring between migratory bird breeding and availability of certain caterpillars or insects. The caterpillars or insects react to the higher temperatures by coming out earlier, but the migratory birds still arrive at the usual time and do not find the regular food for their young[4].

In agriculture changes have already occurred in terms of earlier planting, leading to a longer growing season, but also in the form of crop failures due to changing rainfall patterns. In forest management changes in pest invasions and patterns of forest fires show a clear correlation with the changed climate.

Are human activities responsible for this climate change?

The earth's climate is the result of a number of factors:

- the radiation from the sun and the position of the earth in relation to the sun (the changes in these two are responsible for the ice ages that the earth is experiencing every 100 000 years or so),
- the reflectivity of the earth (called albedo), as influenced by the vegetation and the ice and snow cover (this is influenced by human activities),
- the reflection of sunlight by clouds and fine particles in the atmosphere (from volcanic eruptions, sand storms, but also from coal burning and diesel vehicles), and
- last but not least by the presence of so-called greenhouse gases in the atmosphere, retaining some of the solar radiation.

Some of the factors have a warming, others a cooling effect. Natural greenhouse gases (water, CO_2, methane) are in fact responsible for making planet earth suitable for life (see Box 1.3). The problem arose when human activities (burning of fossil fuel, agriculture, cutting forests, industrial processes for making cement, steel, and other materials) added greenhouse gases to the atmosphere far beyond their natural levels, causing additional warming. So it is the enhanced greenhouse effect that is causing problems.

Box 1.3

The greenhouse effect

The earth is warmed by solar radiation. If no atmosphere would exist, the temperature would be minus 18°C and no life would be possible. But because of the atmosphere that has water vapour, methane, and CO_2, some of the radiation that is sent back into space by the earth is absorbed by the atmosphere and the clouds. This is the natural greenhouse effect. It brings the surface temperature to about 15°C.

Human activities have added greenhouse gases to the atmosphere: CO_2, mainly from deforestation and fossil fuel combustion, methane and nitrous oxides from agriculture and waste, and fluorinated gases from industrial processes. These additional greenhouse gases are responsible for the additional warming of the earth. This is the enhanced greenhouse effect.

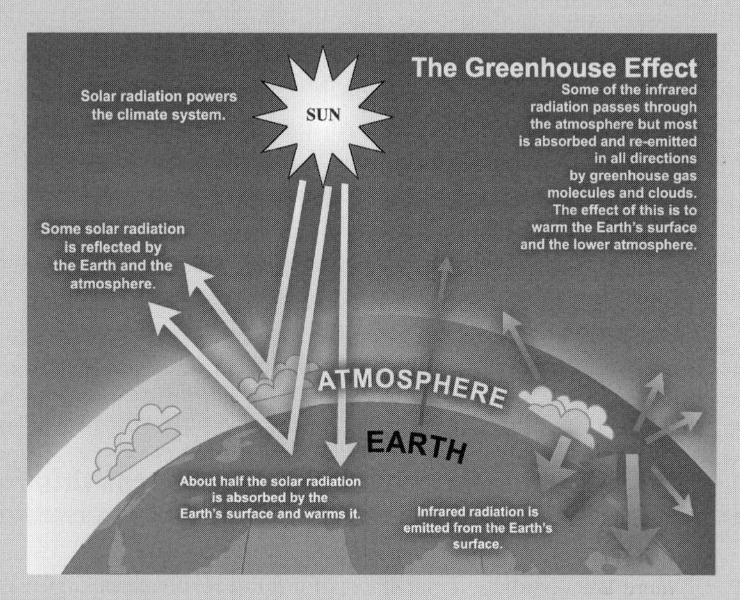

Schematic diagram of the natural greenhouse gas effect.
Source: IPCC Fourth Assessment report, Working Group I, Frequently Asked Questions 1.3, figure 1
(Source: IPCC Fourth Assessment report, Working Group I, Frequently Asked Questions 1.3)

We can measure greenhouse gases in the atmosphere. Carbon dioxide (CO_2), methane (CH_4), and nitrous oxides (N_2O) concentrations have gone up strongly since the beginning of the industrial revolution (see Figure 1.4). CO_2 levels are now about 30% higher than before 1750, N_2O about 50% higher, and CH_4 approximately doubled.

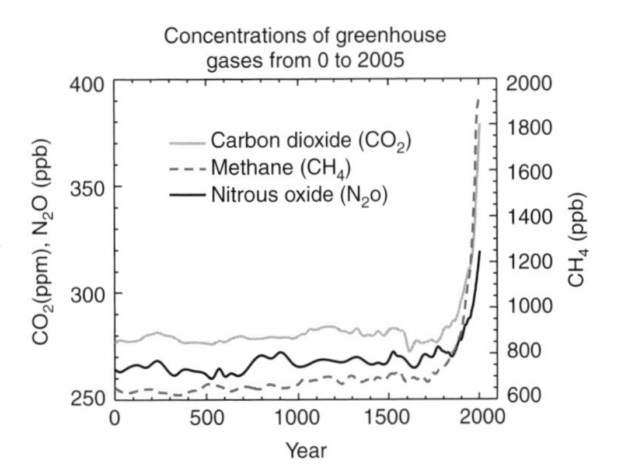

Figure 1.4 **Concentrations of the most important greenhouse gases in the atmosphere over the last 2000 years. Concentration units are parts per million (ppm) or parts per billion (ppb), indicating the number of molecules of a greenhouse gas per million or billion molecules of air.**
Source: IPCC Fourth Assessment report, Working Group I, figure 2.1.

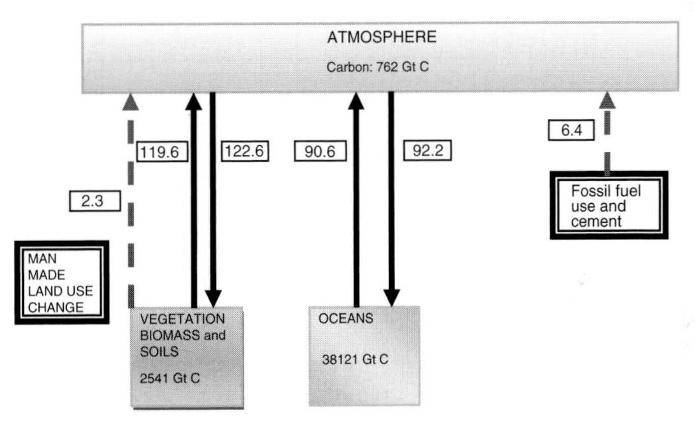

Figure 1.5 **Schematic diagram of global carbon cycle. Shown are stocks of carbon (in GtC) and fluxes (in GtC/yr). Dashed lines represent man-made fluxes of carbon form fossil fuel, cement and land use, change. Land use change numbers are corrected for peatland emissions (see Chapter 9).**
Source: IPCC Fourth Assessment report, Working Group I, figure 7.3 and Working group III, chapter 1.

Concentrations in the atmosphere are a result of emissions of these gases and processes that remove them. This includes natural and man-made emissions and removals. For CO_2 there are very large natural emissions and removals through vegetation and the oceans. Man-made emissions are relatively small compared to these.

Figure 1.5 gives a schematic overview of the natural and man-made emissions and sequestration of CO_2 (sequestration = absorption by growing vegetation). The natural fluxes to and from vegetation and the oceans are typically100 times larger than the man-made fluxes of CO_2. Nevertheless, the man-made fluxes are responsible for the increase in CO_2 emissions in the atmosphere. As the diagram shows, there is a net sequestration of CO_2 in the oceans and in vegetation and soils. That is the reason that only about half of the amount that humans are putting into the atmosphere is staying there.[5]

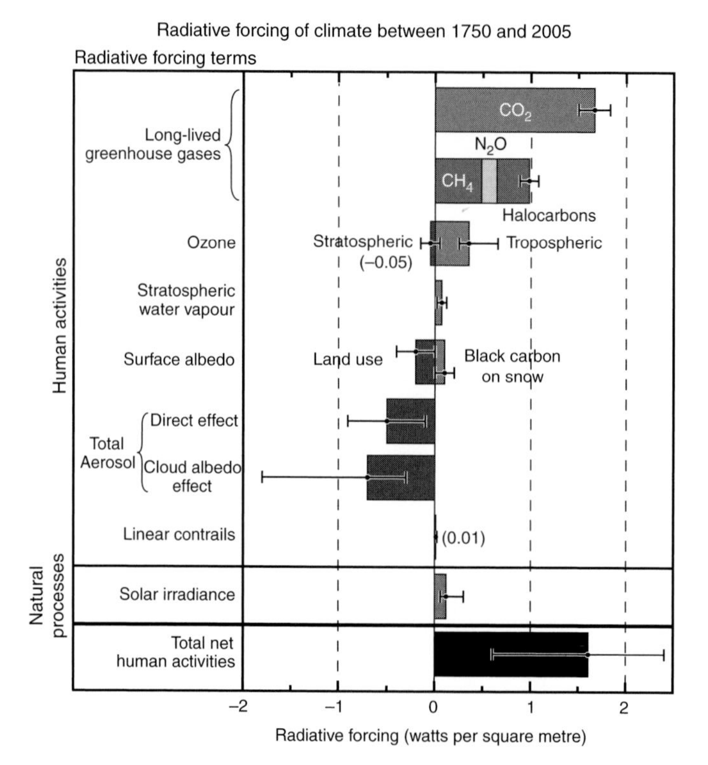

Figure 1.6 **Global average warming and cooling effect of greenhouse gases and other factors in 2005, compared to the situation in 1750. Warming and cooling effect is represented by so-called radiative forcing and expressed in Watt per square meter of the earth surface. Uncertainty ranges are shown with bars.**
Source: IPCC Fourth Assessment report, Working Group I, Frequently asked questions 2.1, figure 2.

The contribution of the various factors mentioned above to warming and cooling is fairly well known, although for some the uncertainties are high. Figure 1.6 gives an overview of the difference between the situation today and that in 1750. It shows significant warming from greenhouse gases (CO_2, CH_4, N_2O, but also fluorinated compounds and ozone) and small contributions to warming from black carbon particles that are deposited on snow and from increased solar radiation. Big cooling effects are caused by a wide range of particles (aerosols), directly through reflection of solar radiation and indirectly because these particles help cloud formation and clouds reflect sunshine. Some cooling has also occurred because the earth has become lighter due to loss of forest cover and reflects more sunshine (increased albedo). Aerosol effects are quite uncertain still. The impact of volcanic eruptions is not visible in Figure 1.6, because the dust and ash blown into the atmosphere by volcanoes disappears within several years. When a big volcanic eruption happens though (the last was Mount Pinatubo in 1991), the average global temperature goes down several tenths of a degree for a few years.

On average there is a clear warming effect. Natural causes (solar radiation, volcanoes) only make a very small contribution. Human beings are responsible; there is no escape from that conclusion. Over time however the relative contributions of human and natural factors changed. Until about 1940 natural forces were playing a big role, but over the last 50 years the human contribution is by far the most important.

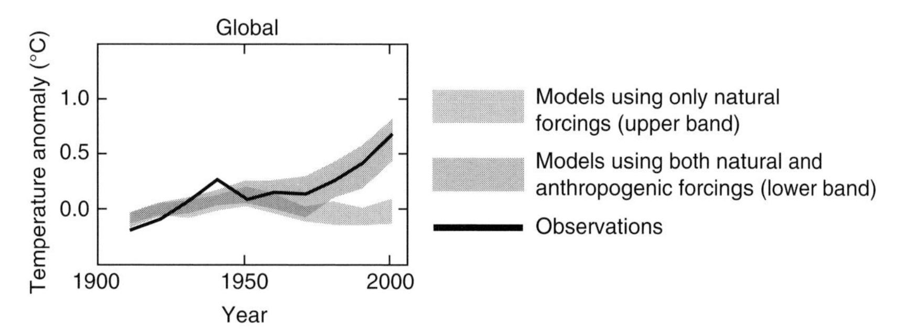

Figure 1.7 **Global average temperature changes compared to the average for 1901–1950. The black line indicates measured temperatures. The lower band indicates the climate model calculation with only natural factors included. The upper band indicates climate model calculations with the effect of greenhouse gases included also.**

Source: IPCC Fourth Assessment report, Working Group I, figure SPM.4.

There is another way to demonstrate that. Climate models have been developed to simulate climatic change, mainly to enable a prediction of future climates. These models have also been used to simulate the climate over the last 150 years. That would allow a comparison with the measurements. If these models are run with only the natural factors included, they do not come close to actual measurements of global average temperatures. When greenhouse gases are added to the calculations they do match the measurements quite well (see Figure 1.7).

How is the climate going to change further in the future?

Greenhouse gases only disappear very slowly from the atmosphere. If we keep adding them to the atmosphere at current rates, concentrations of greenhouse gas in the atmosphere will continue to rise. Without specific policies, emissions of greenhouse gases will continue to increase, so atmospheric concentrations will rise even faster. At the same time concentrations of aerosols tend to go down as a result of policies to clean up air pollution. So cooling forces (from aerosols) decrease and warming forces (from greenhouse gases) increase. As a result, further warming will occur.

Temperatures

Of course it is not precisely known how much warming will increase and by when. It depends on population growth, economic growth, and choices on energy, technology, and agriculture. To deal with this inherent uncertainty, scenarios are used to cover a range of plausible futures. Scenarios are sets of assumptions about the main factors driving emissions. (See Chapter 2 for a more detailed discussion about the causes of greenhouse gas emissions and scenarios.)

Figure 1.8 shows the IPCC SRES scenarios for greenhouse gas emissions (these scenarios also make assumptions on aerosol emissions) and the corresponding increase in the global

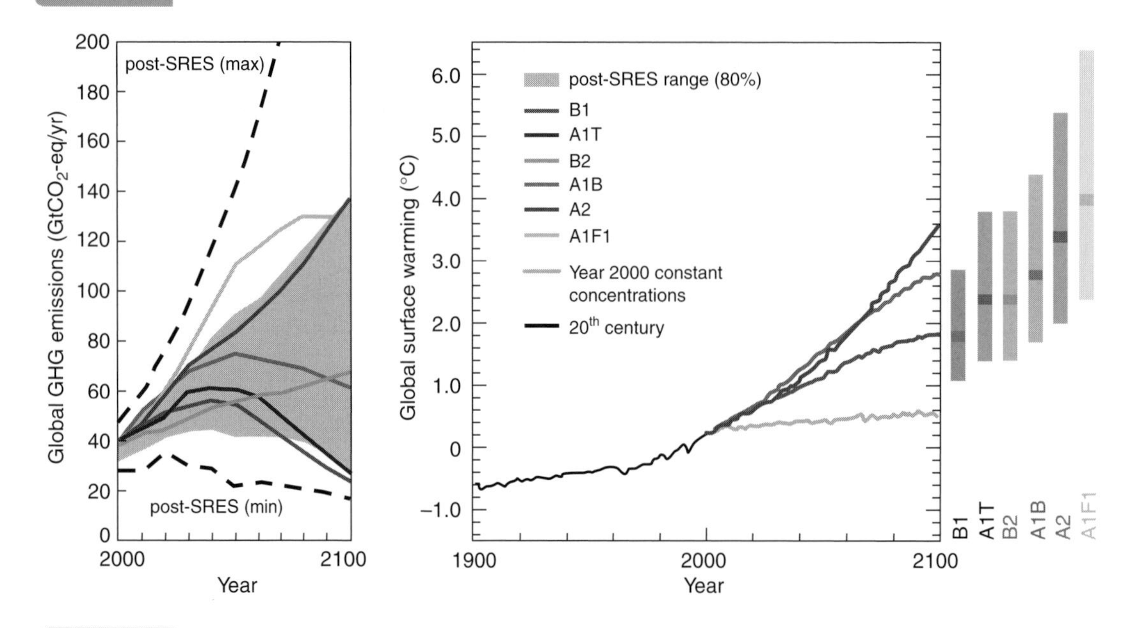

Figure 1.8 **(Left panel) Scenarios for global greenhouse gas emissions, according to IPCC; (right panel) projected global mean temperatures belonging to the scenarios in the left panel.**
Source: IPCC Fourth Assessment report, Synthesis Report, figure SPM.5. See Plate 2 for colour version.

average temperature till the end of this century. By the end of the century global average temperatures will be between 1 and 6.4°C higher than in the period 1980–1999, depending on the scenario (equal to about 1.5 to 6.9°C compared to pre-industrial temperatures).

Temperatures will change differently in different regions. Figure 1.9 shows the pattern that can be expected: stronger warming around the poles (particularly the North Pole) and less warming around the equator. This is caused by the atmospheric circulation patterns that transport heat towards the poles. For the mid range scenario used in this figure temperatures around the North Pole are predicted to be more than 7.5°C higher by the end of the century than in 1990, more than twice as high as the global average.

Other characteristics of the climate by the end of the century include:

- Reduced snow cover
- Widespread increase of summer thaw in permafrost areas
- Strong reduction in summer Arctic sea ice cover (some models predict complete disappearance by the end of the century)
- More heat waves and heavy precipitation
- Stronger tropical cyclones
- Movement of storm tracks towards the North, with changing wind patterns.

Precipitation

For precipitation the general picture is that dry areas will tend to become drier and wet areas wetter. Figure 1.10 shows the changes in precipitation for the December to February

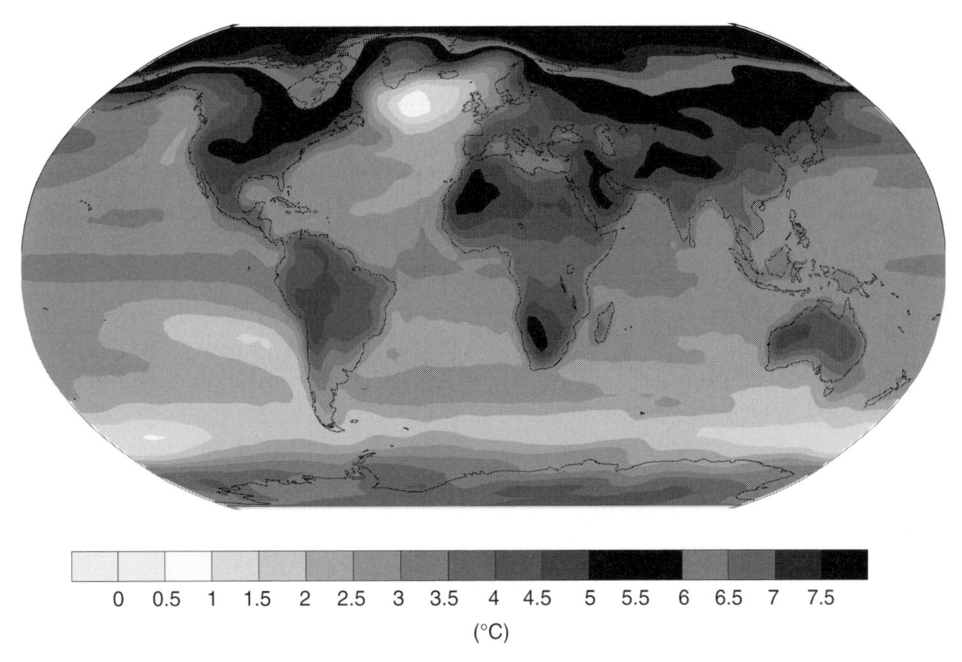

Figure 1.9 **Projected surface temperature changes for the period 2090–2099, compared to 1980–1999. The average of different models is shown for the IPCC SRES A1B scenario (a middle of the range one).**
Source: IPCC Fourth Assessment report, Synthesis Report, figure SPM.6. See Plate 3 for colour version.

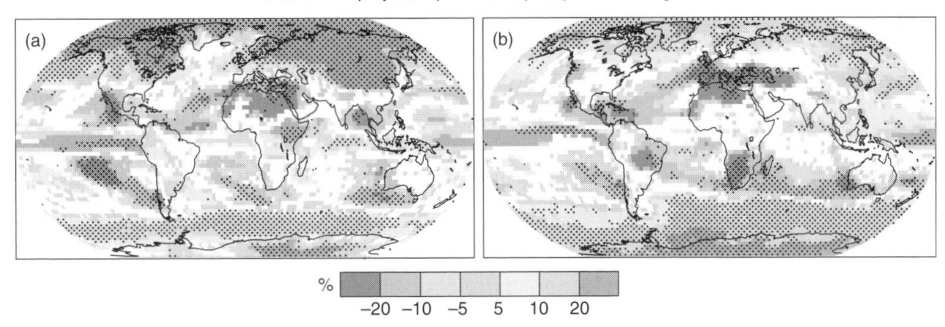

Figure 1.10 **Relative change in precipitation for the period 2090–2099, compared to 1990–1999. (a) December to February. (b) June to August. Model averages are shown for the IPCC SRES A1B scenario (a middle of the road one). White areas are where less than 66% of the models agree about increase or decrease. In stippled areas more than 90% of the models agree.**
Source: IPCC Fourth Assessment report, Synthesis Report, figure 3.3. See Plate 4 for colour version.

and for the June to August period around 2095, compared to the situation in 1990 for a mid range scenario. In most subtropical areas rainfall decreases by up to about 30%, while in high latitudes increases in precipitation of more than 20% can be expected. More pessimistic scenarios regarding greenhouse gas emissions can give even bigger changes.

Sea level rise

Sea levels will rise as a result of warming. As discussed above it is the result of expansion of ocean water and of melting of land ice. It is a slow process. By the end of this century sea level rise is expected to be 20–60cm compared to 1990. But it does not stop there. Even if, through a miracle, concentrations of greenhouse gases in the atmosphere could be kept at the current level, sea level will continue to rise by about 1 meter. And with the sharply increasing concentrations expected over the course of this century, we could be heading for several meters of sea level rise over the next few hundred years. These numbers do not include melting of the large land ice masses in Greenland and Antarctica. The Greenland ice sheet is most at risk of melting. Our current knowledge says that for global temperature increases of 2–4.5°C (well within the range we will see in the course of this century, if no action is taken) the Greenland ice will melt. This will take somewhere between a few hundred to a few thousand years, but total melting means 7 metres of additional sea level rise. This is comparable to the situation on earth about 125000 years ago, when temperatures in Greenland were about 4°C higher than today, a large part of the Greenland ice sheet had melted, and the sea level was 4–6 meters higher than today[6]. Recent observations of ice flow in Greenland suggest that the process of melting could be much faster than climate models have assumed so far. Melt water seems to go down crevices to the base of the ice sheet, where it acts as a lubricant, moving the ice faster out to sea. If this mechanism is the dominant one, complete disintegration and melting could happen in a few hundred years. Our current knowledge is insufficient however to be certain about that[7].

Climate models

Climate change projections are the result of climate change model calculations. Climate models describe in mathematical equations the incoming solar radiation, the retention of energy by greenhouse gases and reflection of radiation by clouds and aerosols, the circulation of air across the globe, the interaction of the atmosphere with the oceans, rainfall, the formation of ice, and many more processes that determine the earth climate. Different models give different outcomes, because of the many assumptions made about the various processes that determine the climate.

One of the most difficult problems is how to describe the different feedback mechanisms in the earth climate system. Water for instance is a powerful greenhouse gas. When the air warms it can contain more water vapour, which strengthens the warming. It also forms clouds that can either have a cooling or a warming effect, depending on the type, height, and structure of the clouds. It is difficult to determine how much additional warming will be the result of this. Other feedback mechanisms are the reflection of ice and snow. More ice and snow means more reflected radiation and cooling. Disappearing snow and ice means warming, which will reduce snow and ice further.

Together these uncertainties are captured in the so-called "climate sensitivity", the warming that occurs for a doubling of CO_2 concentrations in the atmosphere compared to its pre-industrial concentration. The most recent estimate of climate sensitivity is 3°C, with a probability of 66% that it lies between 2 and 4.5°C. That is a big uncertainty. And there is a chance the climate sensitivity is even higher. In other words, when calculating the warming as a result of increases in greenhouse gas concentrations in the atmosphere using the best estimate of climate sensitivity (the 3°C value used in most calculations), actual warming could be twice as small or twice as big. See also Box 1.4.

Box 1.4 **How reliable are current climate model predictions?**

Although models have their limitations, there is considerable confidence in their predictions of future climate change. This confidence is based on the fact that the descriptions of the various processes is based on generally accepted principles of physics, and from their ability to reproduce observed changes in current and past climates. Models can reproduce climate change over the past 150 years pretty well (see Figure 1.6 above), including the short term effect of volcanic eruptions. They also have been able to reproduce the climate over the past 20000 years (including part of the last ice age) reasonably well. Confidence in model predictions is higher for temperatures than for precipitation, although temperature extremes are still difficult. Confidence in model results for large areas is much higher than for smaller regions.

The fact that weather forecasting models become very unreliable beyond periods of a few days does not mean that they cannot be used for climate forecasting. On the contrary, their reliability increases when applied over longer periods of time, when average weather (= climate) is the desired outcome.

(Source: IPCC Fourth Assessment Report, Working Group I, Frequently Asked Questions 8.1)

What will be the impact of future climate change?

Human and natural systems will be exposed to climate change. If the affected people, infrastructures, human activities, or nature are sensitive to climate change, then an impact occurs. We speak about vulnerability to climate change if the capacity to adapt is also taken into account: a system may be sensitive, but if it can adapt easily to the new climatic condition, then it is not vulnerable. The capacity to adapt depends on a range of social and economic factors (see Figure 1.11). In practice the distinction between impact and vulnerability is not always clearly made. In the discussion below the two are separated where possible.

First, the most important impacts will be discussed on a sectoral basis: water, food, nature, health, and infrastructure and human settlements. Then the impacts are grouped together on a regional basis and also vulnerabilities are considered. Finally an overall assessment is made of the most important vulnerabilities.

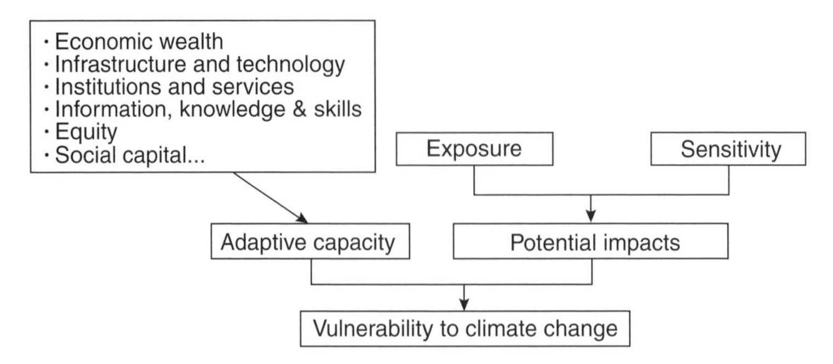

Figure 1.11 **Definition of and factors contributing to vulnerability to climate change.**
Source: Stern review, chapter 4.

Water[8]

Freshwater is vital for human health, food production, forests and ecosystems, cooling water for power plants and industries, and hydropower. Too much water means floods and mud slides, crop failure, and threats to hydropower dams. Too little water means lack of drinking water, crop failure, forest die-back, ecosystem loss, and severe constraints on shipping, industrial production and electricity generation. Sea level rise will mean salt intrusion in fresh groundwater, making it no longer suitable for drinking or agriculture. Increased temperatures and drought periods will make water pollution more serious, so that it can no longer be used for producing drinking water or for irrigation.

Changes in rainfall or snowfall, melting of glaciers, and increasing temperatures (that lead to more evaporation of water) will lead to significant changes in available freshwater. Generally speaking the pattern is similar to the one for changes in precipitation shown in Figure 1.9 above. Wet areas (northern latitudes, tropics) will have more available water, while dry areas (subtropics, in particular Mediterranean and North Africa, Southern Africa, Western Australia, Amazon) will have less water available. By the end of this century water availability in some areas could be 30–50% lower than today, and in others up to 50% higher.

The scale of problems that can be expected with freshwater availability is enormous. Almost 20% of the world's population live in areas where glaciers or snow feed the rivers, which will be affected by glacier melting. Furthermore 20% live in river basins that are likely to have increased risk of flooding by the end of the century. The total number of people that will face increased problems with water availability will be about 2–3 billion by 2080. Pressure on ecosystems will strongly increase. And all of this is in addition to increasing scarcity of freshwater due to population growth, increased wealth, and more agriculture, energy and industry needs.

More specifically, by 2020 75–250 million people in Africa will have to face increased water scarcity. By 2030 water availability in large parts of Australia and New Zealand is going to cause problems. By 2050 freshwater availability in large parts of Asia is going to decline, particularly in large river basins. There is the threat that by mid-century the Amazon forest will be turned into a dry savannah landscape due to drying out of soils. In other parts of Latin America drinking water scarcity and constraints on agriculture and hydropower can be expected. By mid-century fresh groundwater in many small low-lying islands will seriously diminish due to salt water intrusion. In the American West more winter flooding and summer droughts are expected. In Southern Europe water scarcity could reduce hydropower capacity by more than 25% by 2070. And in terms of the impact of flooding, in Bangladesh annual floods will cover 25% more land in the course of this century, where pressure on agricultural land and land for human settlements is already high and the population is still growing.

Food[9]

Increasing temperatures will increase productivity of some crops such as wheat and maize in temperate regions, provided there is enough water. Above a 3°C increase however crop productivity will decline. In tropical and subtropical areas even moderate temperature increases of 1–2°C will reduce crop productivity of most cereal crops that form the basic food of people, such as rice, maize, and wheat. This means increased risk of malnutrition and hunger. Although total global food production could increase up to a warming of 3°C, food scarcity in poor countries is a real danger even at much lower temperature increases. The recent explosion of food prices and the reactions from exporting nations to ban exports of rice and wheat show how fragile the world food market is. Regional scarcity and high prices often mean poor people will not get the food they need. Above 3°C, total food production will decrease, making poor countries and poor people even more vulnerable to hunger. This effect counters the expected improvement in food security as a result of increasing incomes. In addition to the effects of temperature and average rainfall, food production will suffer from irregular rainfall patterns and extreme weather events, such as heat waves, extended droughts, heavy precipitation, and cyclones.

In parts of Africa (Sahel, East Africa, and southern Africa) yields from rain-fed agriculture could be reduced by 50% as early as 2020. Given the strong dependence of many African economies on the agricultural sector (10–70% of GDP earned there), this is not only threatening food security, but will also seriously affect the economy as a whole. In southern and eastern Australia, one of the major exporting areas of wheat, droughts are expected to reduce crop yields. In southern Europe crop yields are primarily affected by water availability. Enhanced irrigation could keep productivity up, but water availability is a problem. In Latin America temperate region crops such as soybeans will do better with moderate temperature increases, but rice production in subtropical and tropical areas is expected to suffer. Productivity in the poultry and cattle industries, which are strong in Latin America, will go down as well. Natural grazing lands will partly suffer from drought, and overgrazing may happen in areas that have water. Both will have a negative influence on pastoral populations.

Fishery will also be affected. Sensitivity to ocean temperatures is well known. The El Nino cyclical movement of warm water across the Pacific Ocean towards South America strongly reduces fish catch along the coast of Peru. With a changing climate movement of fish to different areas, changes in available food, and spreading of diseases to new areas and to fish farms (now about half of the total wild catch) off the coast all may have negative consequences. Higher fish growth rates are unlikely to compensate for this. Climate change impacts come on top of heavy overfishing in many parts of the oceans as well as disturbance or loss of breeding grounds in mangrove forest, coral reefs, and tidal areas. Local fishing communities will suffer from these changes.

Nature[10]

Nature is formed by all the ecosystems on earth. Ecosystems form the backbone of the earth's ability to provide habitable conditions. They consist of webs of plants, animals, insects, and bacteria that interact with each other to create a living system. They provide invaluable services to the global economy via the provision of clean water, shelter, food, building materials, medicines, recreation, and tourism[11]. They also capture large amounts of CO_2 from the atmosphere and reduce the amounts that stay in the atmosphere (they are 'net sinks' of CO_2). Ecosystems are under enormous pressure already. Over the past 50 years, humans have changed ecosystems more rapidly and extensively than in any comparable period in human history. This has resulted in a substantial and largely irreversible loss in the diversity of life on earth. Climate change impacts are going to be added to this.

Ecosystems are very much adapted to climatic conditions. They are optimized to the climate we had for the last thousands of years. Slow changes, i.e. changes over periods of thousands of years, can usually be accommodated by natural ecosystem adaptation. Rapid changes as we are facing now, i.e. where we see significant changes over a period of 100 years or so, are too fast for many species to adapt to. Within this century ecosystems will see the highest CO_2 concentrations in at least the last 650 000 years, the highest temperatures in at least the last 740 000 years, and the most acid ocean waters in more than 20 million years. Natural adaptation is further threatened by man-made and natural obstacles for migration of animals and plants (roads, towns, rivers, mountains).

The threats from climate change do not only come in the form of higher temperatures, heat waves, and changes in precipitation. Wildfires as a result of drought, explosions in insect numbers as a result of changing climate, and more acid ocean water as a result of CO_2 dissolving in sea water, all contribute to the impacts on ecosystems. Impacts on ecosystems are often of the so-called 'threshold' type. Above a certain level of temperature or acidity or drought one or more species can no longer survive (which can easily lead to extinction for species that are unique to certain areas) and with the decline in those species ecosystems as a whole may collapse.

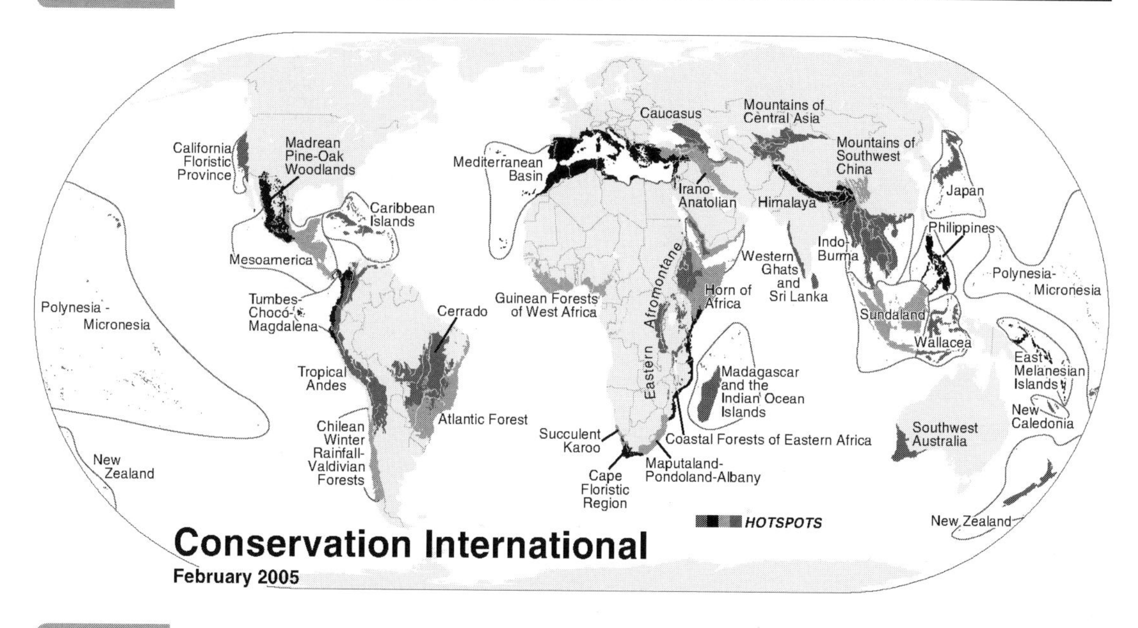

Figure 1.12 **Biodiversity hotspots.**
Source: Conservation International, www.biodiversityhotspots.org. See Plate 5 for colour version.

The scale of the threat of climate change to nature is extremely large. For global average temperature increases of 1–2°C above what they were around 1990 (1.5–2.5 °C above the pre-industrial era) many sensitive and unique ecosystems are threatened with extinction. This includes ecosystems in many of the 34 areas with high concentrations of unique species that are seriously threatened (so-called 'biodiversity hotspots') that have been identified[12] (see Figure 1.12). At 1–3°C warming of sea water, coral reefs, currently already affected by warming, will face widespread bleaching and die off. At 1.5 to 2.5°C warming 20–30% of all species for which research was done are likely to be faced with extinction. For temperature increases beyond that many more species will face extinction.

In addition, some ocean organisms that form shells are going to be negatively affected by the increasing acidity of ocean water. Some form an important part of plankton and their disappearance could have major impacts on the food chain and on reflection at the surface and cloud formation above oceans. With higher temperatures the decomposition of plant material increases and as a result the capture and sequestration of CO_2 (the 'net carbon sink') has a good chance of changing into a net source of CO_2 before the end of this century.

Latin America is particularly under threat when it comes to ecosystems. All of its seven biodiversity hotspots are impacted by climate change, in addition to the threat to the Amazon forest. In Australia a number of very important ecosystems, including the Great Barrier Reef, are going to be negatively affected. In the Alps in Europe up to 60% of mountain plants face extinction by the end of the century, because it is no longer possible for these plants to move up the mountain to cooler areas. Polar bears have been declared a "threatened species" by the US government recently[13] because the continuing loss of

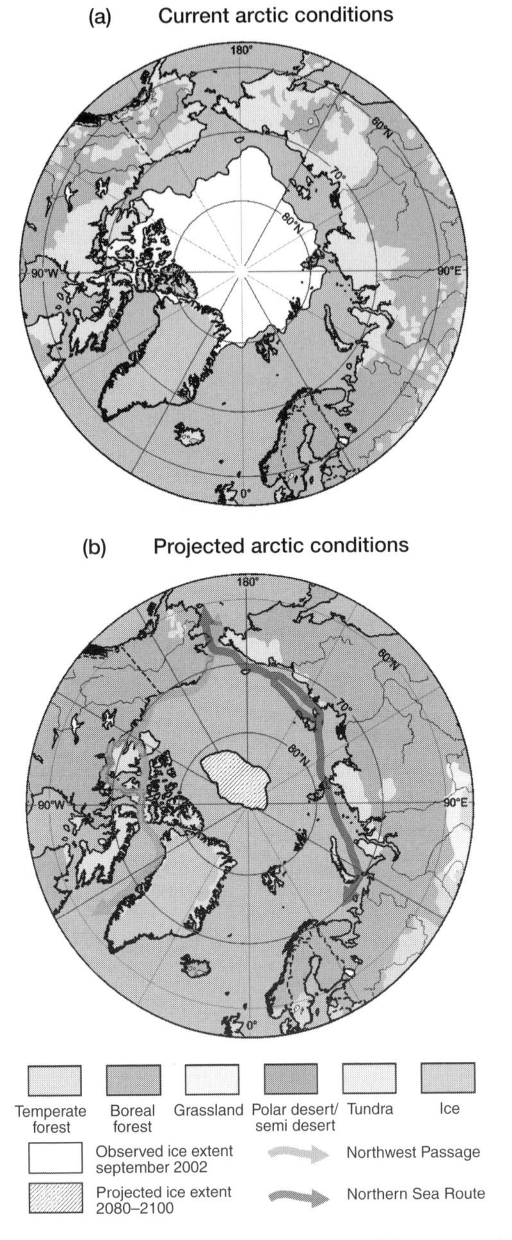

Figure 1.13 **Arctic sea ice and vegetation of Arctic and neighbouring regions. (a) 2005 conditions. (b) Projected for 2090–2100 under an IPCC IS92a scenario. Note the sharp decline in sea ice and tundra area.**
Source: IPCC Fourth Assessment Report, Working group II, figure TS.16. See Plate 7 for colour version.

sea ice as a result of climate change is a serious threat to their survival (see Figure 1.13 and Box 1.5). Forest fires are likely to increase strongly in the big Northern forests of Siberia and Canada. Tundra areas will shrink causing loss of nesting areas for the several hundred million migratory birds that fly to the Arctic in summer[14] (see also Figure 1.13).

Box 1.5

Polar bears threatened with extinction

Polar bears are good swimmers, but they need sea ice to hunt for seals that live on the ice. Sea ice also gives them a way to move from one area to another. With more and more sea ice melting in summer their food supply will be in danger. Given that they give birth in winter and do not eat for 5 to 7 months in that period, having ample food in spring and summer is critical to them and their cubs. And that depends completely on the ice conditions. Signs of declining conditions for bears are already visible in the southern most ranges in Canada. Breeding success is already going down. Only if polar bears can adapt to a land-based summer lifestyle is there hope for their survival. Competition with brown and grizzly bears will be a big obstacle however.

Polar bear in arctic, 1999. Greenpeace/Daniel Beltrá.
Source: Greenpeace, http://www.greenpeace.org/international/photosvideos/photos/polar-bear-in-arctic.
(Source: Arctic Climate Impact Assessment, Synthesis Report, and IPCC Fourth Assessment Report, Working group II, box 4.5)

Health

Human health will be affected negatively by climate change. As a result of food scarcity malnutrition will increase. More frequent and intense extreme weather events (floods, storms, fires, droughts, heat waves) will lead to injuries and death. The heat wave of 2003 in Europe that caused an additional 35000 deaths could be a regular phenomenon by the end of the century. Higher temperatures and longer droughts will increase the risk of water pollution and food poisoning, leading to an increase in diarrhoea. Some mosquito-borne diseases like dengue fever and malaria will spread to some areas where they are not occurring now.

There are some positive effects as well, mainly the reduction of cold-related deaths and disappearance of malaria from some areas. On balance the negative effects clearly dominate (see Figure 1.14).

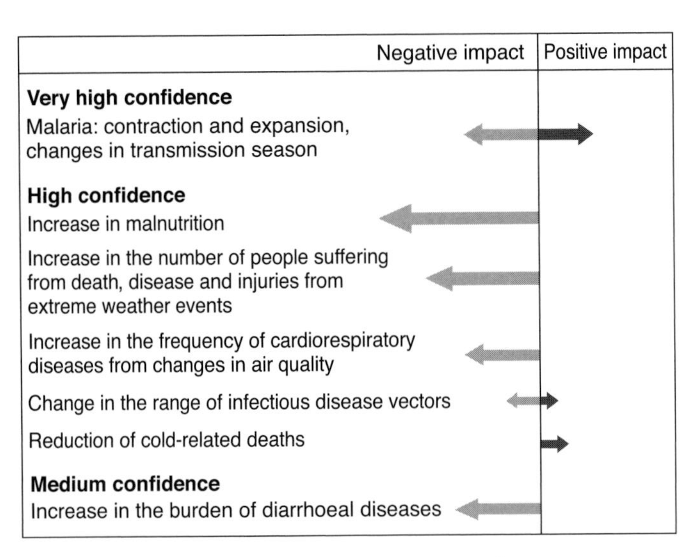

Figure 1.14 Selected health impacts of climate change.
Source: IPCC Fourth Assessment Report, Working group II, figure TS.9.

Poor and elderly people, children, small farmers, and people in coastal areas will face the biggest impacts of these health risks because they are more exposed to climate change impacts, and they also do not have access to good health services.

Infrastructure and human settlements

The biggest risk for infrastructure and human settlements comes from coastal and river flooding as a result of increased precipitation and sea level rise[15]. With river deltas and coastal areas having enormous concentrations of people, impacts can be very serious, particularly in South-East Asia and Africa. Without adaptation more than 100 million people could face coastal flooding every year by the end of the century[16]. Millions of people could be permanently displaced in major coastal delta regions as a result of sea level rise alone (see Figure 1.15). Low-lying coastal urban areas, especially those that also undergo natural land subsidence and are in cyclone prone areas, such as Bangkok, New Orleans and Shanghai, face the risk of great damage. Low-lying small islands may even have to be abandoned. In the Caribbean and Pacific more than 50% of people live within 1.5 km from the shore.

At present about 120 million people are exposed annually to tropical cyclones and on average more than 10000 people were killed every year between 1980 and 2000. Impacts from more intense tropical cyclones will lead to greater damage, particularly in coastal areas.

Very different risks are faced in permafrost (permanently frozen ground) areas. Melting of frozen ground in summer will happen in many areas where the ground is frozen year round (see also Figure 1.13). Roads, buildings, pipelines, and high voltage electricity lines may be seriously damaged and expensive reconstruction needed[17].

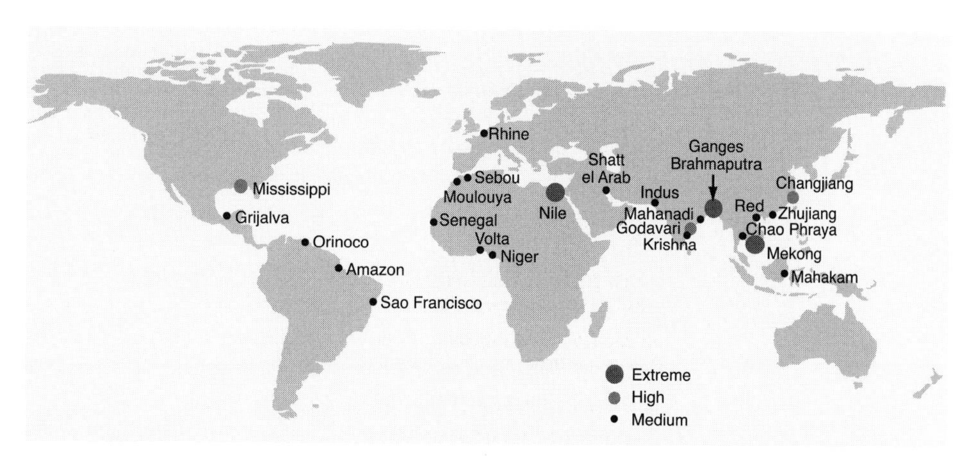

Figure 1.15 **Vulnerability of coastal river delta areas. Dots represent potential number of displaced persons due to sea level rise in combination with erosion and reduced sediment deposition from rivers in 2050. Extreme = more than 1 million; high = 50000 to 1 million; medium = 5000 to 50000.**
Source: IPCC Fourth Assessment Report, Working Group II.

What is the combined effect of these impacts regionally?

The magnitude of climate change varies from region to region. Impacts from climate change are manifold and they come on top of other stresses. Table 1.1 gives an overview of the most prominent risks for each region.

In some areas the capacity to adapt to climate change is more limited than in others, usually because of poverty. That brings us to vulnerability: low adaptive capacity means a high vulnerability. In terms of the most vulnerable regions, the Arctic stands out because of the high rates of change and the big impact on ecosystems and human communities. Africa is particularly vulnerable because of its lack of adaptive capacity. Small islands and low-lying river deltas in Asia and Africa face the biggest risk of mass migration.

How can we characterize the overall vulnerability to climate change?

An overall picture of vulnerability to climate change cannot be captured in one indicator. The IPCC has used a set of five indicators that reflect the range of vulnerabilities that are relevant:

Table 1.1. Examples of some projected regional impacts

Region	Projected impacts
Africa	• By 2020, between 75 and 250 million people are projected to be exposed to increased water stress due to climate change • By 2020, in some countries, yields from rain-fed agriculture could be reduced by up to 50%. Agricultural production, including access to food, in many African countries is projected to be severely compromised. This would further adversely affect food security and exacerbate malnutrition • Towards the end of the 21st century, projected sea level rise will affect low-lying coastal areas with large populations. The cost of adaptation could amount to at least 5–10% of gross domestic product (GDP) • By 2080, an increase of 5–8% of arid and semi-arid land in Africa is projected under a range of climate scenarios (TS)
Asia	• By the 2050s, freshwater availability in Central, South, East and South-East Asia, particularly in large river basins, is projected to decrease • Coastal areas, especially heavily populated megadelta regions in South, East, and South-East Asia, will be at greatest risk due to increased flooding from the sea and, in some megadeltas, flooding from the rivers • Climate change is projected to add to the pressures on natural resources and the environment, associated with rapid urbanization, industrialization, and economic development • Disease and death due to diarrhoeal disease associated with floods and droughts are expected to rise in East, South, and South-East Asia due to projected changes in the hydrological cycle
Australia and New Zealand	• By 2020, significant loss of biodiversity is projected to occur in some ecologically rich sites including the Great Barrier Reef and Queensland Wet Tropics • By 2030, water security problems are projected to intensify in southern and eastern Australia and in northern and eastern New Zealand • By 2030, production from agriculture and forestry is projected to decline over much of southern and eastern Australia, and over parts of eastern New Zealand, due to increased drought and fire. However, in New Zealand, initial benefits are projected in some other regions • By 2050, ongoing coastal development and population growth in some areas of Australia and New Zealand are projected to exacerbate risks from sea level rise and increases in the severity and frequency of storms and coastal flooding
Europe	• Climate change is expected to magnify regional differences in Europe's natural resources and assets. Negative impacts will include increased risk of inland flash floods, and more frequent coastal flooding and increased erosion (due to storminess and sea level rise) • Mountainous areas will face glacier retreat, reduced snow cover, and winter tourism, and extensive species losses (in some areas up to 60% under high emissions scenarios by 2080)

	• In Southern Europe, climate change is projected to worsen conditions (high temperatures and drought) in a region already vulnerable to climate variability, and to reduce water availability, hydropower potential, summer tourism, and, in general, crop productivity • Climate change is also projected to increase the health risks due to heat waves, and the frequency of wildfires
Latin America	• By mid century, increases in temperature and associated decreases in soil water are projected to lead to gradual replacement of tropical forest by savanna in eastern Amazonia. Semi-arid vegetation will tend to be replaced by arid land vegetation • There is a risk of significant biodiversity loss through species extinction in many areas of tropical Latin America • Productivity of some important crops is projected to decrease and livestock productivity to decline, with adverse consequences for food security. In temperate zones soybean yields are projected to increase. Overall, the number of people at risk of hunger is projected to increase (TS; *medium confidence*) • Changes in precipitation patterns and the disappearance of glaciers are projected to significantly affect water availability for human consumption, agriculture, and energy generation
North America	• Warming in western mountains is projected to cause decreased snowpack, more winter flooding, and reduced summer flows, exacerbating competition for over-allocated water resources • In the early decades of the century, moderate climate change is projected to increase aggregate yields of rain-fed agriculture by 5–20%, but with important variability among regions. Major challenges are projected for crops that are near the warm end of their suitable range or which depend on highly utilized water resources • During the course of this century, cities that currently experience heat waves are expected to be further challenged by an increased number, intensity, and duration of heat waves during the course of the century, with potential for adverse health impacts • Coastal communities and habitats will be increasingly stressed by climate change impacts interacting with development and pollution
Polar regions	• The main projected biophysical effects are reductions in thickness and extent of glaciers and ice sheets and sea ice, and changes in natural ecosystems with detrimental effects on many organisms including migratory birds, mammals, and higher predators • For human communities in the Arctic, impacts, particularly those resulting from changing snow and ice conditions, are projected to be mixed • Detrimental impacts would include those on infrastructure and traditional indigenous ways of life • In both polar regions, specific ecosystems and habitats are projected to be vulnerable, as climatic barriers to species invasions are lowered
Small Islands	• Sea level rise is expected to exacerbate inundation, storm surge, erosion, and other coastal hazards, thus threatening vital infrastructure,

settlements, and facilities that support the livelihood of island communities
- Deterioration in coastal conditions, for example, through erosion of beaches and coral bleaching is expected to affect local resources
- By mid century, climate change is expected to reduce water resources in many small islands, e.g. in the Caribbean and Pacific, to the point where they become insufficient to meet demand during low rainfall periods
- With higher temperatures, increased invasion by non-native species is expected to occur, particularly on mid- and high-latitude islands

Unless stated explicitly, all entries are from WGII SPM text, and are either very high confidence or high confidence statements reflecting different sectors (Agriculture, Ecosystems, Water, Coasts, Health, Industry, and Settlements). The WGII SPM refers to the source of the statements, timelines, and temperatures. The magnitude and timing of impacts that will ultimately be realised will vary with the amount and rate of climate change, emission scenarios, development pathways, and adaptation.
Source: IPCC Fourth Assessment Report, Synthesis Report, Table SPM.2.

Unique and threatened systems

As discussed extensively above, many ecosystems (coral, polar, mountain, and other systems) are vulnerable to increases in temperature of a few degrees centigrade. They typically have very little room to adapt to new circumstances.

Extreme weather events

Many of the more serious impacts are not caused by a gradual change of the climate, but by the extremes in temperature, precipitation, or wind speed. Table 1.2 gives an overview of these.

Distribution of impacts and vulnerabilities

Vulnerability is not about global averages. It is about the weakest people, the most sensitive coastal areas, and the most vulnerable infrastructure. When low-lying islands and coastal areas have to be abandoned, while other parts of the world face only minor problems, this cannot be averaged out. A chain is as strong as the weakest link. Africa, polar regions, and Asian megadelta regions were identified above as the most vulnerable.

Aggregate impacts

An obvious aggregation of impacts is the value of lost food production and damage to infrastructure and other market systems. Aggregating other non-market impacts such as species loss or loss of human life is much more difficult to quantify in monetary terms. As

Table 1.2. Examples of possible impacts of climate change due to changes in extreme weather and climate events, based on projections to the mid to late 21st century. These do not take into account any changes or developments in adaptive capacity. The likelihood estimates in column 2 relate to the phenomena listed in column 1

Phenomenon and direction of trend	Likelihood of future trends based on projections for 21st century using SRES scenarios	Examples of major projected impacts by sector			
		Agriculture, forestry and ecosystems	Water resources	Human health	Industry, settlement, and society
Over most land areas, warmer and fewer cold days and nights, warmer and more frequent hot days and nights	Virtually certain	Increased yields in colder environ-ments; decreased yields in warmer environments; increased insect outbreaks	Effects on water resources relying on snowmelt; effects on some water supplies	Reduced human mortality from decreased cold exposure	Reduced energy demand for heating; increased demand for cooling; declining air quality in cities; reduced disruption to transport due to snow, ice; effects on winter tourism
Warm spells/heat waves. Frequency increased over most land areas	Very likely	Reduced yields in warmer regions due to heat stress; increased danger of wildfire	Increased water demand; water quality problems, e.g. algal blooms	Increased risk of heat-related deaths, especially for the elderly, chronically sick, very young, and socially isolated	Reduction in quality of life for people in warm areas without appropriate housing; impacts on the elderly, very young, and poor
Heavy precipitation events. Frequency increases over most areas	Very likely	Damage to crops; soil erosion, inability to cultivate land	Adverse effects on quality of surface and groundwater; contamination of water supply; water	Increased risk of deaths, injuries, and infectious, respiratory and skin diseases	Disruption of settlements, commerce, transport, and societies due to flooding: pressures on urban and

Table 1.2. (*cont.*)

Phenomenon and direction of trend	Likelihood of future trends based on projections for 21st century using SRES scenarios	Examples of major projected impacts by sector			
		Agriculture, forestry and ecosystems	Water resources	Human health	Industry, settlement, and society
			scarcity may be relieved		rural infrastructures; loss of property
Area affected by drought increases	Likely	Land degradation; lower yields/crop damage and failure; increased livestock deaths; increased risk of wildfire	More widespread water stress	Increased risk of food and water shortage; increased risk of malnutrition; increased risk of water- and food-borne diseases	Water shortage for settlements, industry, and societies; reduced hydropower generation potentials; potential for population migration
Intense tropical cyclone activity increases	Likely	Damage to crops and trees; damage to coral reefs	Power outages causing disruption of public water supply	Increased risk of deaths, injuries, water- and food-borne diseases	Disruption by flood and high winds; loss of insurance in vulnerable areas; potential for population migrations, loss of property
Increased incidence of extreme high sea level (excludes tsunamis)	Likely	Salt water intrusion in irrigation water, estuaries, and freshwater systems	Decreased fresh-water availability due to saltwater intrusion	Increased risk of deaths and injuries; migration-related health effects	Costs of coastal protection versus costs of relocation; potential for movement of populations; also see tropical cyclones above

Source: IPCC Fourth Assessment Report, Synthesis Report, Table SPM.3.

will be discussed below, total costs can go up to 5–20% of global GDP, with higher costs for very vulnerable countries. Other quantifications, for instance millions of people affected by certain impacts, are often even more useful to reflect the magnitude of the aggregated impacts.

Large-scale irreversible events

Some impacts are irreversible. When a plant or animal species becomes extinct, it is gone forever. When glaciers and ice sheets melt, they cannot be reconstructed. When sea levels rise metres as a result of that, it means much of the inundated land can no longer be reclaimed. When ocean circulation changes, there is no way to bring it back to its original state and it will have large impacts on fisheries and ocean ecosystems. It could trigger big changes in regional climate.

When melting permafrost generates large methane emissions from the new swamps that develop, that leads to a self-reinforcing warming. As indicated above, species extinction is likely to happen on a large scale in the course of this century if no action is taken. The other big and irreversible impacts probably will not occur this century, because more drastic warming is needed than what can be expected this century. With uncontrolled climate change however these conditions may well exist beyond 2100. The level of warming above which such irreversible processes occur are called ‘tipping points’.

Overall picture

Compared to the previous IPCC assessment, when knowledge about vulnerabilities was more limited, the picture has become more pessimistic. Vulnerabilities are more serious now for a given level of warming and the confidence in the findings is higher[18].

What does this mean for development?

From the discussions above it is clear that climate change impacts are wide ranged and affect many people and important economic sectors, directly or indirectly. The total impact will be bigger than the sum of the individual impacts. Developing countries are the most vulnerable because of their much greater reliance on climate sensitive economic sectors such as agriculture, their low incomes, poor health systems, rapid population growth, and limited protection against extreme weather events. In addition, for many developing countries these impacts come on top of many other problems like poverty, lack of food security, environmental degradation, loss of biodiversity, and natural hazards. Achieving progress in key development areas, as for instance covered under the Millennium Development Goals, will be made much more difficult. Climate change can become a major obstacle for the elimination of poverty. Achieving real sustainable development will become impossible for countries facing large climate change impacts, while this development is critical to increasing their adaptive capacity to reduce vulnerability to climate change.

How would this affect the economy? The broader social, economic, and environmental impacts that are discussed above reflect quite well the seriousness of the threats. However, economic aspects are so prominent in decision making that the impact on a country's GDP is providing useful information.

Calculating economic costs of impacts is very problematic (see Chapter 3 for a more in-depth discussion). A whole range of impacts needs to be quantified in monetary terms. Some affect goods that have a market value, in which case calculating the costs is relatively straightforward. Others affect health or nature, which does not have a market value and must therefore be quantified in indirect ways, for instance by asking people how much they are willing to spend to protect endangered species. Quantifying the value of a human life is particularly sensitive. This is the first source of uncertainty and of differences in outcomes from different studies.

Then there are other problems: how to compare costs made in the short term with costs to be made in the distant future when people might have become much richer? This is the so-called discount rate problem. And there is the problem of adding up all costs across countries with very different incomes. And how far do we go in looking into the future? Do we stop at impacts that are likely to happen this century or are we looking much further into the future with more corresponding serious climate change impacts? And do we use only average impacts or do we take into account low probability events that may have catastrophic consequences? All these things matter enormously when calculating the damages from climate change[19].

How much does climate change cost? The range of costs found is 1–5% of global GDP for global average temperature increases of about 4°C, going up in some studies to 10% for 6°C warming. In developing countries costs are generally above average. The Stern review got even higher numbers when looking at per capita consumption: 5–20% of GDP, when the future is weighed heavily, impacts on poor people are counting more than impacts on rich people and by taking the low probability, high risks into proper account.

The conclusion must be that development, particularly in poorer countries, can be seriously undermined by climate change impacts.

Notes

1. For a more detailed discussion see IPCC Fourth Assessment Report, Synthesis Report, chapter 1 and Working Group I, chapters 3 and 4.
2. IPCC Fourth Assessment Report, Working Group I, box 3.6.
3. IPCC Fourth Assessment Report, Working Group II, chapter 1.4.
4. IPCC Fourth Assessment Report, Working Group II, chapter 4.4.10.
5. See IPCC Fourth Assessment Report, Working Group I, chapter 7.3 for a more detailed description of the carbon cycle.
6. IPCC Fourth Assessment Report, Working Group I, chapter 6.4.1.2.
7. IPCC Fourth Assessment Report, Working Group I, box 4.1.

8. See for an in-depth discussion IPCC Technical Paper on Water, 2008.
9. See for an in-depth discussion IPCC Fourth Assessment Report, Working Group II, chapter 5.
10. See for an in-depth discussion IPCC Fourth Assessment Report, Working Group II, chapter 4.
11. Millennium Ecosystem Assessment, Synthesis Report, http://www.millenniumassessment.org/en/synthesis.aspx.
12. A biological hotspot is defined by two criteria: (1) it is a region that contains at least 1500 plant species, unique for that area or ecosystem, and (2) it has already lost 70% or more of its original area. Collectively, 25 areas in 1999 held no less than 44% of the world's plants and 35% of terrestrial vertebrates in an area that formerly covered only 12% of the planet's land surface. The extent of this land area had been reduced by almost 90% of its original extent, such that this wealth of biodiversity was restricted to only 1.4% of land surface. There are now 34 of these hotspots identified. See http://www.biodiversityhotspots.org.
13. On 15 May 2008 the polar bear was declared a "threatened species" under the US Endangered Species Act; see http://alaska.fws.gov/fisheries/mmm/polarbear/issues.htm.
14. See Arctic Climate Impact Assessment, Synthesis report, page 45.
15. See for an in-depth discussion IPCC Fourth Assessment Report, Working Group II, chapter 6.
16. IPCC Fourth Assessment Report, Working Group II, chapter 6.2 and box 6.3.
17. Arctic Climate Impact Assessment, Overview Report.
18. IPCC Fourth Assessment Report, Synthesis Report, chapter 5.2.
19. See IPCC WG II, 2007, chapter 20.6 for an in-depth discussion.

2 Greenhouse gas emissions

What is covered in this chapter?

This is a book about controlling man-made climate change. Therefore we start with the man-made gases and aerosols that are responsible for climate change. They fall into two categories: (1) the six gases covered under the Kyoto Protocol: carbon dioxide, methane, nitrous oxide, sulphur hexafluoride, perfluorinated compounds, and hydrofluorocarbons; (2) ozone, chlorofluorocarbons, and aerosols. Their emissions sources are discussed in terms of the processes and the sectors of the economy where they emerge and the contributions of different countries. The strong increase and continuing upward trends of greenhouse gas emissions form a big challenge for emission reduction.

Contributions to warming

As discussed in Chapter 1, the contribution of gases and aerosols to warming depends on their effectiveness to retain solar radiation (called radiative properties) and their concentration in the atmosphere.

Controlling climate change therefore requires control of these concentrations. And concentrations are the combined result of input (emissions) and disappearance of gases. It is like filling a bath. If we want to control the water level, and the drain is closed, it means the tap has to be shut. And so it works with greenhouse gases (see Figure 2.1). Greenhouse gases disappear very slowly from the atmosphere. It takes 100 years before half of an amount of carbon dioxide (CO_2) put into the atmosphere has disappeared, but about 20% stays in the atmosphere for thousands of years. For methane (CH_4) it takes 12 years for two-thirds of it to disappear (called the 'lifetime'). For nitrous oxide (N_2O) this takes 110 years. For fluorinated gases the lifetime of the most common gases ranges from about 10 to several thousand years. Aerosols that contribute to cooling by comparison have a short residence time of several years. Given the slow disappearance of the most important greenhouse gases, emissions have to be reduced to very low levels if we want to prevent concentrations from rising above a certain level.

Table 2.1. **Global warming potentials**

Gas	Global warming potential 20 years	100 years	500 years
CO_2	1	1	1
CH_4	72	25	7.6
N_2O	289	298	153
HFC23	12000	14800	12200
SF6	16300	22800	32600

Source: IPCC Fourth Assessment Report, Working Group I, table TS.2

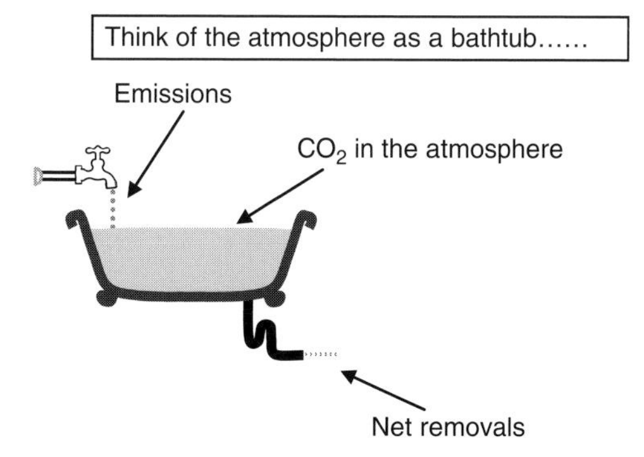

Figure 2.1 **Schematic representation of the atmosphere as a bath tub.**

The radiative properties of greenhouse gases are expressed in the global warming potential (GWP). This is the warming of an amount of that gas released into the atmosphere, compared to the warming of the same amount of CO_2 over a period of time. It captures both the radiative property of the molecules and the residence time in the atmosphere. For the most important greenhouse gases the GWPs are given in Table 2.1. The GWP for CO_2 is thus by definition equal to 1. Depending on the period of time chosen to compare the gases, the GWP of other greenhouse gases changes. For CH_4, a relatively short lived gas with powerful radiative properties, the GWP is 75 for a 20 year period, but for a 100 year period it drops to 25, because much of the CH_4 has disappeared in that period. This is of course more pronounced for a period of 500 years, which explains the GWP of 7.6. Some greenhouse gases like sulphur hexafluoride (SF6) and hydrofluorocarbon 23 (HFC23) are very powerful, with a radiative effect more than 10000 times that of CO_2.

GWPs are handy to add up the effect of different gases. If the quantities of each gas are multiplied with their respective GWPs, then adding them up gives you the total, so-called CO_2 equivalent emission. This CO_2 equivalent measure is used frequently throughout this book.

Kyoto greenhouse gases

Emission trends

The 1997 Kyoto Protocol agreement[1] focused on the major man-made contributors, carbon dioxide (CO_2), methane (CH_4), nitrous oxide (N_2O), sulphur hexafluoride (SF_6), perfluorinated fluorocarbons (PFCs), and hydrofluorocarbons (HFCs), the reason being that other gases and aerosols are much harder to control. With 75% of the warming caused by the Kyoto gases, it was also a good start at controlling climate change.

Emissions of the Kyoto gases have risen sharply over the last 35 years. The total[2] went up 70% between 1970 and 2004 with CO_2, the largest contributor, increasing by 80%.

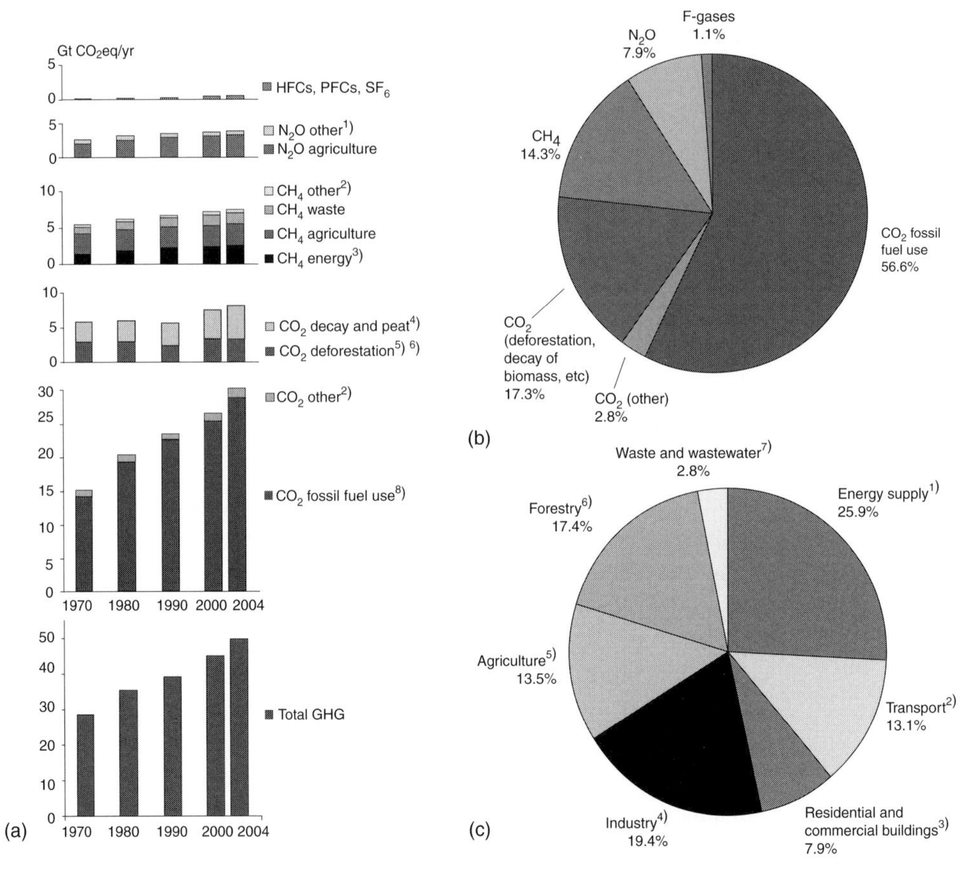

Figure 2.2 (a) Global annual man-made emissions of greenhouse gases from 1970 to 2004. (b) Share of different gases in total emissions in 2004. (c) Share of different sectors in total emissions in 2004. Gases are weighted according to their GWP and expressed in terms of CO_2-eq.
Source: IPCC Fourth Assessment Report, Synthesis Report, figure SPM.3.

The proportion of CO_2 in the 2004 emissions is slightly more than 75%, for CH_4 it is 15%, and for N_2O 8% (see Figure 2.2b).

Where are these emissions coming from?[3]

CO_2 comes mainly from burning coal, oil, and gas (75%). Smaller amounts are produced from turning oil and gas into plastics and other compounds that eventually are decomposed into CO_2 again (3%) as well as from manufacture of cement through decomposition of one of the main ingredients, limestone (3%). About 20% of the total CO_2 emissions comes from deforestation and decomposition of peat lands, crop residues, and organic materials in agricultural soils.

CH_4 comes from a variety of sources, the largest being livestock, particularly cattle and sheep (25%). This is followed by leaks from extraction, processing, and distribution of natural gas (15%). Other important sources are rice cultivation (12%), associated gas from coal production (10%), and decomposition of organic waste in waste water treatment (9%) and landfills (7%).

N_2O mainly comes from fertilized grasslands and croplands, where nitrogen fertilizers are decomposed in the soil (35%), followed by animal waste (26%). Surface water polluted with nitrogen accounts for about 15%. Small amounts come from chemical factories, such as those for nylon production (5%) and waste water treatment (2%). Cars with catalytic converters produce small quantities of N_2O (about 1% of the total).

Fluorinated gases (mostly HFCs) are emitted mainly from air conditioners in cars and refrigerators, as well as from the production of industrial chemicals. SF_6 is mainly used as an insulator in electrical equipment.

Economic sectors

If we organize the main sources of greenhouse gas emissions according to the sectors of the economy, we see that energy supply is the largest (26%), followed by industry (19%), the forest sector (17%), agriculture (14%), transport (13%), the building sector (8%), and waste management (3%)[4]. Emissions from electricity supply and transport are growing fastest. Figure 2.2c gives the global distribution in 2004.

Confusion can arise around sector contributions, because emissions can be counted in different ways. The numbers given above are based on emissions at the point where they enter the atmosphere (so-called 'point of emission allocation'). So emissions from electricity generation are counted under the energy supply sector. However, it can be more useful to count such emissions under the sector where that electricity is used (so-called "end-use allocation"). That can give a better picture of how electricity emissions

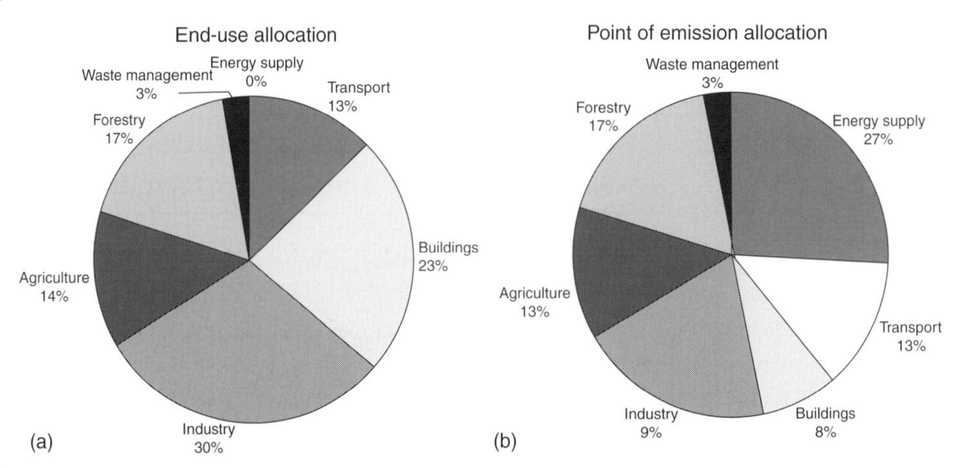

Figure 2.3 Comparison of sector shares of global greenhouse gas emissions, according to two different allocation methods.

can be reduced through energy savings. Counting emissions according to these so-called end-use sectors gives a very different picture, with the building sector being the largest, followed by forestry (see Figure 2.3).

Another complicating issue is exported goods. The current accounting system for emissions, as adopted under the Climate Convention, allocates emissions of exported coal, oil, and gas to the user country, but emissions of manufacturing goods to the exporting country. The argument is that for manufactured goods such a system is simpler, because no calculations have to be made about the emissions contained in exported goods, but also that the exporting country has the economic benefits of that export. For many countries there is not a big difference between the two systems, because they are also importers, so that the effects more or less cancel out. For a country like China however, with a huge export surplus, it does matter (see Box 2.1).

Box 2.1

Greenhouse gas emissions embedded in China's exported goods

Due to China's large export of manufactured goods, about one third of its domestic CO_2 emissions are in fact related to exports (see figure). Emissions from importing countries of course would go up if the emissions from imported goods were counted there. For instance the emissions from the UK would have been about 11% higher had all imported goods from China been produced domestically. For other countries the picture can be very different of course.

If these exported goods had been manufactured in the importing countries instead, global CO_2 emissions would have been lower. The reason is the relatively high carbon intensity of China's energy supply.

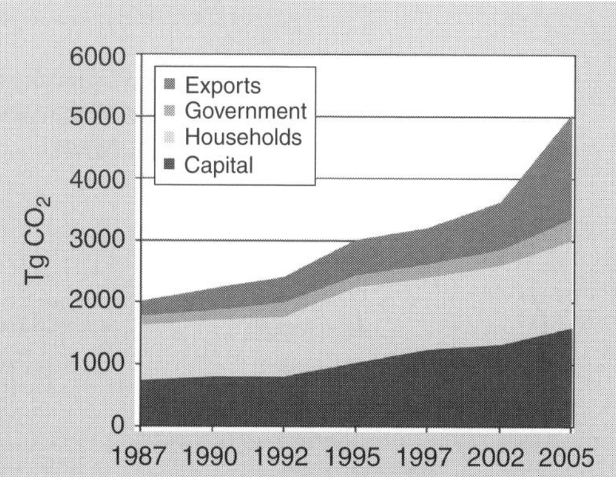

China's total domestic CO_2 emissions, divided by driving demand: exports, Governmental consumption, household consumption, and capital investment

CO_2 emissions from China. Exports means the emissions generated by the manufacturing of exported goods.

Source: Weber C. L. et al. The contribution of Chinese exports to climate change, Energy Policy, 2008, doi:10.1016/j.enpol.2008.06.009.

(Source: US Congressional Research Service, China's greenhouse gas emissions and mitigation policies, September 2008; You Li and C.N. Hewitt, The effect of trade between China and the UK on national and global carbon dioxide emissions, Energy Policy vol. 36, Issue 6, June 2008, pp. 1907–1914)

Which countries are responsible for greenhouse gas emissions?

The largest emitter of greenhouse gases is China, followed by the USA, the European Union, Indonesia, and India. This is the ranking for all greenhouse gases together, including land use change[5]. Leaving out emissions from land use change, which is often done when presenting country data, does change the picture significantly. Without land use change emissions, Indonesia for instance drops from place 4 to 12 and Brazil from place 7 to 13[6] (see Figure 2.4).

It is more illustrative and fairer to compare countries on the basis of average emissions per person[7], and this changes the ranking dramatically (see Figure 2.5). An average American citizen emits about 5 times as much as an average Chinese citizen and about 8 times as much as an average Indian citizen.

However, average citizens do not exist. A relatively poor country like India has a considerable number of rich people, whose consumption pattern causes much higher emissions than the average for the country, and is comparable to citizens in developed countries. And relatively wealthy countries do have poor people who produce low emissions. Out of the 6.5 billion people on earth, about 750 million have high emissions (more than 10t CO_2/yr) and a billion people very low emissions (less than 0.1 t CO_2/yr)[8]. This brings us to the issue of lifestyle.

Personal emissions and lifestyle

It is obvious that personal emissions of greenhouse gases depend on lifestyle. And that means consumption of electricity for home appliances, gas for heating and cooking, and

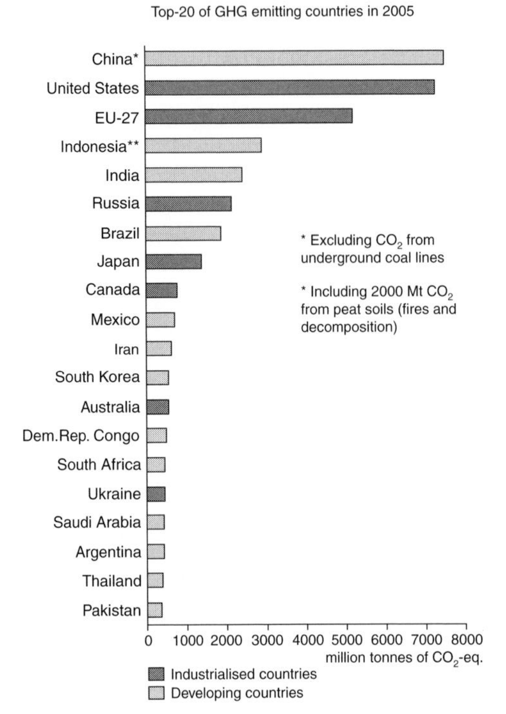

Figure 2.4 **Total greenhouse gas emissions including all Kyoto gases and including land use change in 2005 for the 20 largest emitters.**
Source: Netherlands Environmental Assessment Agency[6].

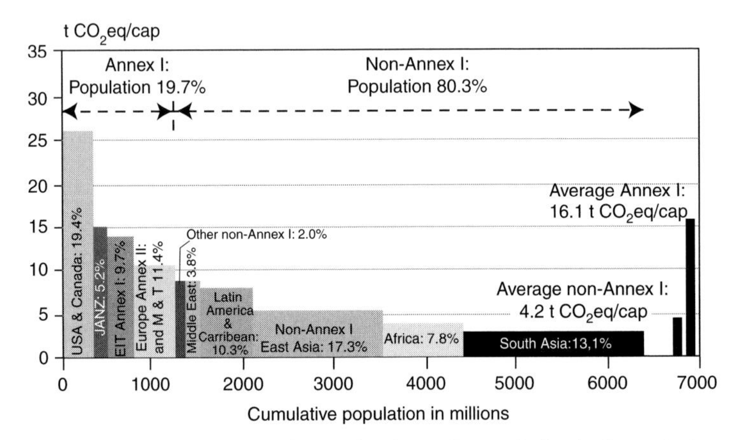

Figure 2.5 **Emission per person for various regions (2004 data). On the vertical axis the emissions per person of all greenhouse gases, expressed as t CO_2-eq, are shown; on the horizontal axis the number of people in the various regions is given. The surface area of the rectangles for each region is proportional to the total emissions. Annex-I is a term from the Climate Convention (see Chapter 12), covering 36 industrialized countries; non-Annex-I covers all other (developing) countries. The world average emission per capita is 6.5 tCO_2 eq/cap.**
Source: IPCC Fourth Assessment Report, Working Group III, figure SPM.3.a.

fuel (motorcycle, bus, car, plane), or electricity (train, tram) for transport. Food consumption also contributes, as well as the indirect emissions of consumer products and emissions generated at the workplace. What are the emissions due to these various activities?

An average UK citizen consumes about 1600 kWh of electricity per year, 740 m^3 of gas for home heating, and drives roughly 6500 km in a car. Add to that a substantial amount of hot and cold water, train and bus rides, several flights, and lavish consumption of food and other consumer goods and you get a rather greenhouse gas intensive consumption pattern. To convert these consumption data to emissions the average CO_2 emission per kilowatthour electricity and the average fuel consumption of cars, buses, and airplanes need to be used (see Box 2.2). This adds up to about 12.5 tonne of CO_2 per average UK person per year (see Table 2.2[9]). Of course, individual lifestyles vary considerably and so do personal emissions. There are many personal CO_2 calculators available online[10].

Box 2.2 CO_2 emissions per unit of energy or activity

Per unit of energy (GJ):
- Coal: 90 (kg/GJ)
- Oil: 70 (kg/GJ)
- Gas: 50 (kg/GJ)

Per unit of fuel:
- Litre of gasoline: 2.3 kg/l
- Litre of diesel: 2.6 kg/l
- Cubic metre of natural gas: 1.6 kg/m^3

Per km driven:
- Efficient car (1 l on 20 km; gasoline): 115 g/km; with 2 people: 57 g/km/person
- Inefficient car (1 l on 8 km; gasoline): 287 g/km with 2 people: 143 g/km/person
- Diesel vehicle (1 l on 15 km; diesel): 173 g/km
- Truck (1 l on 3 km; diesel): 870 g/km
- Bus: (1 l on 5 km; diesel): 520 g/km; with 20 people: 13 g/km/person

Per km in airplane (see note):
- Short flight: 150 g/km/person
- Long flight: 110 g/km/person

Per kilowatt hour electricity:
- From coal: 0.85–1.35 kg/kWh
- From gas: 0.4–0.52 kg/kWh
- From hydropower: 0.01–0.08 kg/kWh
- From nuclear: 0.04–0.012 kg/kWh
- From wind: 0–0.03 kg/kWh

Note: Data from UK DEFRA Company Greenhouse Gas Reporting Manual. The aviation emissions are not corrected for the multiplier effect due to release of emissions at high altitude. This multiplier is about a factor 2–4 according to the IPCC Special Report on Aviation, 2005.

Table 2.2. Personal greenhouse gas emissions for an average UK person

Activity	Average consumption per person	Emissions (tonne CO_2/person/ year)
House heating	Gas: 740 m^3/person/year	1.2
Hot water, cooking		0.4
Lighting, appliances	Electricity: 1600 kWh/person/year	0.7
Transport: motorcycle, car	6525 km/person/year	1.2
Transport: bus, rail		0.1
Transport: air		1.8
Other direct		0.6
Indirect emissions from food		2.1
Indirect emissions from consumer goods		3.1
Indirect emissions from workplace		1.3
TOTAL		12.5

Source: Goodall C. How to live a low-carbon life, Earthscan, London, 2007.

Emission intensity of the economy

Emissions can also be related to the size of the economy. Normally the size of the economy is expressed as gross domestic product (GDP). The higher the GDP, the higher the energy use and greenhouse gas emissions. But there are differences among countries. If a country has an economy that is energy and fossil fuel intensive, emissions per unit of GDP will be higher than for a country whose economy is not so dependent on energy use (see Chapter 3 for more detail).

Comparing countries' economies does entail some complexities. There are basically two ways to do it. One is to compare the GDPs by converting the local currency into a standard currency, say the US dollar. This is then called GDP at market exchange rates (GDP_{mer}). Such a comparison does not take into account the differences in local prices. People can have relatively low incomes, but with low prices for food, housing, etc. they can be better off than people in another country with higher incomes. If those things are taken into account, a corrected GDP can be calculated before it is converted to an international currency. That is the so-called GDP at purchasing power parity (GDP_{ppp}), which will be used here.

Figure 2.6 shows that industrialized countries generally have a more energy efficient economy than former communist countries in Eastern Europe and Asia (so-called economies-in-transition) and developing countries. However, the US economy is only slightly more efficient than that of India (the South Asia region) and only about 30%

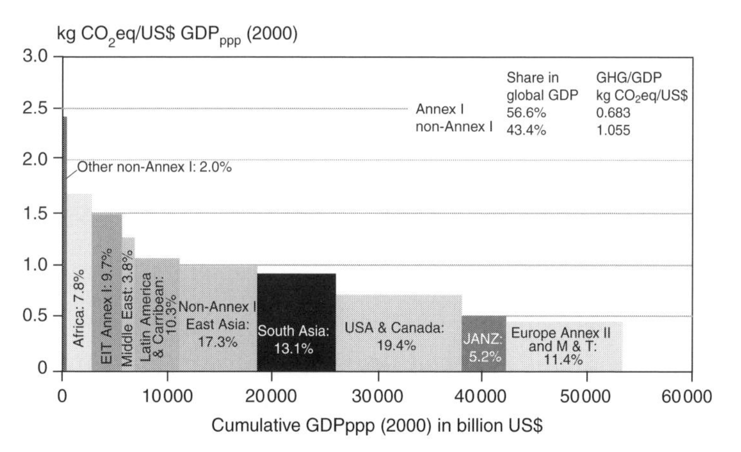

Figure 2.6 **Emission per unit of GDP_{ppp} for various regions (2004 data). On the vertical axis the emissions of all greenhouse gases per unit of GDP_{ppp}, are shown; on the horizontal axis the number of people in the various regions is given. The surface area of the rectangles for each region is proportional to the total emissions. The world average emission per unit of GDP is 0.84 kg CO_2eq/US.**
Source: IPCC Fourth Assessment Report, Working Group III, figure SPM.3.b.

more efficient than that of China. Japan and Europe are the most efficient economies, being about 25% more efficient than the US and more than three times as efficient as the so-called economies-in-transition.

For better understanding of differences between countries it can be useful to look at emissions per unit of product. For instance, steel plants differ with respect to the amount of CO_2 emissions per tonne of steel produced due to differences in processes and in the efficiency of energy use. Knowing these numbers is particularly useful in finding ways to reduce emissions (see Chapter 8 for more detail). The same approach can be followed for a whole range of energy efficient products, such as cement, glass, aluminium, paper, and others. When comparing energy use and greenhouse gas emissions in buildings it is often helpful to express energy or emissions per unit of floor space. In all cases the numbers are of course influenced by the type of fuel used or the carbon emissions of the electricity used.

Finally, it can also be enlightening to look at cumulative emissions. Due to the long life time of greenhouse gases the cumulative amount emitted to the atmosphere is directly correlated with the concentration. That means for instance that responsibility for the increased concentrations of CO_2, CH_4, N_2O, and other greenhouse gases that we see today lies predominantly with the industrialized countries, which started to emit CO_2 150 years ago. When we look at the cumulative emissions between 1950 and 2002 of CO_2 from energy only, developed countries are responsible for 71% and developing countries 29%. But when we also include CO_2 from deforestation, the shares become almost equal: developed 51%, developing 49%. Extending the period over which cumulative emissions are looked at to 1850–2002 would add the emissions from deforestation in developed countries that mostly happened before 1950. As a result the share of developed versus developing countries goes back to about 70:30 again[11].

Other gases and aerosols

As indicated above, the Kyoto gases are only responsible for 75% of the warming effect due to greenhouse gases and aerosols and they do not cause any cooling, as some other gases and many aerosols are doing[12]. So what are these other gases and aerosols and where do they come from?

Gases covered under the Montreal Protocol on protecting the ozone layer[13]

The Montreal Protocol, established in 1988, controls gases that damage the ozone layer that protects the earth against ultraviolet radiation. This layer sits in the stratosphere, 10–50 km above the earth. Most of these gases are also greenhouse gases, in particular chlorofluorocarbons (CFCs), hydrochlorofluorocarbons (HCFCs), halons, and gases like methylchloroform, methylbromide and carbontetrachloride. CFCs and HCFCs are by far the most important. They were not included under the Kyoto Protocol because they were already regulated under the Montreal Protocol. Together they are responsible for 10% of the warming and 3% of the cooling effects.

Emissions mainly come from refrigeration and air conditioning, insulating and packaging foams, fire extinguishers, and industrial cleaning agents. Total emissions of CFCs and HCFCs have been declining strongly since the Montreal Protocol came into force in 1988. As a result of the ban on CFCs, emissions went sharply down, while those of HCFCs and HFCs went up (see Figure 2.7).

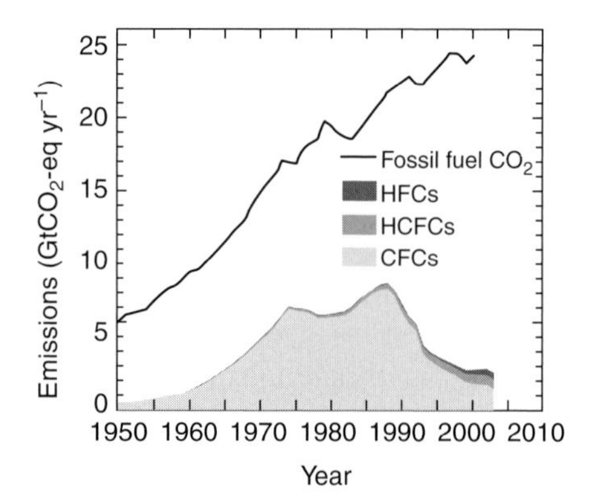

Figure 2.7 **Emissions of fluorinated gases (GWP weighted), in comparison with fossil fuel CO_2 emissions.**

Source: IPCC Special Report on Protecting the Ozone Layer and the Global Climate System, 2005.

Ozone

Ozone in the troposphere (the first 10km of the atmosphere) is responsible for slightly more than 10% of the warming (see Figure 1.6). It is not emitted as such, but formed by reactions of other pollutants in the atmosphere under the influence of sunlight. These so-called precursor gases are carbon monoxide (CO), nitrogen oxides (NOx), and hydrocarbons (methane and others). These gases are emitted by fossil fuel burning engines (cars, generators, etc.) and industrial furnaces, as well as from a variety of industrial sources. None of these gases are controlled under the Kyoto Protocol, but since they are well known air pollutants, there are regulatory measures in many countries affecting their emissions. Ozone itself is a primary concern from the point of view of impacts on human health, crops, and ecosystems. Average tropospheric ozone concentrations have increased by about 50% since 1860[14], but there are strong regional variations, mostly as a result of air pollution policies in Europe, North America, and Japan that led to lower emissions of precursor gases.

Ozone in the stratosphere (10–50 km above the earth) acts as a filter for harmful UV radiation. It is formed under the influence of sunlight, but is also disappearing due to reactions with so-called ozone depleting substances. CFCs are the most prominent of these ozone depleters, which is the reason they are being phased out under the Montreal protocol. Ozone depletion is most visible during September to December above the South Pole, the so-called 'Antarctic ozone hole'. As a result of this depletion process average stratospheric ozone concentrations have declined compared to 1750, contributing a little bit to cooling (see Figure 1.6).

Aerosols

There are many different aerosols with different properties that affect the extent to which they absorb or scatter solar radiation. Most aerosols have a cooling effect, but some contribute to warming. Aerosols are responsible for more than 80% of the total cooling. The effect is partially direct (solar radiation directly affected by the particles), partially indirect because aerosols enhance clouds that then reflect sunlight. The most important cooling aerosols are sulphates (formed in the atmosphere from sulphur dioxide emissions as a result of burning coal and oil; responsible for about 60% of the total cooling), nitrates (also formed in the atmosphere from nitrogen oxide; 15%), dust (from soils and roads; 15%), and organic carbon (formed due to incomplete combustion in industry, power generation, traffic, and homes as well as from agricultural waste burning; 12%). Black carbon (different from organic carbon because it originates from the burning of fossil fuel only, but is formed in the same way as organic carbon) has a warming effect that takes away about 30% of the overall aerosol cooling.

Emissions of these aerosols are not known very precisely. Historically sulphur dioxide emissions have been proportional to growing fossil fuel use. Since 1970 however air pollution abatement policies in Europe and North America have slowed down this growth considerably, even though emissions in Asia have grown with increasing fossil fuel use. The most recent global trend is a more or less stable emission of around 55–60 million tonnes of sulphur per year. Nitrogen oxide emissions showed a similar pattern, with current annual emissions around 30 million tonnes of nitrogen.

Emissions of organic and black carbon are particularly uncertain due to limited inventory studies. They have increased with increasing fossil fuel use, agriculture, and deforestation. Current estimates are 3–10 million tonnes/year for black carbon and 5–17 million tonnes/year for organic carbon. No recent trend can be identified.

None of the aerosol emissions have been regulated under any international agreement so far, because of the large uncertainty in emissions that would make agreed policy intervention very difficult.

How will emissions develop in the future?

Future emissions will of course depend on what we do about climate change. If worldwide action is taken to curb greenhouse gas emissions, the situation will be very different from a 'business as usual' future. Let us first look at this 'business as usual' or 'no action' situation.

Driving forces

In order to come up with plausible estimates of future emissions it is important to understand the forces that influence them, the so-called driving forces.

In its simplest form we can say:

$$\text{Emissions} = \text{number of people} \times \text{income per person} \times \text{emissions per unit of income}$$

Number of people or population is straightforward. Income per person (expressed as GDP[15]/capita) reflects economic development. The emissions per unit of income depend mainly on the amount and type of energy used, technology choices, land use and land use change, and lifestyle (what the money is spent on). The various driving forces will be discussed here briefly.

Population

Population projections for this century have been lowered since the early 1990s, based on falling birth rates in many parts of the world. The most recent projections suggest a world population of 8–9 billion in the year 2100, but with a fairly large uncertainty range

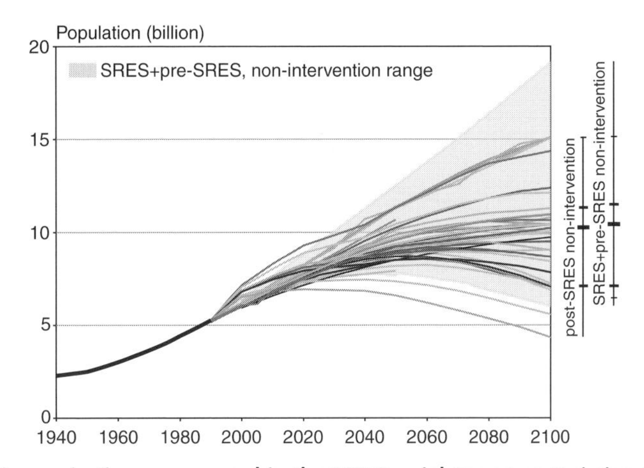

Figure 2.8 **Global population projections as reported in the IPCC Special Report on Emission Scenarios (SRES+ pre-SRES; light shaded area) and in the IPCC Fourth Assessment Report, Working Group III, chapter 3 (post-SRES non-intervention, dark shaded area).**

Source: IPCC Fourth Assessment Report, Working Group III, figure 3.1. See Plate 6 for colour version.

of 5–15 billion, caused by uncertainty about future birth rates (see Figure 2.8). Some of the lower projections even show a decline in the population after the middle of the century. To deal with this uncertainty scenarios are used. Scenarios are certain combinations of assumptions belonging to a possible future situation. A scenario for a future with high economic growth and ample attention for education and social justice would give a relatively low population growth, because birth rates are likely to go down faster in such a situation. In a low economic growth scenario without strong social policies population growth would be high.

Economic development

Economic development, expressed as global GDP, is projected to increase strongly in the future[16]. In light of the large number of people still living in poverty, this is a necessity and a matter of social justice. Overall economic growth of course does not say anything about income differences, but we leave that aside for this discussion.

In light of historic development, assumptions for future global average economic growth rates vary between 1% and 3% per year. By the end of this century that could lead to large differences in global GDP (4 to 20 times the current global income). Growth rates in different parts of the world will show even bigger differences. To deal with uncertainties scenarios are used, in which growth rates are chosen to be consistent with the kind of economic and social policies assumed. Figure 2.9 shows the range of the IPCC SRES scenarios for the period until 2030, together with some other projections from the Worldbank, the International Energy Agency, and the US Department of Energy.

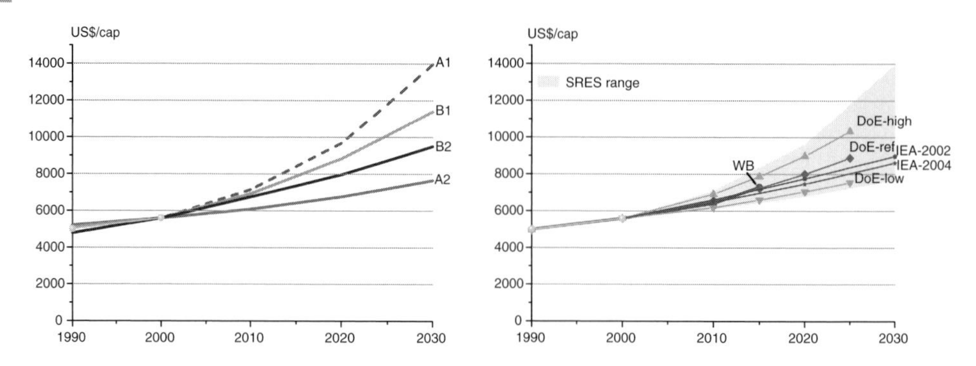

Figure 2.9 **Scenarios for development of global income (GDP) till 2030. DoE= US Dept of Energy; IEA = International Energy Agency; WB= Worldbank; shaded area is from IPCC-SRES = scenarios from IPCC Special Report on Emission Scenarios.**
Source: IPCC Fourth Assessment Report, Working Group III, figure 3.3.

Emissions per unit of income

Emissions per unit of income are driven by four important factors:

- Energy use
- Technology
- Land use
- Lifestyle

These are discussed below.

Energy use

One of the most important factors that drives emissions per unit of income is energy use[17]. Historically there has been a strong correlation between income and energy use (see Figure 2.10).

This figure tells us that energy is essential for development. It also shows that there is a fairly large spread in energy use at any given income level, meaning that certain countries managed to develop with relatively low energy use compared to others. Or, in other words, some countries have a much lower energy intensity (energy per unit of GDP) of their economy than others. Historically global energy intensity has been declining since the 1960s due to a shift towards a more service based economy and improved technology (the amount of energy used by cars, appliances, buildings, manufacturing processes, etc.). It is now about 25% lower than in 1960. Scenarios for this century estimate it will further decline by about 1% per year, leading to something like a 75% reduction by the end of the century compared to 1960. Technology and lifestyle (what people prefer to do with their time and money) make a difference. So energy intensity could be even lower.

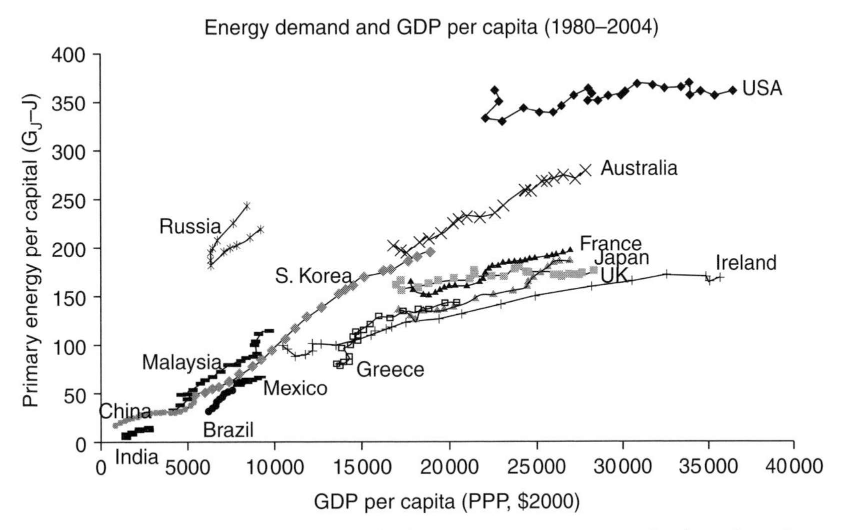

Figure 2.10 **Relationship between GDP_{PPP} per capita and primary energy use per capita for selected countries over the period 1980–2004. Russian data only 1992–2004.**

Source: Steven Koonin, Energy Trends and Technologies: facts, challenges and responses, iis-db.stanford.edu/evnts/5153/Drell_Lecture_0208.pdf ; data from UN, DOE/EIA.

The type of energy used, in particular the carbon content of it, also matters a lot. The dominant energy source since the start of the industrial revolution has been fossil fuel. First there was coal, later came oil, which was followed by natural gas. Gas produces about half the CO_2 of coal for a given amount of useful energy (see also Box 2.2). As a result the average carbon content of the world's energy use is now about 30% lower than in the year 1900. However, since the year 2000, this trend seems to have reversed. Global carbon intensity is going up due to a shift to coal because of sharply increased prices of natural gas and heavy use of coal in fast growing developing countries such as China and India[18].

Scenarios for this century still show a decline of about 0.4% per year, but with a high uncertainty. This could halve carbon intensity by the end of the century compared to 1960; however a small increase also is possible. Again, technology is playing a big role, because large scale use of nuclear power or renewable energy could make a great difference.

Figure 2.11 shows the historic CO_2 emissions as a function of income, an analogous picture to the one on energy. Note the relatively low per capita emissions of France, caused by a conscious decision after the 1970 oil crisis to develop a nuclear power based electricity sector (currently about 80% of electricity in France is nuclear).

Land use[19]

Over the past centuries human civilization has changed land cover dramatically, especially by converting forest and wilderness areas into agricultural land. This process is continuing, particularly in developing countries. Land use change is responsible for about

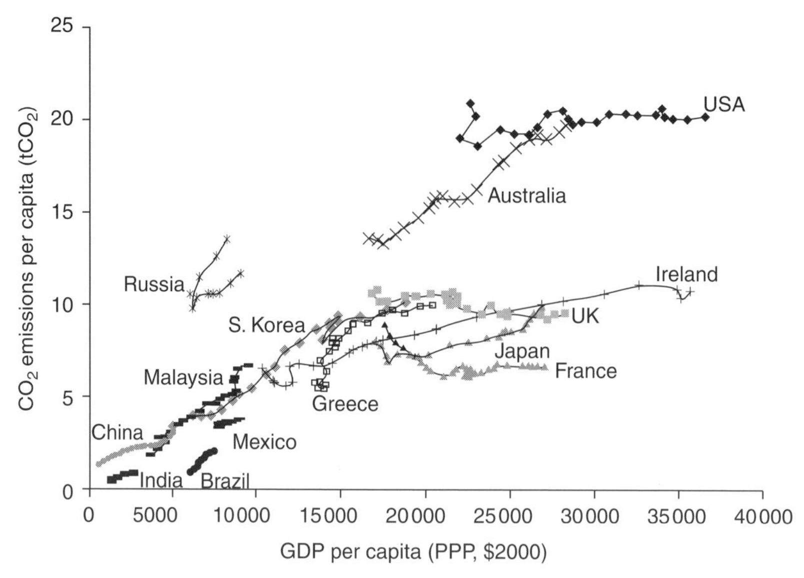

Figure 2.11 **Relationship between GDP_{PPP} per capita and CO_2 emission per capita for selected countries over the period 1980–2004. Russian data only 1992–2004.**
Source: Steven Koonin, BP, iis-db.stanford.edu/evnts/5153/Drell_Lecture_0208.pdf ; data from UN, DOE/EIA.

one-third of all of the CO_2 that was put in the atmosphere during the entire industrial era. It contributes about 20% to current global emissions. So future emissions will also depend heavily on how land use is going to develop.

Unfortunately future land use is difficult to project. Demand for food and timber and the land needed for that heavily depends on population, productivity of agriculture, and lifestyle. For a vegetarian diet about 80% less land is required to feed one person than for a meat based diet. Preservation of land for nature protection is another factor determining future land use. Scenarios for this century generally show cropland and grassland increasing and forests declining, but the spread is large. Some scenarios assume strong productivity growth in combination with lower population growth and strong forest protection policies, leading to an increase in forest land and maintenance of cropland and grassland areas. Other scenarios project increases of 40–50% in cropland and grassland areas, with up to 20% further loss of forest areas.

Emission projections

To project greenhouse gas emissions for this century all driving forces have to be combined into emissions scenarios. Since there is in principle an unlimited number of combinations of the various assumptions of all the relevant drivers, a sort of 'standardized' set of scenarios was developed by the IPCC[20], the so-called SRES scenarios. They defined four different

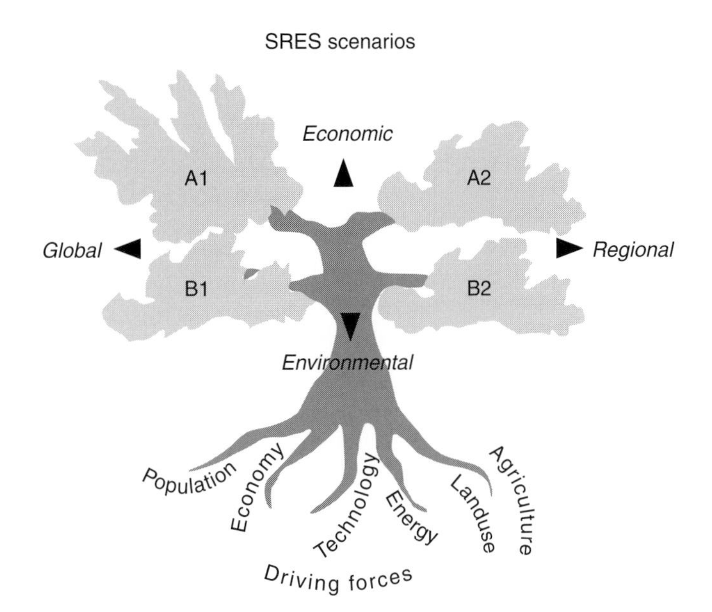

Figure 2.12 Schematic illustration of SRES scenarios. The four scenario 'families' are shown, very simplistically, as branches of a two-dimensional tree. The value orientation (economic versus environmental) is shown on a vertical axis, the geographical orientation (global versus regional) on a horizontal one. The schematic diagram illustrates that the scenarios build on the main driving forces of GHG emissions. Each scenario family is based on a combination of value and geographical orientation. Each scenario family has a common specification of some of the main driving forces. The A1 storyline branches out into four groups of scenarios to illustrate that alternative development paths are possible within one scenario family.

Source: IPCC Special Report on Emission Scenarios, figure TS-2.

'worlds' (or scenario families) by looking at two dimensions: (1) the value systems of societies: economic or environmental; (2) the orientation: global or regional, and then forming four different combinations (see Figure 2.12). They did *not* assume any specific policy to reduce greenhouse gas emissions.

For each of the 'worlds' consistent assumptions for the main drivers were made. For instance, high economic growth goes together with lower population growth and a faster introduction of new technology, whereas a strong environmental value system and low energy life styles are consistent. The characteristics of these four different worlds are summarized in Box 2.3.

Box 2.3

The main characteristics of the four SRES storylines and scenario families

By 2100 the world will have changed in ways that are hard to imagine – as hard as it would have been at the end of the 19th century to imagine the changes of the 100 years since. Each storyline assumes a distinctly different direction for future developments, such that the four storylines differ in increasingly irreversible ways. Together they describe divergent futures that encompass a significant portion of the underlying uncertainties in the main driving forces. They cover a wide range of key 'future' characteristics such as population growth, economic development, and technological change. For this reason, their plausibility or feasibility should not be considered solely on the basis of an extrapolation of *current* economic, technological, and social trends.

- The A1 storyline and scenario family describes a future world of very rapid economic growth, low population growth, and the rapid introduction of new and more efficient technologies. Major underlying themes are convergence among regions, capacity building, and increased cultural and social interactions, with a substantial reduction in regional differences in per capita income. The A1 scenario family develops into four groups that describe alternative directions of technological change in the energy system. Two of the fossil-intensive groups were merged in the SPM.
- The A2 storyline and scenario family describes a very heterogeneous world. The underlying theme is self-reliance and preservation of local identities. Fertility patterns across regions converge very slowly, which results in high population growth. Economic development is primarily regionally oriented and per capita economic growth and technological change are more fragmented and slower than in other storylines.
- The B1 storyline and scenario family describes a convergent world with the same low population growth as in the A1 storyline, but with rapid changes in economic structures toward a service and information economy, with reductions in material intensity, and the introduction of clean and resource-efficient technologies. The emphasis is on global solutions to economic, social, and environmental sustainability, including improved equity, but without additional climate initiatives.
- The B2 storyline and scenario family describes a world in which the emphasis is on local solutions to economic, social, and environmental sustainability. It is a world with moderate population growth, intermediate levels of economic development, and less rapid and more diverse technological change than in the B1 and A1 storylines. While the scenario is also oriented toward environmental protection and social equity, it focuses on local and regional levels.

Based on these four different worlds, scenarios were developed to cover a wide range of possible outcomes. In addition to the four worlds described, for one world two variants were identified, to cover the range of technologies and energy choices, the so-called A1 High Tech (A1T) and A1 Fossil Intensive (A1FI) scenarios. For each of these six scenario families a representative scenario was chosen, together generally called the SRES scenarios.

Projections of greenhouse gas emissions for this century with these scenarios span a wide range, as illustrated in Figure 1.8. By the end of the century there could be up to a fourfold increase or a slight reduction compared to the year 2000. The decline of emissions in the second half of the century happens in scenarios that assume a stabilization and decline in global population. In the medium term however all scenarios show a strong increase of emissions. To bring it a bit closer to home, for the year 2030 the projected increase of all greenhouse gas emissions is somewhere between 25% and 90%.

Two-thirds to three-quarters of this increase will come from developing countries, in line with their economic development.

One important lesson from this is that socio-economic development matters a lot. The different scenarios reflect different socio-economic development paths that can in principle be influenced. So choices made in economic and social policy can make a huge difference in terms of future greenhouse gas emissions.

Are actual emissions higher than what scenarios project?

New scenarios developed after the SRES scenarios do not show a significantly different picture[21]. They lie within the range covered by the SRES scenarios. Although the newest insights lead to some differences in assumptions for population, other drivers, in particular economic growth and the carbon intensity of energy, have compensated this.

Comparing the scenarios with actual emissions over the past few years shows that they are around the high end of the scenario range. Some scientists have argued they are even significantly higher, but careful analysis of the data used shows this is probably not the case[22]. In addition, it is dangerous to draw conclusions about long term trends on the basis of data for only a few years. It is likely the years 2008 and 2009 will show a lower increase of emissions due to the worldwide economic recession. Nevertheless these findings are worrisome. It means that the necessary emission reductions to avoid major climate change damages will be more difficult to realize and the risk of short-term climate change impacts increases.

So what does this mean?

From the perspective of controlling climate change the emission trends outlined above are bad news. While drastic reductions of emissions are required to stop the atmospheric concentrations from rising, current emissions have a strong upward trend and without action projections for the future also are strongly upwards. On top of that the cooling effect from aerosols may go down, when air pollution in developing countries is addressed. So controlling climate change is an uphill battle: population increase, increasing incomes, and higher demand for energy to improve well-being in poor countries all point in the opposite direction. Action to reduce emissions has to overcome that and then bring emissions down drastically. Chapters 5 to 9 will discuss this for the most important economic sectors.

Notes

1. See for more detail on the Kyoto Protocol chapter 12.
2. Weighted according to the Global Warming Potential of each gas.

3. Ibid.
4. Tourism is covered under transport and buildings.
5. Emissions from bunker fuels used in international shipping and aviation is excluded, because no international agreement has been reached on how to allocate these emissions over the various countries.
6. http://www.mnp.nl/en/dossiers/Climatechange/FAQs/index.html?vraag=10title=Which%20are%20the%20top-20%20CO2%20or%20GHG%20emitting%20countries%3F#10
7. This includes also emissions that are not caused by individual consumption, such as everything related to export industries and international transport.
8. Chakravarty S et al. PNAS, 2009, doi:10.1073/pnas.0905232106.
9. Goodall C. How to live a low-carbon life, Earthscan, London, 2007.
10. See for instance Act-on at http://actonco2.direct.gov.uk/index.html; Conservation International at http://www.conservation.org/act/live_green/carboncalc/Pages/methodology.aspx
11. Baumert K et al. Navigating the numbers, chapter 6, WRI,Washington DC, 2005.
12. The combined cooling effect of aerosols, damage to the stratospheric ozone layer, and land use change is about half of the total warming effect of the other compounds.
13. IPCC Special Report on protecting the ozone layer and the global climate system, 2005.
14. Horowitz L. Past, present, and future concentrations of tropospheric ozone and aerosols: Methodology, ozone evaluation, and sensitivity to aerosol wet removal. *Journal of Geophysical Research*, 111, 2006, D22211, doi:10.1029/2005JD006937.
15. GDP = gross domestic product, a measure of the total income of a country.
16. IPCC Fourth Assessment Report, Working Group III, chapter 3.2.1.2.
17. IPCC Fourth Assessment Report, Working Group III, chapter 3.2.1.5.
18. See chapter 5, figure 5.1.
19. IPCC Fourth Assessment Report, Working Group III, chapter 3.2.1.6.
20. IPCC Special Report on Emissions Scenarios, Cambridge University Press, 2000.
21. IPCC Fourth Assessment Report, Working Group III, chapter 3.2.2.
22. Van Vuuren D, Riahi K. Do recent emission trends imply higher emissions forever? Climatic Change, 2008, pp 1–12; doi:10.1007/s10584–008–9485-y.

3 Keeping climate change within sustainable limits: where to draw the line?

What is covered in this chapter?

One of the big questions in controlling climate change is "how far do we go in limiting climate change?" The climate has already changed and greenhouse gases in the atmosphere today will lead to further change, even if emissions were completely stopped overnight. Emissions are increasing strongly. Social and technical change is slow, and so is political decision making. There are also costs to be incurred. So where to draw the line? This chapter will look at the normative clauses that are part of the Climate Convention, the role of science in decision making, and some of the political judgements that have been made. It will explore the emission reduction implications of stabilization of greenhouse gas concentrations in the atmosphere. It will investigate how such reductions can be realised. It will look into the role of adapting to a changed climate as part of the approach to manage the risk of climate change. Finally, costs of doing nothing will be compared to the costs of taking action.

What does the Climate Convention say about it?

The United Nations Framework Convention on Climate Change[1] (to be referred to as UNFCCC or Climate Convention), signed at the 1992 Rio Summit on Environment and Development and effective since 1994, has an article that specifies the 'ultimate objective' of this agreement[2]. It says:

> *The ultimate objective of this Convention and any related legal instruments that the Conference of the Parties may adopt is to achieve, in accordance with the relevant provisions of the Convention, stabilization of greenhouse gas concentrations in the atmosphere at a level that would prevent dangerous anthropogenic interference with the climate system. Such a level should be achieved within a time-frame sufficient to allow ecosystems to adapt naturally to climate change, to ensure that food production is not threatened and to enable economic development to proceed in a sustainable manner.*

This text has far reaching implications. It mandates stabilization of greenhouse gas concentrations in the atmosphere, which will require eventually bringing emissions of greenhouse gases down to very low levels (see below). It also specifies explicit criteria for what that concentration level ought to be:

- the level should be chosen so as to avoid 'dangerous man-made interference' with the climate system, meaning as a minimum that:
  - ecosystems can still adapt naturally
  - food production is not threatened
  - economic development is still sustainable
- the speed at which concentration levels (and therefore the climate) are allowed to change should also be limited.

What risks and whose risks?

Most of these criteria are about the negative impacts of climate change on ecosystems and the economy (see Chapter 1). The point about sustainable economic development however also implies concern about the response to climate change. In theory a radical response in cutting emissions or spending a fortune on protective measures to cope with climate change could threaten sustainable development. So there are two sides to this problem of choice: the risks of climate change impacts on the one hand and the risks of responding to it on the other. Balancing those two risks is an essential element of making decisions on what is 'dangerous'.

The other important dimension is *whose risk* we are looking at. Climate change impact will be very unevenly spread. Within countries and between countries there will be huge differences in vulnerability of people. Low lying island nations will be threatened in their very existence, long before sea level rise is going to be a major issue for many other countries. Livelihoods of poor people in drought prone rural areas will be endangered long before most people in rich countries begin noticing serious local climate impacts (see more detailed discussion in Chapter 1). In general this requires an attitude of protecting the weakest. What is no longer tolerable for the most vulnerable groups ought to be taken as the limit for the world.

The multimillion dollar question is of course what that 'dangerous' level precisely is. At the time the UNFCCC was agreed there was no way that countries could agree on a specific concentration level. And after 14 years of further discussion that is still the case.

Should science give us the answer?

Control of climate change can be achieved through stabilizing concentrations in the atmosphere. This limits global mean temperatures and that reduces climate change impacts. To stabilize concentrations requires emissions to go down to very low levels. The lower the stabilization level, the earlier these low emissions levels should be reached. Figure 3.1 shows these relationships in a simple manner.

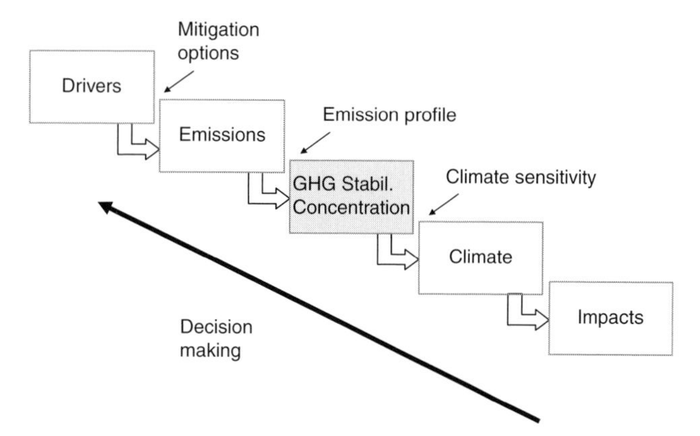

Figure 3.1 Schematic drawing of stabilizing concentrations of GHGs in the atmosphere and the upstream and downstream relationships with emissions, temperatures, and impacts.

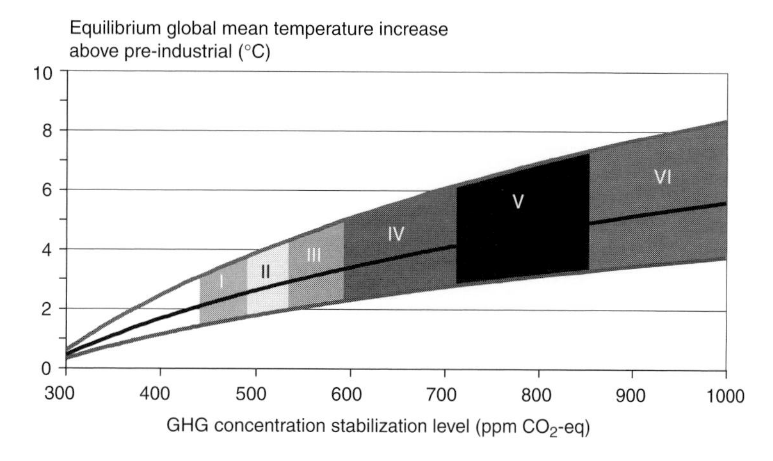

Figure 3.2 Concentrations of CO_2 and other greenhouse gases (expressed as CO_2-equivalent) and equilibrium temperature increases for a range of stabilization levels. Temperatures given on the y-axis are equilibrium temperatures, i.e. temperatures that belong to a stabilization level after the earth system has come to a steady state. These temperatures are higher than the temperature at the time the stabilization level is reached initially. Category numbering (roman numbers I to VI) represent categories of stabilization levels as used in the IPCC assessment.
Source: IPCC Fourth Assessment Report, Working Group III, figure SPM.8.

As summarized in Chapter 1, there is a fairly straightforward relationship between the mean global temperature and the impacts that can be expected, even if there are still significant uncertainties and gaps in our knowledge. Figure 3.2 shows in a nutshell how greenhouse gas concentration levels relate to global mean temperature and Figure 3.3 how climate change impacts relate to global mean temperature increase (above the pre-industrial temperature level).

What is striking in Figure 3.2 is the large uncertainty about global mean temperatures corresponding to a certain stabilization level of greenhouse gas concentrations (for instance at a concentration of 600 ppm CO_2-eq the corresponding temperature lies between 2.5 and

5°C).Why is this? It is caused by the uncertainty in the so-called 'climate sensitivity'. As explained in Chapter 1, climate sensitivity is defined as the warming for a doubling of CO_2 concentrations in the atmosphere. The best estimate of that climate sensitivity at the moment is 3°C (the black line in the middle of the band in Figure 3.2), but with an uncertainty range of 2–4.5°C (reflected in the range shown in the figure) and the possibility that it is even higher. This means there is a 50% probability that temperatures will be 2°C or less above those in the pre-industrial era at a concentration level of 450ppm CO_2-equivalent, but there is also a 50% probability that they will be 2°C or higher. There is even a 17% probability that temperatures at that concentration will be above 3°C.

Another important point is that the equilibrium temperatures referred to in Figure 3.2 are slightly different (several tenths of a degree) from the temperatures at the time concentrations have been stabilized. This is caused by the time it takes for oceans to get into equilibrium with the atmosphere. For the higher stabilization levels it can take centuries before the equilibrium temperature is reached.

We now have a lot of scientific information on the impacts of climate change, as summarized in Chapter 1. Figure 3.3 shows what kinds of climate change impacts can be avoided when limiting global mean temperatures.

GMT range relative to pre-industrial	Geophysical systems Example: Greenland ice sheet	Global biological systems Example: terrestrial ecosystems	Global social systems Example: water	Global social systems Example: food supply	Regional systems Example: Polar Regions	Extreme events Example: fire risk
>4–6	Near-total deglaciation	Large-scale transformation of ecosystems and ecosystem services More than 35% of species committed to extinction	Severity of floods, droughts, erosion, water quality deterioration will increase with increasing climate change	Further declines in global food production	Continued warming likely to lead to further loss of ice cover and permafrost. Arctic ecosystems further threatened, although net ecosystem productivity estimated to increase	Frequency and intensity likely to be greater, especially in boreal forests and dry peat lands after melting of permafrost
3.6–4.6	Commitment to widespread to near-total deglaciation 2–7 m sea level rise over centuries to millennia	Global vegetation becomes net source of C	Sea level rise will extend areas of salinization of ground water, decreasing freshwater availability in coastal areas		While some economic opportunities will open up (e.g. shipping), traditional ways of life will be disrupted	
2.6–3.6	*Lowers risk of near-total deglaciation*	Widespread disturbance, sensitive to rate of climate change and land use; 20–50% species committed to extinction. *Avoids widespread disturbance to ecosystems and their services, and constrains species losses*	Hundreds of millions of people would face reduced water supplies	Global food production peaks and begins to decrease. Lowers risk or further declines in global food production associated with higher temperatures		
1.6–2.6	Localized deglaciation (already observed due to local warming), extent would increase with temperature	10–40% of species committed to extinction *Reduces extinction to below 20–50%, prevents vegetation becoming carbon source* Many ecosystems already affected	Increased flooding and drought severity *Lowers risk of floods, droughts, deteriorating water quality and reduced water supply for hundreds of millions of people*	Reduced low latitude production. Increase high latitude production	Climate change is already having substantial impacts on societal and ecological systems	Increased fire frequency and intensity in many areas, particularly where drought increases
0.6–1.6	*Lowers risk of widespread to near-total degiaciation*	*Reduces extinctions to below 10–30%; reduces disturbance levels*		Increased global food production *Lowers risk of decrease in global food production and reduces regional losses (or gains)*	*Reduced loss of ice cover and permafrost; limits risk to Arctic ecosystems and limits disruption of traditional ways of life*	*Lowers risk of more frequent and more intense fires in many areas*

Figure 3.3 **Relationship between concentration stabilization levels and the impacts that can be expected at the respective equilibrium temperatures. Text in italics indicates reduction of risks.**
Source: IPCC Fourth Assessment Report, Working Group III, figure 3.38 and table 3.11.

Still, a choice on where to draw the line regarding what level of climate change would constitute a 'dangerous' situation is a matter of value judgement. Science and scientists are not supposed to make such value judgements. These kinds of decisions should be left to political processes, because they involve the weighing of various risks, involve ethical questions, and are inherently subjective.

Scientists are just human beings, they have certain personal convictions and perspectives and so it happens that some of them make statements about what ought to be done. As a citizen they of course have every right to speak out. As scientists they should limit themselves however to showing the implications of different degrees of climate change and of the costs of taking action. Their role is to inform decision makers, not to step into their shoes. Even then it is difficult to completely eliminate personal perspectives.

This attitude has not always prevailed. In 1987, the UN Advisory Group on Greenhouse Gases proposed limits to climate change: not more than 1–2°C above the pre-industrial era temperature, a change in global mean temperature of not more than 0.1°C per 10 years, and a sea level rise of not more than 0.2–0.5 m above the 1990 level. These proposals were based on the then available scientific information about impacts on ecosystems and the risk of melting of large ice masses on Greenland and the Antarctic. Nevertheless they were pure value judgements. Fortunately, after the establishment of the UN Intergovernmental Panel on Climate Change (IPCC) in 1988, science returned to a more objective and informative role. The 'mantra' of IPCC for its assessment reports is 'policy-relevant, but not policy prescriptive'. It means the IPCC is not making recommendations. It lays out the implications of different choices, but does not make a judgement of what is right or what is wrong.

In the next section you can read how emission reductions are connected to stabilization levels and temperature limits.

What are the implications of stabilizing greenhouse gas concentrations in the atmosphere?

The relationship between increase of global mean temperature and concentrations of greenhouse gases in the atmosphere has been discussed above. But what are the implications for global emissions? As outlined in Chapter 2, for any level of stabilization of concentrations, emissions have to go down to very low levels. The lower the concentration level, the sooner this has to happen. Figure 3.4 shows what this means for emissions of CO_2 for stabilization levels between 450 and 650 ppm CO_2 equivalent (see Box 3.1 for explanation of these units and where we are now).

Calculations like this are done with the help of global carbon cycle models, factoring in all natural and man-made emission sources and fixation of CO_2 in land, vegetation, and oceans. Comparable models for other greenhouse gases are also used. See Box 3.2 for a description.

Box 3.1

How to express concentration levels and where are we now?

Concentrations of greenhouse gases in the atmosphere are expressed in parts per million by volume (ppm). In order to capture the cumulative effect of the various greenhouse gases (and aerosols) and have a simple unit, their combined contributions are expressed in ppm CO_2-equivalent; in other words, the CO_2 concentration that would give the same warming effect as the sum of the individual concentrations of the individual gases (and aerosols).

The 2005 atmospheric concentration levels, expressed as CO_2-equivalent concentrations, were as follows:

CO_2: 379ppm
All Kyoto gases (see Chapter 2): 430ppm CO_2 equivalent
All greenhouse gases (incl. gases with ozone depleting potential (ODP)): 455ppm CO_2 equivalent
All greenhouse gases and aerosols: 375ppm CO_2 equivalent

(Source: IPCC Fourth Assessment Report, Synthesis Report, p 20, notes to table SPM.6)

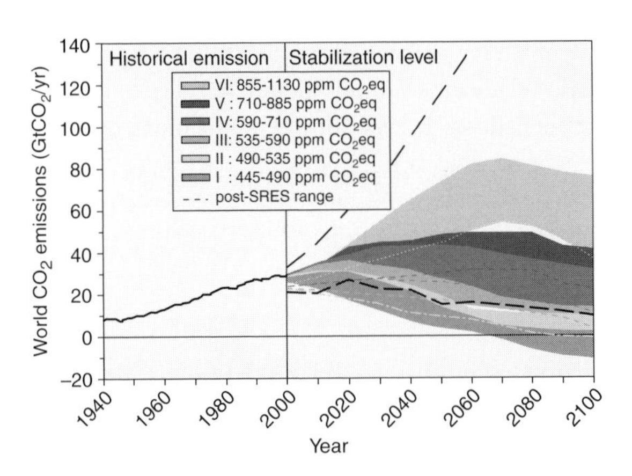

Figure 3.4 **CO_2 emission reductions required to achieve stabilization of greenhouse gas concentrations in the atmosphere at different levels, compared to 2000 emission levels. The wide bands are caused by different assumption about the emission trends without action (so-called 'baselines') and different assumptions about timing of reductions. Emission trajectories are calculated with various models (see also Box 3.2).**

Source: IPCC Fourth Assessment Report, Synthesis Report, figure SPM 11. See Plate 8 for colour version.

As is obvious from Figure 3.4, the emission reductions required for certain stabilization levels are not precisely known. This is caused by different assumptions in the calculations about 'no-action' emission trends (baselines) and timing of reductions. In other words, there are different ways to get to a specific stabilization level.

If we look a bit closer at required emission reductions, it is clear the implications are enormous. For a stabilization level of 450ppm CO_2 equivalent (roughly what is

Table 3.1. **Characteristics of stabilization scenarios, the resulting long term equilibrium global average temperature, the sea level rise component from thermal expansion, and the required emission reductions**

CO_2 concentration at stabilization (2005 = 379 ppm)	CO_2eq concentration at stabilization including GHGs and aerosols (2005 = 375 ppm)	Peaking year for CO_2 emissions[a,b]	Change in global CO_2 emissions in 2050 (% of 2000 emissions)[a,b]	Global average temperature increase above pre-industrial at equilibrium, using 'best estimate' climate sensitivity[c,d]	Global average sea level rise above pre-industrial at equilibrium from thermal expansion only[e]
Ppm	Ppm	Year	Percent	°C	metres
350 – 400	445 – 490	2000 – 2015	−85 to −50	2.0 – 2.4	0.4 – 1.4
400 – 440	490 – 535	2000 – 2020	−60 to −30	2.4 – 2.8	0.5 – 1.7
440 – 485	535 – 590	2010 – 2030	−30 to +5	2.8 – 3.2	0.6 – 1.9
485 – 570	590 – 710	2020 – 2060	+10 to +60	3.2 – 4.0	0.6 – 2.4
570 – 660	710 – 855	2050 – 2080	+25 to +85	4.0 – 4.9	0.8 – 2.9
660 – 790	855 – 1130	2060 – 2090	+90 to +140	4.9 – 6.1	1.0 – 3.7

[a] The emission reductions to meet a particular stabilization level reported in the mitigation studies assessed here might be underestimated due to missing carbon cycle feedbacks.
[b] Ranges correspond to the 15th to 85th percentile of the post-TAR scenario distribution. CO_2 emissions are shown so multi-gas scenarios can be compared with CO_2-only scenarios (see Figure SPM.3).
[c] The best estimate of climate sensitivity is 3°C.
[d] Note that global average temperature at equilibrium is different from expected global average temperature at the time of stabilization of GHG concentrations due to the inertia of the climate system. For the majority of scenarios assessed, stabilization of GHG concentrations occurs between 2100 and 2150.
[e] Equilibrium sea level rise is for the contribution from ocean thermal expansion only and does not reach equilibrium for at least many centuries.
Source: IPCC Fourth Assessment Report, Synthesis Report, Table SPM.6.

required to keep global mean temperature rise to 2°C), global CO_2 emissions would have to start coming down by about 2015 and by 2050 should be around 50–85% below the year 2000 levels. For a 550ppm CO_2 equivalent stabilization level (leading to about 3°C warming), global CO_2 emissions should start declining no later than about 2030 and should be 5–30% below 2000 levels by 2050. In light of the expected upward trend of global CO_2 emissions (40–110% increase between 2000 and 2030) and the time it takes for countries to agree about the required action and to implement reduction measures, these reduction challenges are staggering. The impact of the Kyoto Protocol is a drop in the ocean compared to this. It is expected to lead, by 2012, to a slight slowdown of the increase in emissions, but is insufficient to stop the increase in emissions. Table 3.1 lists the required emission reductions for different stabilization levels.

How can drastic emission reductions be realized?

Options for reducing greenhouse gas emissions fall into five categories:

- more efficient use of energy and energy conservation (= not using energy)
- using lower carbon energy sources (switching from coal to gas, renewable energy, nuclear)
- capturing of CO_2 from fossil fuels and CO_2 emitting processes and storing that in geologically stable reservoirs
- reducing emissions of non-CO_2 gases from industrial and agricultural processes
- fixing CO_2 in vegetation by reducing deforestation, forest degradation, protecting peat lands, and planting new forests.

Many technologies are commercially available today to reduce emissions at reasonable costs. These technologies will be further improved and their costs will come down. By 2030 several other low carbon technologies that are currently under development will have reached the commercial stage (Table 3.2). Chapters 5, 6, 7, 8, and 9 will discuss these options in detail for the major economic sectors.

Knowledge of the available technologies is not enough to answer the question of whether substantial reduction of emissions can be achieved in the long term. What is also needed is the expected development in the absence of climate change action (the baselines). Furthermore there are limitations to the speed with which power plants and other infrastructure can be replaced by low carbon alternatives. By putting the information about technologies, their cost over time, the rate at which they can be implemented, and the baseline into computer models, the resulting emissions over time can be calculated for any assumed scenario of climate change action. Calculations can also be done in a ‘reverse mode’. Then the desired emission reduction profile is determined first, based on a carbon cycle model of the earth system. The emission reduction options are then applied until the required reductions from a baseline are met. The cheapest options are applied first, followed by the more expensive ones. Box 3.2 describes the calculation process for one of these models.

Table 3.2. Selected examples of key sectoral mitigation technologies, policies and measures, constraints and opportunities

Sector	Key mitigation technologies and practices currently commercially available.	Key mitigation technologies and practices projected to be commercialized before 2030
Energy Supply	Improved supply and distribution efficiency; fuel switching from coal to gas; nuclear power; renewable heat and power (hydropower, solar, wind, geothermal, and bioenergy); combined heat and power; early applications of CCS (e.g. storage of removed CO_2 from natural gas)	Carbon capture and storage (CCS) for gas, biomass, and coal-fired electricity generating facilities; advanced nuclear power; advanced renewable energy, including tidal and waves energy, concentrating solar, and solar PV
Transport	More fuel efficient vehicles; hybrid vehicles; cleaner diesel vehicles; biofuels; modal shifts from road transport to rail and public transport systems; non-motorized transport (cycling, walking); land use and transport planning	Second generation biofuels; higher efficiency aircraft; advanced electric and hybrid vehicles with more powerful and reliable batteries
Buildings	Efficient lighting and daylighting; more efficient electrical appliances and heating and cooling devices; improved cooking stoves; improved insulation; passive and active solar design for heating and cooling; alternative refrigeration fluids, recovery and recycle of fluorinated gases	Integrated design of commercial buildings including technologies, such as intelligent meters that provide feedback and control; solar PV integrated in buildings
Industry	More efficient end-use electrical equipment; heat and power recovery; material recycling and substitution; control of non-CO_2 gas emissions; a wide array of process-specific technologies	Advanced energy efficiency; CCS for cement, ammonia, and iron manufacture; inert electrodes for aluminium manufacture
Agriculture	Improved crop and grazing land management to increase soil carbon storage; restoration of cultivated peaty soils and degraded lands; improved rice cultivation techniques and livestock and manure management to reduce CH_4 emissions; improved nitrogen fertilizer application techniques to reduce N_2O emissions; dedicated energy crops to replace fossil fuel use; improved energy efficiency	Improvements of crop yields
Forestry/ forests	Afforestation; reforestation; forest management; reduced deforestation; harvested wood product management; use	Tree species improvement to increase biomass productivity and carbon sequestration. Improved remote sensing

Table 3.2. *(cont.)*

Sector	Key mitigation technologies and practices currently commercially available.	Key mitigation technologies and practices projected to be commercialized before 2030
	of forestry products for bioenergy to replace fossil fuel use	technologies for analysis of vegetation/soil carbon sequestration potential and mapping land use change
Waste	Landfill methane recovery; waste incineration with energy recovery; composting of organic waste; controlled waste water treatment; recycling and waste minimization	Biocovers and biofilters to optimize CH_4 oxidation

Source: IPCC Fourth Assessment Report, Working Group III, Table SPM.3.

Box 3.2 The IMAGE-TIMER-FAIR Integrated Modelling Framework

Calculations of how to achieve deep reductions consist of the following steps:

- Make an assumption about a baseline of emissions without action
- Set an atmospheric concentration objective
- Define clusters of emissions pathways for a period of 50–100 years or longer that match the concentration objectives with the help of a built-in model of the global carbon cycle. In determining those emission pathways limitations are set for the speed at which global emissions can be reduced (usually 2–3% per year globally)
- Then a set of measures is sought from a built-in database of reduction options and costs that, from a global viewpoint, achieve the required emission reductions. The selection is done so that costs are kept to a minimum, i.e. the cheapest options are used first. In substituting baseline energy supply options with low carbon ones the economic lifetime of existing installations is taken into account and so are other limitations to using the full potential of reduction options.

All calculations are performed for 17 world regions. For calculating regional reductions and costs, the global reduction objectives are first divided between these regions using a pre-defined differentiation of commitments. The resulting regional reduction objectives can then be realized via measures both inside and outside the region. Emissions trading systems allow these reductions to be traded between the various regions.

The model can produce calculations of the cost of the reduction measures. The costs always concern the direct costs of climate policy, i.e. the tonnes reduced times the cost per tonne. No macroeconomic impacts in terms of lower GDP, moving of industrial activity to other countries, or the loss of fossil fuel exports can be calculated. Reference is made to other analyses. Co-benefits, such as lower costs for air pollution policy, are not included in the calculations. (Source: Van Vuuren et al. Stabilising greenhouse gas concentrations at low levels: an assessment of options and costs. Netherlands Environmental Assessment Agency, Report 500114002/2006)

Table 3.3. Estimate of the total cumulative technical potential of options to reduce greenhouse gas emissions during the period 2000–2100 (in $GtCO_2$-eq)

Option	Cumulative technical potential ($GtCO_2eq$)
Energy savings	>1000
Carbon capture and storage	>2000
Nuclear energy	>300
Renewable	>3000
Carbon sinks	>350
Non-CO_2 greenhouse gases	>500

Source: From climate objectives to emission reduction, Netherlands Environmental Assessment Agency, 2006, http://www.mnp.nl/en/publications/2006/FromClimateobjectivestoemissionsreduction.Insightsintotheopportunitiesformitigatingclimatechange.html

Technical potential of reduction options

The emission reduction that can be obtained from a specific reduction option is of course limited by the technical potential of that option. The technical potential is what can technically be achieved based on our current understanding of the technology, without considering costs. However, part of the technical potential could have very high costs. A first check of the viability of scenarios for drastic emission reduction is to compare the overall need for reduction with the total technical potential for all options considered. Table 3.3 shows estimates of the cumulative technical potential of the most important reduction options for the period 2000 to 2100. These are so-called conservative estimates, i.e. they give the minimum potential that is available.

The total technical potential for all options combined for this whole century is of the order of 7000 billion tonnes CO_2-equivalent[3]. This can then be compared with required cumulative reductions of 2600, 3600, and 4300 billion tonnes CO_2-equivalent for stabilization at 650, 550, and 450ppm CO_2-equivalent, respectively[4]. This first order comparison thus shows even the lowest stabilization scenario considered (450 ppm CO_2-equivalent) to be technically feasible.

Replacement of existing installations

The next step in the calculations is to combine introduction of reduction options in specific regions to a portfolio in such a way as to minimize costs. This means taking the cheaper options first. But it also means that reduction technologies are introduced to the extent they can be absorbed in the respective sector. For example, most models assume existing electric power plants are not replaced until their economic lifetime is reached. Low carbon energy supply options (e.g. wind power) in the model calculations thus only are used for replacing outdated power plants and for additional capacity needed.

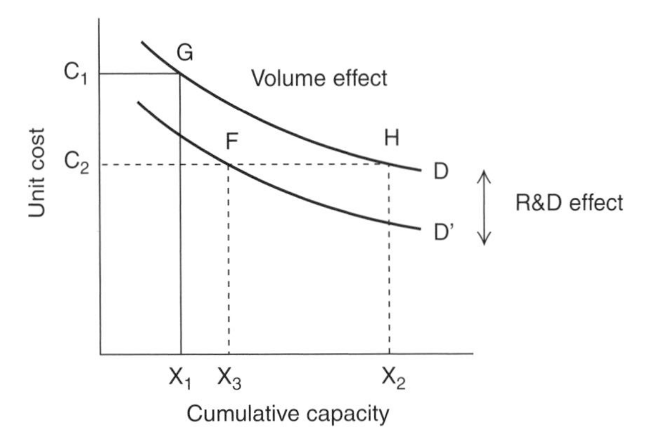

Figure 3.5 Cost reduction of technologies as a result of learning by doing (costs go down proportional to the cumulative capacity built, as in line D) and as a result of Research and Development (costs go down when R&D delivers results, as in line D′). Cost reduction from c_1 to c_2 can be obtained by expanding capacity from x_1 to x_2, or, alternatively, by R&D investments and increasing capacity from x_1 to x_3. R&D is usually more important in the early stages of development of a technology. When a technology is more mature the capacity effect usually dominates.

Source: Tooraj Jamasb, Technical Change Theory and Learning Curves: Patterns of Progress in Energy Technologies, Working paper EPRG, Cambridge University, March 2006.

Technological learning

Over time technologies become cheaper because of improvements in research and development and cost savings due to the scale of production. This is called 'technological learning' (see Figure 3.5). As an example, the price of solar (PV) energy units over the period 1976–2001 dropped 20% for each doubling of the amount produced.

These technological improvements and cost reductions are explicitly incorporated in the 'no action' case (the so-called baseline): efficiency of energy use increases; costs of renewable energy come down; and new technologies enter the market, even without specific climate change action. Traditional fossil fuel technologies also improve and costs come down unless fossil fuel prices go up (which happened recently). The effect of specific climate change action leading to increased deployment of technologies with lower emissions comes on top of this.

How important the baseline improvements are is shown in Figure 3.6. If technology had been frozen at the 2001 level, emissions in the baseline would have been twice as high by 2100.

Emission reductions

Figure 3.7 shows the outcome of calculations with one particular model[5] for a stabilization level of 450 ppm CO_2-equivalent. These results are typical for model calculations aiming at stabilization at this level. The right hand panel shows the contribution of the various

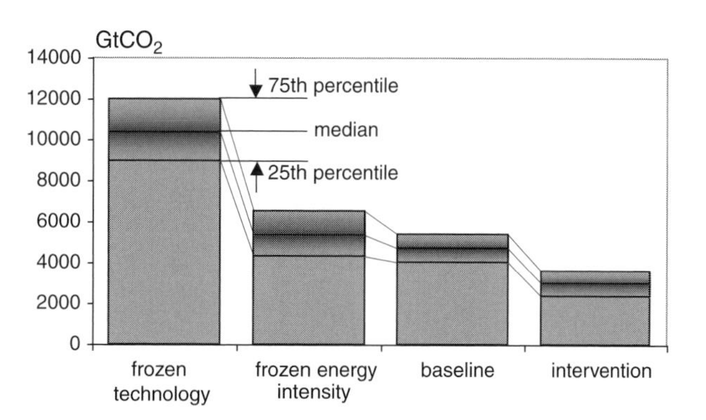

Figure 3.6 Cumulative emission of greenhouse gases over the period 2000–2100 for different assumptions about technological learning. 'Frozen technology' assumes technologies do not improve beyond their current status; in the baseline normal technological learning is assumed as happened in the past; the intervention case assumes additional climate change action to reduce emissions. Note the large difference between 'frozen technology' and the baseline, showing the importance of technological learning.

Source: IPCC Fourth Assessment Report, Working Group III, fig 3.32.

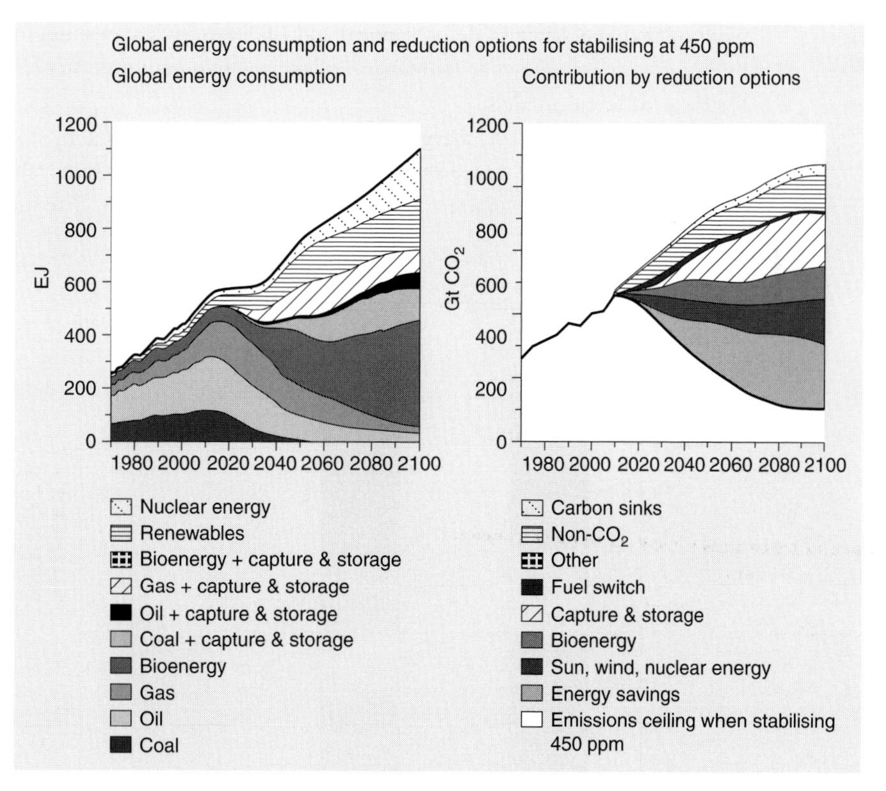

Figure 3.7 Contribution of reduction options to the overall emissions reduction (right hand panel) and changes in the energy supply system for stabilization at 450ppm CO_2-eq.

Source: From climate objectives to emission reduction, Netherlands Environmental Assessment Agency, 2006, http://www.mnp.nl/en/publications/2006/FromClimateobjectivestoemissionsreduction.Insightsintotheopportunitiesformitigatingclimatechange.html.

reduction options ('wedges'). One thing that stands out in this figure is the large contribution of energy efficiency improvement and CO_2 capture and storage (CCS). Energy efficiency is a relatively cheap option with a lot of potential. When the cost of achieving substantial reductions increases, CCS is more attractive than other more expensive options. The contributions of non-CO_2 gas reductions take place at an early stage, reflecting the relatively low cost of these options.

The left hand panel of Figure 3.7 shows the changes in the energy supply system as a result of implementing reduction options. The energy supply system continues to rely on fossil fuels (about 80% in 2005 and about 50% in 2050, but half of it will be 'clean fossil' (with CO_2 capture and storage). Fuel switching (from coal and oil to gas) and additional forest planting (so-called 'carbon sinks') play a very modest role. Biomass energy however gets a major share in the energy supply system after the middle of the century.

Different models give different results. An important reason for this is the different assumptions about the cost of reduction options, leading to a different order in which these options are introduced. Omission of certain options from the model and assumptions about availability of options, economic lifetimes of power plants or industrial installations, and economic growth also contribute to these model differences. Forest measures (forest planting and avoidance of deforestation) and CCS are for instance not included in the AIM model. Figure 3.8 shows a comparison of the relative contribution of reduction measures for three different models.

Similar differences in energy supply options are produced by the various models. The AIM model for instance shows a very high proportion of renewable energy by 2100 in the low level stabilization case, while other models do not. The main reason for this

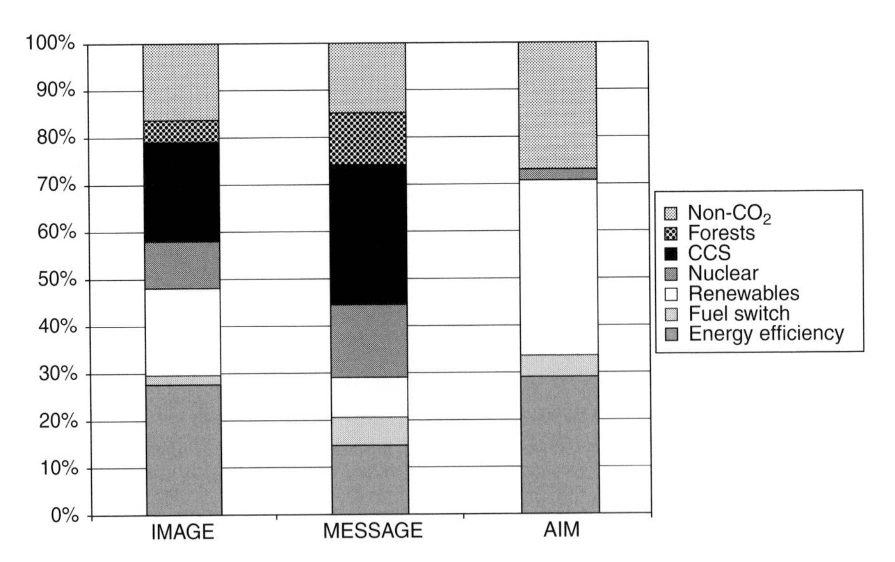

Figure 3.8 **Relative contribution of reduction measures to cumulative reductions in the period 2000–2100 in three models for stabilization at about 500ppm CO_2-eq.**

Source: IPCC Fourth Assessment Report, Working Group III, figure 3.23.

variation of course is the absence of CCS from the available reduction options. There are also significant differences in total energy consumption in the various model outcomes as a result of different assumptions on the cost and potential of energy efficiency improvements.

Better to adapt to climate change than to avoid it?

The task of restructuring the energy system in order to achieve the lower stabilization levels is enormous. The political complications of getting global support for it are huge. Therefore the suggestion is sometimes made to focus efforts on adaptation to climate change as a way to manage the climate change risks. Is this a sensible approach? Let us investigate what adaptation means.

Societies have adapted to climate variability and climate change for a long time: building of dikes and putting buildings on raised foundations against floods, water storage and irrigation systems to cope with lack of precipitation, adjustment of crop varieties and planting dates in agriculture, and relocation of people in areas where living conditions have deteriorated greatly. Planned adaptation, i.e. adaptation in anticipation of future climate change, is beginning to happen[6]. In the Netherlands, for instance, management plans for coping with increased river flows and higher sea levels have been adjusted and substantial investment in overflow areas for river water and strengthened coastal protection against sea level rise are being made[7]. In Nepal adaptation projects have been implemented to deal with the risk of glacial lake outburst floods caused by melting glaciers[8].

There are many possible ways in which to adapt to future climate change. Table 3.4 lists some typical examples for a range of economic sectors, together with relevant policy actions and problems or opportunities.

Many adaptation options are serving other important objectives, such as protecting and conserving water (through forest conservation and efficient irrigation), improving productivity of agriculture (moisture management of soils), improving the protection of biological diversity (protection of mangrove forests, marshes), and creating jobs (infrastructural works)[9].

Most climate change impacts will occur in the future. Developing countries that are the most vulnerable to climate change need to develop their infrastructure and economic activity to improve the living conditions of their people and to create jobs. This means there are enormous opportunities to integrate climate change into development decisions right now. There is no need to wait until climate change impacts manifest themselves. In other words, development can be organized so that societies become less vulnerable to climate change impacts: development can be made 'climate-proof' (see Chapter 4).

There are serious limitations to adaptation. Adaptation will be impossible in some cases, such as melting of big ice sheets and subsequent large sea level rise, loss of ecosystems and species, and loss of mountain glaciers that are vital to the water supply of large areas. And even where adaptation is technically possible, it must be realised that the

Table 3.4. Selected examples of planned adaptation by sector

Sector	Adaptation option/ strategy	Underlying policy framework	Key constraints and opportunities to implementation (Normal font = constraints; *italics* = *opportunities*)
Water	Expanded rainwater harvesting; water storage and conservation techniques; water re-use; desalination; water-use and irrigation efficiency	National water policies and integrated water resources management; water-related hazards management	Financial, human resources and physical barriers; *integrated water resources management*; *synergies with other sectors*
Agriculture	Adjustment of planting dates and crop variety; crop relocation; improved land management, e.g. erosion control and soil protection through tree planting	R&D policies; institutional reform; land tenure and land reform; training; capacity building; crop insurance; financial incentives, e.g. subsidies and tax credits	Technological & financial constraints; access to new varieties; markets; *longer growing season in higher latitudes*; *revenues from 'new' products*
Infrastructure/ settlement (including coastal zones)	Relocation; seawalls and storm surge barriers; dune reinforcement; land acquisition and creation of marshlands/wetlands as buffer against sea level rise and flooding; protection of existing natural barriers	Standards and regulations that integrate climate change considerations into design; land use policies; building codes; insurance	Financial and technological barriers; availability of relocation space; *integrated policies and managements*; *synergies with sustainable development goals*
Human health	Heat-health action plans; emergency medical services; improved climate-sensitive disease surveillance and control; safe water and improved sanitation	Public health policies that recognize climate risk; strengthened health services; regional and international cooperation	Limits to human tolerance (vulnerable groups); knowledge limitations; financial capacity; *upgraded health services*; *improved quality of life*
Tourism	Diversification of tourism attractions and revenues; shifting ski slopes to higher	Integrated planning (e.g. carrying capacity; linkages with other sectors); financial incentives,	Appeal/marketing of new attractions; financial and logistical challenges; potential adverse

	altitudes and glaciers; artificial snow making	e.g. subsidies and tax credits	impact on other sectors (e.g. artificial snow making may increase energy use); *revenues from 'new' attractions*; *involvement of wider group of stakeholders*
Transport	Realignment/relocation; design standards and planning for roads, rail, and other infrastructure to cope with warming and drainage	Integrating climate change considerations into national transport policy; investment in R&D for special situations, e.g. permafrost areas	Financial and technological barriers; availability of less vulnerable routes; *improved technologies and integration with key sectors (e.g. energy)*
Energy	Strengthening of overhead transmission and distribution infrastructure; underground cabling for utilities; energy efficiency; use of renewable sources; reduced dependence on single sources of energy	National energy policies, regulations, and fiscal and financial incentives to encourage use of alternative sources; incorporating climate change in design standards	Access to viable alternatives; financial and technological barriers; acceptance of new technologies; *stimulation of new technologies*; *use of local resources*

Source: IPCC Fourth Assessment Report, Synthesis Report, table SPM.4.

capacity required to implement it and the costs of doing it might be prohibitive. Think of people on low lying islands, poor farmers in drought prone rural areas in Africa, people in large low lying river delta regions, or on vulnerable flood plains in densely populated parts of Asia. But also in highly developed areas there are serious limitations to adaptation as the huge impacts of hurricane Katrina in New Orleans in 2005 and the heat wave in Europe in 2003 showed[10].

Given the limitations of adaptation, it does not appear to be a good strategy to rely only on adaptation. Limiting climate change through emission reductions (mitigation) can avoid the biggest risks that cannot realistically be adapted to. Mitigation does not eliminate all risks however. Even with the most ambitious efforts that would keep global average temperature rise within 2°C above the pre-industrial level, there is going to be substantial additional climate change. Adaptation to manage the risks of that is needed anyway. Adaptation is also needed to manage the changes in climate that are already visible today. Adaptation and mitigation are thus both needed. It is not a question of 'either-or' but of 'and-and', or, in other words, 'avoiding the unmanageable and managing the unavoidable'[11].

What are the costs?

The first question is: 'costs of what?' Too often only the costs of controlling climate change are considered. The costs of inevitable adaptation to a changed climate are usually forgotten, although it is clear that the less is done on emissions reductions, the more needs to be done on adaptation. But what is worse is that the 'costs of doing nothing', i.e. of the impacts of uncontrolled climate change, are often completely ignored. That distorts the picture. In other words, the only sensible way to look at costs is to look at both sides of the balance sheet: the cost of reducing emissions on the one hand and the costs of adaptation and the costs of the remaining climate change impacts on the other. In fact, to get a realistic picture of the true costs, the indirect costs and the benefits of taking action need to be included also as a correction to the mitigation costs. Many actions to reduce emissions have other benefits. A good example is the avoidance of air pollution when coal is replaced by natural gas in order to reduce CO_2 emissions.

Mitigation costs

Costs of mitigation can be expressed in several ways. One is the cost of avoiding 1 tonne of CO_2 (or a mixture of gases expressed as CO_2-equivalent). Knowing how many tonnes you need to avoid under a specific mitigation programme, and multiplying that number with the cost per tonne, gives you the total costs of that programme (in fact investment and operational costs, but often called 'abatement costs').

There is also another cost perspective: the cost to the economy as a whole, or how much the overall 'wealth' (expressed for instance in the GDP of a country) is affected by mitigation policies. There no simple relationship between the two cost measures. Expenditures as such do not reduce wealth. In fact the opposite is true: more economic activity (expenditures) means a higher GDP. However, spending money on reducing greenhouse gases normally means that money is not spent on something else. Many other economic (but not all) activities produce more wealth than reducing greenhouse gas emissions and therefore overall wealth could be reduced as a result of mitigation action. The 'foregone increase of wealth' (by choosing mitigation instead of more productive activities) is then the macro-economic cost of that mitigation action. Note that the cost of the damages due to climate change is not included. Nor are the effects of adaptation[12].

Expenditures for mitigation in long-term mitigation strategies leading to stabilization of greenhouse gas concentrations in the atmosphere can be substantial. As outlined above, over time more and more costly reduction options need to be implemented in order to drive down emission to very low levels. The deeper the cuts in emissions, the higher the cost of the last tonne avoided will be (called the 'marginal cost'). Figure 3.9a shows how the marginal cost develops over time for different stabilization scenarios.

The marginal cost is shown for a typical set of stabilization calculations as a function of time for different stabilization scenarios. Ambitious scenarios lead to a stronger increase of marginal costs. Total abatement costs are determined by the average costs and the volume of the required reductions. These costs are shown in Figure 3.9b. To put cost

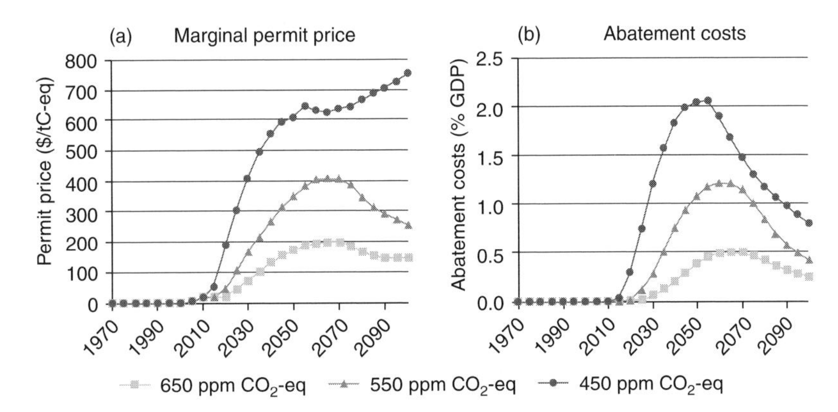

Figure 3.9 **(a) Development of the marginal cost of a tonne of CO_2-eq avoided for different scenarios. (b) Abatement costs expressed as % of global GDP for the same stabilization scenarios as in (a). For all scenarios an IPCC SRES B2 baseline was used.**

Source: Van Vuuren et al. Stabilising greenhouse gas concentrations at low levels: an assessment of options and costs, Netherlands Environmental Assessment Agency, Report 500114002/2006.

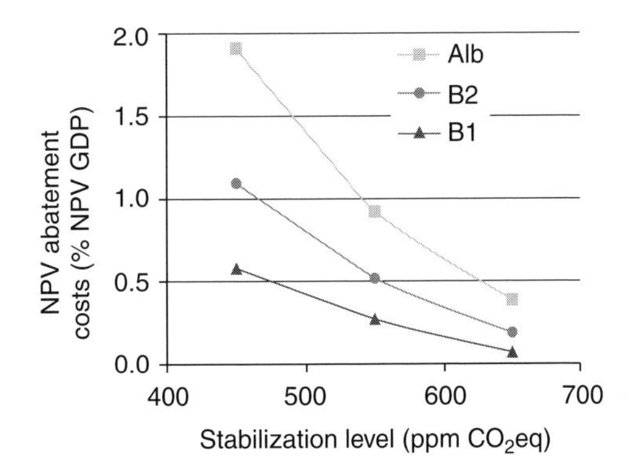

Figure 3.10 **Discounted (Net Present Value) of cumulative abatement costs for different stabilization levels, expressed as % of discounted GDP, for different baselines.**

Source: Van Vuuren et al. Stabilising greenhouse gas concentrations at low levels: an assessment of options and costs, Netherlands Environmental Assessment Agency, Report 500114002/2006.

numbers in perspective, they are usually expressed as % of global GDP. For low level stabilization they can go up to about 1–2% of GDP in the period around 2050. This means that for ambitious stabilization scenarios expenditures for emission reduction would be of the same order of magnitude as those for all environmental measures taken today in most industrialized countries.

To make cost comparisons easier, costs can be accumulated over the century and expressed as the so-called 'net present value' (future costs discounted to the present). Typical numbers found for this cumulative cost are 2–3% of global GDP for the most stringent scenarios (leading to low stabilization levels) and most pessimistic assumptions, to less than 1% for higher stabilization levels and more optimistic assumptions[13].

Table 3.5. Coverage of sectoral estimates of adaptation costs and benefits in the literature (size of check mark indicates degree of coverage)

Sector	Coverage	Cost estimates	Benefit estimates
Coastal zones	Comprehensive – covers most coastlines	√	√
Agriculture	Comprehensive – covers most crops and growing regions	–	√
Water	Isolated case studies in specific river basins	√	√
Energy (demand for space cooling and heating)	Primarily North America	√	√
Infrastructure	Cross-cutting issue – covered partly in coastal zones and water resources. Also isolated studies of infrastructure in permafrost areas	√	–
Heath	Very limited	√	–
Tourism	Very limited – winter tourism	√	–

Source: OECD, Economic aspects of adaptation to climate change, 2008.

Costs are not only affected by different stabilization levels, but also by assumptions about the trends without action (so-called baselines). Baselines that reflect high growth economies, heavily relying on fossil fuels, lead to higher abatement costs (see Figure 3.10).

Adaptation costs

Adaptation is a local issue. Building a picture of global adaptation costs therefore requires summing up a large number of local and regional studies, which is where the problem lies. Studies are limited. In terms of coastal defence and agriculture the coverage of studies is reasonable. Beyond that, coverage is poor. Where coverage is reasonable, studies are not harmonized, making it very difficult to get an aggregate cost number. Avoided risks are not clearly defined, so that it is unclear what precisely is achieved for a certain additional investment (see Table 3.5).

For coastal defence in response to rising sea levels many studies were undertaken in all parts of the world. Cost estimates for small low lying island states are the highest: for most countries close to 1% of GDP per year, with much higher numbers for the Marshall Islands, Micronesia, and Palau. The costs are somewhat lower for coastal countries[14]. But studies have not been standardized regarding the sea level rise to which adaptation is tailored.

Studies on adaptation in agriculture have focused on minimizing productivity loss. Outcomes show that productivity loss can be reduced by at least 35% and sometimes can be avoided completely or additional yields can be obtained (meaning current practices are not optimal). Reliable data on cost are not available, and usually rough estimates are made on increasing R&D (something like 10%), agricultural extension (also 10% or so), and investment (2% increase)[15].

For the water sector only one rough estimate is available currently and that one only looks to 2030. It suggests at least US$10 billion per year is needed in that timeframe, which is small compared to the US$50 trillion annual world GDP. For infrastructure, health, and tourism there are only a few isolated studies available.

Notwithstanding a poor knowledge base, attempts have been made to estimate global adaptation costs across all sectors. The Worldbank did a study based on investments that are sensitive to climate change. Others followed this method. The numbers for global adaptation costs range from about 10 to 100 billion US$/year. Studies undertaken by the Climate Change Convention are based on sectoral data and are higher: 30–170 billion US$/year. Later studies arrive at higher numbers, with the highest being about 0.3% of global GDP. These numbers are very uncertain however and could easily be proven wrong by more detailed studies undertaken in the future.

Co-benefits

Strong reductions in greenhouse gas emissions can help address other problems, air pollution being one of them. Reducing fossil fuel use does not only reduce emissions of CO_2, but also of small particles, SO_2, and NOx that cause serious health problems. When the reduction in health problem is quantified in dollar terms (although that is tricky because of the assumptions that have to be made), this covers a significant part of the mitigation costs. Or, in other words, net mitigation costs are much smaller. When avoidance of crop damage and damage to ecosystems due to better air quality is added, net mitigation costs go down further still[16].

There are other co-benefits. Energy efficiency measures and a shift to renewable energy sources will reduce imports of oil and gas, improving the energy security of many countries. Employment can be generated through labour intensive energy efficiency improvements in existing buildings and production and installation of renewable energy installations[17]. Figure 3.11 shows the magnitude of some of these co-benefits for different stabilization scenarios. For the most stringent 450 ppm CO_2equivalent scenario, reduction in loss of life due to air pollution and oil imports is of the order of 30%.

Costs of climate change damages

Attempts have been made to express the damages from climate change in monetary terms. This is an extremely difficult exercise. There are large uncertainties about regional

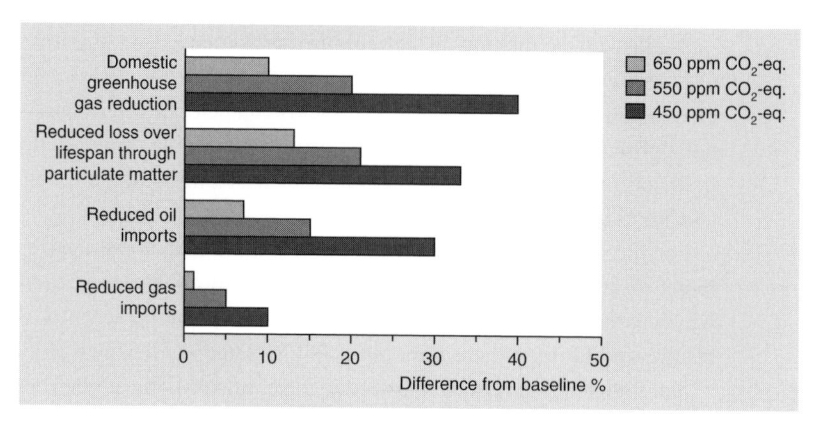

Figure 3.11 **Co-benefits of climate policy for air quality and energy security in 2030 for different stabilization scenarios. Improvements are shown as % of the baseline.**

Source: Van Vuuren et al. Stabilising greenhouse gas concentrations at low levels: an assessment of options and costs, Netherlands Environmental Assessment Agency, Report 500114002/2006.

impacts, because not every region has been studied enough and, more importantly, because climate change at regional and local scale cannot yet be predicted with certainty. In many areas it is, for instance, not yet known if the climate will get wetter or dryer. Climate models are not sophisticated enough yet.

But even knowing the impacts does not mean these impacts can be translated into costs. For things that have a market value, such as food, it is relatively simple: loss of production can be converted into a loss of income for the farmer. Lack of drinking water can be translated into costs by calculating, for instance, the costs of building pipelines to bring drinking water from other regions or of producing drinking water from sea water. Costs of building sea walls and dikes to protect against sea level rise can also be calculated. But how can a value be placed on the loss of land that can no longer be protected or abandonment of islands and relocation of people? When it comes to human disease or death or loss of species and ecosystems, it becomes even more problematic to attach a monetary value[18]. This is the first source of uncertainty and of differences in outcomes from different studies.

Then there is the choice of which impacts to include. Should low probability, high consequence events such as slowing of the ocean circulation or melting of the Greenland ice cap be included or not? And if so, how is the impact quantified? Is mass migration as a result of an area being no longer suitable for habitation covered? If climate change is going to be worse than the current best estimate, are the impacts of that evaluated or not? This is the second reason for the large uncertainties and differences between study outcomes.

A third major factor in the uncertainty of cost calculations is the so-called discount factor. This reflects the value attached to impacts in the future versus those happening today. In most economic calculations future costs are given a lower value, the argument being that future generations will have higher incomes and more options. Costs go down by a certain percentage per year they lie in the future. This is the discount rate. In fact a discount rate is the inverse of an interest rate on an investment made today. Just as a capital grows over time with a certain interest rate, so a future cost is reduced to a present cost with a certain discount rate. For costs of regular economic activities these discount rates can range from a few per cent to 10–15% or more.

What does this mean? For a discount rate of 5%, the cost counted today will only be less than 50% of the cost that is incurred 10 years into the future. This implies that costs of climate change impacts that may happen 50 or 100 years into the future in fact count as almost zero today. In such a situation, and certainly when it comes to impacts over a period of 100–200 years that may to some extent be irreversible and potentially catastrophic, such discount factors are widely seen as ethically unacceptable. Very low or even zero discount rates are then advocated for such situations. The Stern review[19] for instance chose a very low discount rate on exactly those grounds. However, there is no general consensus on the exact value of the different discount rates for such situations, explaining why outcomes of cost estimates can vary widely.

The fourth cause of uncertainty in calculations of cost of impacts is the relative weight. Do we simply add up costs or give them a weighting based on the size of the population that is affected? And do we weigh all costs equally, or are costs in poor countries or for poor people given more weight from a fairness point of view? The loss of a certain amount of money has a much bigger impact on a poor person than on a rich one. This is called equity weighting. In some calculations this is applied, in others not, which makes a big difference.

Notwithstanding this myriad of problems, calculations have been made. It will be no surprise that they span a wide range as a result of different assumptions and smaller or larger coverage of potential impacts. They are also very likely underestimating the real costs. Costs can be expressed in different ways: as a percentage of GDP or as the costs per tonne of CO_2 or CO_2-equivalent emitted today.

Many estimates express the total costs as a percentage loss of GDP, for a given degree of climate change. GDP is a measure of economic output. For a global average temperature increase of about 4°C, most estimates show a global average loss that varies from 1% to 5% of global GDP, with some studies going up to about 10% loss for about 6° C warming. Developing countries are facing higher than average losses[20]. Even a single catastrophic event, such as a major tropical cyclone, can cause enormous damage in poor countries. The drought in Southern Africa in 1991–1992 for instance caused a drop in income in Malawi of over 8%. Hurricane Mitch caused damages in Honduras totalling about $1250 per inhabitant, 50% more than the per capita annual income.[21]

The Stern review used similar numbers to express the damages from climate change impacts as referred to for the Honduras case, namely the loss of consumption or income per person. This gives a more direct idea of the economic impact as felt by people, because it does not include the economic output generated by 'clean-up' activities as a result of climate change impact damages. Stern came to much higher estimates of losses than most of the other estimates mentioned above, i.e. 5–20% of GDP for temperature increases of 7–9°C. This stronger warming assumption, which by the way is well within the range of estimates for the next 200 years or so, is of course one explanation. They also used a very low discount rate, applied equity weighting, and included the risk of much stronger climate change than the best estimate available today. These assumptions are not unrealistic however.

Costs can also be expressed in a different way by calculating the total future damages that are caused by 1 tonne of CO_2 emitted today and to discount these future costs to today. That produces the so-called 'social cost of carbon (SCC)'. Estimates for this SCC vary widely, for the reasons given above. Based on available studies the estimate is US\$5–95 per tonne of CO_2-equivalent emitted today. Some studies give lower or higher numbers. The advantage of using this SCC is that it can easily be compared with the costs of avoiding this amount of CO_2-equivalent emissions. Since most emission reduction technologies have a cost of less than US\$100 per tonne today, avoidance becomes attractive. For emissions in the future the SCC will be higher, because damages increase at higher greenhouse gas concentrations in the atmosphere. For a tonne emitted in 2030 for instance the SCC is estimated to be something like US\$10–190 per tonne of CO_2-equivalent. This is of the same order of magnitude or higher than the expected costs of drastic reductions of emissions, leading to stabilization at very low concentrations (of the order of US\$30–120/tCO_2-eq). This does not yet take into account the fact that the SCC is very likely underestimated because of the limitations of the current studies.

Risk management

How should all these factors be weighed up when deciding where to draw the line on climate change? The answer is 'risk management'. That means considering the risks of climate change impacts, how reduction of greenhouse gas emissions, increasing forest carbon reservoirs and adaptation could reduce those risks, what the costs and co-benefits of those actions are, and what policy actions would be needed to realize these actions. This is not a simple process.

There are basically two different approaches to this risk management problem:

- determine what a 'tolerable' risk of climate change impacts is (political judgement based on scientific evidence), determine how this can be achieved at the lowest possible costs, and then consider if this is practicable from a policy point of view
- do a cost–benefit analysis to compare the monetized climate change damages with the cost of taking action, ensuring the costs are not higher than the benefits

Political judgement: the EU's 2 degree target

The first approach has been chosen by the European Union. At the political level the European Union formulated its 'two degree target' in 1996. Based on the then available scientific information, as summarized in the IPCC's Second Assessment Report, the EU proposed a limit of 2°C above the pre-industrial level as the 'maximum tolerable level' of climate change for global use and adopted it as guidance for its own policies. It was reconfirmed at the highest level of heads of state and prime ministers of the EU Member

States in 2007[22]. This 2°C target has been the basis for the EU's negotiating position for the Kyoto Protocol, its unilateral policy, adopted in 2007, to reduce EU greenhouse gas emissions to 20% below 1990 levels and its position on a new agreement to follow the Kyoto Protocol (30% reduction below 1990 levels by 2020 for all industrialized countries). It has been endorsed by a few other countries and many non-governmental environmental organizations.

When setting the 2 degree target the EU kept an eye on the costs, co-benefits, and required policy for achieving this target (but not in the form of a cost–benefit analysis), although the available scientific and technical information was limited at the time. Since 1996 much more information has become available, which shows that staying below 2°C of warming compared to pre-industrial levels is going to be tough, although not impossible and not very costly. In fact no studies so far show specific reduction scenarios that achieve a lower temperature increase without early retirement of power plants and industrial installations.

Most countries responsible for the biggest share of global greenhouse gas emissions have been reluctant to state a long term goal for controlling greenhouse gases and climate change. Japan came the closest with its proposal of reducing global emissions to half their 2005 level by the year 2050[23]. This was subsequently endorsed by the G8 leaders in 2008 in Japan, but with a significant weakening: the base year was omitted[24]. The reasons behind that are that the formulation as proposed by Japan is significantly weaker than what the EU 2 degrees target requires (a 50–85% reduction compared to 1990). At the other end of the spectrum the USA was not even ready to subscribe to the Japanese proposal. More recently, leaders of the major economies have expressed support for a 2°C limit.

Cost–benefit comparison

When applying a traditional cost–benefit analysis, monetized costs of climate change impacts are compared with the costs of mitigation, adaptation, and co-benefits. Unfortunately, a reasonable estimate of the global costs of adaptation cannot be given, nor can the co-benefits be quantified. This leaves us with a comparison between the costs of impacts (without adaptation) and the cost of stabilizing greenhouse gas concentrations at specific levels. Even that comparison is problematic, particularly due to the huge uncertainty in the costs of the climate change damages (see above). When we take the lowest level of stabilization that was assessed by the IPCC (i.e. 445–490 ppm CO_2-eq) and we look at the cost of the last tonne avoided (the so-called marginal cost) in 2030, we see a range of something like US\$30–120/tCO_2-eq (see above). This is of the same order of magnitude as the damages of a tonne of greenhouse gases emitted, expressed as the 'social costs of carbon' (US\$10–190/tCO_2-eq, see above). In light of the underestimation of the cost of impacts and the co-benefits of mitigation action (positive, but not quantified), it seems to make sense to take aggressive action, because benefits are higher than the costs[25].

The Stern review came to the same conclusion, in a much more unambiguous way. They compared the costs of aggressive actions (1–2% of GDP) to the costs of the damages without controls (5–20% of GDP) and concluded that taking aggressive action is

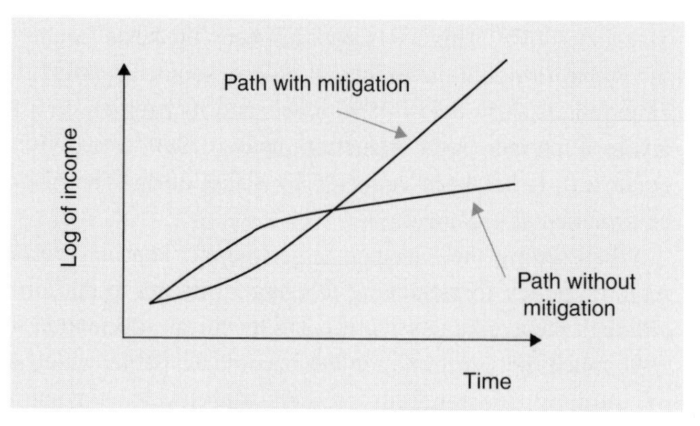

Figure 3.12 **Schematic drawing comparing economic growth paths for a situation with and without mitigation.**
Source: Stern review on the economics of climate change, Figure 2.3.

much cheaper than doing nothing. As indicated above the difference with the IPCC results comes from the assumptions the Stern review made on the discount rate, the inclusion of low probability, high consequence impacts and the equity weighting they applied[26]. The schematic drawing in Figure 3.12 illustrates nicely that in the short term mitigation costs lead to lower economic growth, but that this is compensated later by the negative impacts of climate change on the economy.

So what do we know now?

This chapter shows how the critical question on the appropriate level of stabilization of greenhouse gas concentrations in the atmosphere can be approached. It indicates that global average temperature increase can be limited to 2°C compared to pre-industrial and reasonable costs. This is only the case if aggressive action in the short term is taken. It also clarifies that many risks of climate change, particularly the most serious ones, can be avoided in this way. Adaptation remains important however, because at 2°C warming there will still be many negative impacts, affecting many people, particularly in developing countries. From a risk management perspective taking this aggressive action is justified.

Notes

1. See also Chapter 12.
2. See UNFCCC, http://www.unfccc.int.
3. The technical potentials of the various options cannot be simply added, because there could be overlaps and competition between some of the options for certain parts of the economy.
4. From climate objectives to emission reduction, Netherlands Environmental Assessment Agency, 2006, table 1, http://www.mnp.nl/en/publications/2006/FromClimateobjectives-toemissionsreduction.Insightsintotheopportunitiesformitigatingclimatechange.html.

5. IMAGE-TIMER 2.3; more information can be found in Van Vuuren et al. Stabilising greenhouse gas concentrations at low levels: an assessment of options and costs, Netherlands Environmental Assessment Agency, Report 500114002/2006.
6. IPCC Fourth Assessment Report, Working group II, ch 17.2.
7. http://www.verkeerenwaterstaat.nl/english/topics/water/water_and_the_future/water_vision/.
8. IPCC Fourth Assessment Report, Working Group II, box 17.1.
9. IPCC Fourth Assessment Report, Working Group II, ch 17.2.2.
10. IPCC Fourth Assessment Report, Working Group II, ch 17.4.2.
11. Confronting climate change: avoiding the unmanageable and managing the unavoidable, United Nations Foundation, 2007.
12. IPCC Fourth Assessment Report, Working Group III, ch 2.4.
13. IPCC Fourth Assessment Report, Working Group III, ch 3.3.5.3.
14. Agrawala S, Frankhauser S (eds) Economic Aspects of Adaptation to Climate Change, OECD, Paris, 2008.
15. See note 14.
16. IPCC Fourth Assessment Report, Working Group III, ch 11.8.
17. See also Chapter 10.
18. Economic methods for attaching a monetary value to a human life consider for instance the life-time 'earning power' (income over life-time). This automatically leads to a much higher value of a human life in rich countries than in poor countries. This raises serious ethical problems. An intense political debate on this issue started when the 1995 IPCC Second Assessment Report, Working Group III was released in which such calculations featured.
19. Stern review: the economics of climate change, chapter 2.
20. IPCC Fourth Assessment Report, Working Group II, ch 20.
21. Stern review, chapter 4.3.
22. European Council conclusions, March 2007.
23. See the 'Cool Earth 50' proposal as outlined in the speech of Prime Minister Abe, http://www.mofa.go.jp/policy/environment/warm/coolearth50/speech0705.html.
24. http://www.khou.com/sharedcontent/projectgreen/greenarticles/stories/070808kvueG8climate-cb.35fb6506.html.
25. IPCC Fourth Assessment Report, Working Group III, ch 3.5; the discussion there takes into account the wide range of estimates for damages in the literature and the full range of outcomes of stabilization studies. The conclusion of the IPCC did not take into account the co-benefits.
26. Stern review: the economics of climate change, ch 13.

4 Development first

What is covered in this chapter?

Development drives greenhouse gas emissions through increased consumption of fossil energy, increasing populations, industrial production, and increasing consumption. It also shapes the way societies are able to respond to climate change and other challenges. Climate change as caused by greenhouse gases emitted as a result of development can undermine that same development. This chapter looks at how to get out of this vicious circle. The solution can only be found if development objectives are seen as the starting point. Alleviating poverty and providing people with decent living conditions has to remain central. The way to get there can and must be changed. Integrating climate change into development decisions and making development more sustainable is the way to go. But what does that mean in practice and how easy is it to reconcile conflicting priorities? These are some of the issues that this chapter will investigate.

Development and climate change

Chapter 2 pointed out that greenhouse gas emissions are driven by development: population growth, economic development, technology choices, consumption patterns, and energy and land use. Building a society around unlimited availability of cheap fossil fuels, leading to poorly insulated buildings, gas-guzzling cars, and inefficient industries (what most countries have done so far), makes it very hard to adjust to a situation that requires efficient use of energy and low greenhouse gas emissions. A transport infrastructure centred on the car and urban sprawl can only be changed over a long period of time. Every day a wide array of investments in energy systems, buildings, factories, and infrastructure further shape a society and its greenhouse gas emissions.

Vulnerability to climate change also depends on development. As explained in Chapter 1, poor countries and poor people are the most vulnerable. They are more dependent on agriculture, which is most sensitive to changes in rainfall patterns and temperatures. Irrigation is not available in many places. Drinking water is scarce in large areas. There is little protection against flooding. Health services are inadequate. And worse, development is often making societies more vulnerable by building houses in flood plains, destroying mangrove forests that used to protect coasts against storm surges and hurricanes, and cutting down forests that retain water.

Rich countries can take measures to protect people against drought by building irrigation systems, against flooding by building dikes, and against hurricanes by building strong houses and providing shelter. They can also provide good health services to counter increased exposure to infectious diseases. That can be costly however. And that does not mean rich countries are not vulnerable. The enormous damage done to New Orleans by hurricane Katrina in 2005 shows that neglect of appropriate investments to deal with the risk of hurricanes and storm surges can have dire consequences.

So the relationship between development and climate change is a 'two-way street' (see Figure 4.1). Development is the driver and also the recipient of climate change.

Can or even should development policy be used in controlling climate change? Is dealing with climate change not distracting from or even endangering development? Industrialized countries developed without taking care of the environment and cleaned up when they could afford it. So why would developing countries develop differently?

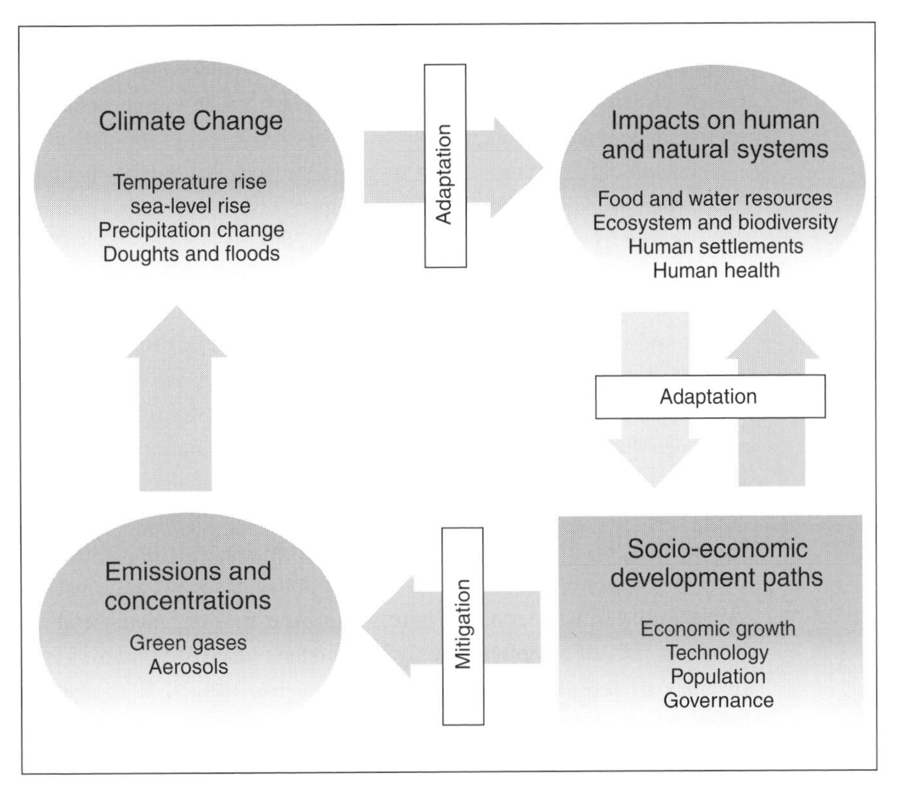

Figure 4.1 **Interrelationship between development and climate change. Development is both the driver and the recipient of climate change.**

Source: IPCC Third Assessment report, Synthesis Report, 2001.

What does climate change mean for development?

In Chapter 1 the impacts of climate change were discussed extensively. Water, ecosystems, food production, coastal areas, and human health all face impacts of climate change, leading to increasingly negative consequences as climate change progresses. Africa, Asia, and small island states are particularly vulnerable. These impacts affect development. In countries that rely heavily on agriculture changing rainfall patterns, higher temperatures, heat waves, droughts, and floods present a great threat to agricultural production and can negatively affect economic growth. Availability of drinking water, already problematic in some regions, will be further threatened, directly affecting the livelihoods of rural people. Sea level rise and increasing frequency of heavy rainfall will lead to more flooding, particularly in densely populated river deltas, negatively affecting social and economic development.

The basic elements of poverty eradication, as summarized in the Millennium Development Goals (see Table 4.1), will be much harder to reach with the climate change projected, particularly in the period beyond 2015 when the eradication of poverty should be finally achieved.

Impacts of climate change can go beyond specific damages that are repairable. It can become impossible for people to survive extreme drought or repeated flooding in which case abandoning their lands and migrating to safer grounds is the only way out. There are several potentially explosive areas in the world where that may happen, increasing the risk of violent conflict (see Figure 4.2). That brings climate change into the realm of international security concerns.

Development assistance given to poor countries is targeted to a considerable extent at sectors that are vulnerable to climate change. For instance, in Tanzania the percentage of all aid going to vulnerable sectors is about 20%, in Bangladesh 30%, and in Nepal about 60%[1]. Overall, the World Bank estimates that about one quarter of its loans portfolio is subject to significant risk from climate change[2].

Costs of climate change damages were discussed in Chapters 1 and 3, leading to the conclusion that extreme events like hurricanes or floods can have a disastrous impact on the economy of poor countries. Looking at the overall impact of gradual climate change it was concluded that losses of income of 5–20% in poor countries can be expected for 7–9°C of warming, something that is well within the range of uncontrolled warming if we look beyond the year 2100. The conclusion must be that climate change can really undermine development, so it cannot be neglected when making the main socio-economic decisions.

As far as controlling emissions of greenhouse gases is concerned developing countries cannot afford to wait either. It is true that industrialized countries have been responsible for about 70% of all greenhouse gas emissions since 1850 and therefore are responsible for most of the climate change we are seeing today[3]. It is also true that per capita emissions in developing countries are still much lower than in industrialized countries. But without a policy on climate change two-thirds to three-quarters of all additional emissions for the period until 2030 will come from developing countries and for the period thereafter the share will be even higher. Or, to put it in different terms, even if industrialized countries were to reduce their emissions to zero, emissions from developing countries would be too high to keep climate change to tolerable levels.

Table 4.1. Potential impacts of climate change on Millennium Development Goals

Millennium Development Goals: Climate Changes as a Cross-Cutting Issue	
Millennium Development Goal	Examples of Links with Climate Change
Eradicate extreme poverty and hunger **(Goal 1)**	• Climate change is projected to reduce poor people's livelihood assets, for example, health, access to water, homes, and infrastructure • Climate change is expected to alter the path and rate of economic growth due to changes in natural systems and resources, infrastructure, and labour productivity. A reduction in economic growth directly impacts poverty through reduced income opportunities • Climate change is projected to alter regional food security. In particular in Africa, food security is expected to worsen
Health related goals: • combat major diseases • reduce infant mortality • improve maternal health **(Goals 4, 5, and 6)**	• Direct effects of climate change include increases in heat related mortality and illness associated with heat waves (which may be balanced by less winter cold related deaths in some regions) • Climate change may increase the prevalence of some vector borne diseases (for example malaria and dengue fever), and vulnerability to water, food, or person-to-person borne diseases (for example cholera and dysentery) • Children and pregnant women are particularly susceptible to vector and water borne diseases. Anemia – resulting from malaria – is responsible for a quarter of maternal mortality • Climate change will likely result in declining quantity and quality of drinking water, which is a prerequisite for good health, and exacerbate malnutrition – an important source of ill health among children – by reducing natural resource productivity and threatening food security, particularly in Sub-Saharan Africa
Achieve universal primary education **(Goal 2)**	• Links to climate change are less direct, but loss of livelihood assets (social, natural, physical, human, and financial capital) may reduce opportunities for full-time education in numerous ways. Natural disasters and drought reduce children's available time (which may be diverted to household tasks), while displacement and migration can reduce access to education opportunities
Promote gender equality and empower women **(Goal 3)**	• Climate change is expected to exacerbate current gender inequalities. Depletion of natural resources and decreasing agricultural productivity may place additional burdens on women's health and reduce time available to participate in decision making processes and income generating activities • Climate related disasters have been found to impact more severely on female-headed households, particularly where they have fewer assets to start with

Table 4.1. (*cont.*)

Millennium Development Goals: Climate Changes as a Cross-Cutting Issue

Millennium Development Goal	Examples of Links with Climate Change
Ensure environmental sustain ability **(Goal 7)**	• Climate change will alter the quality and productivity of natural resources and ecosystems, some of which may be irreversibly damaged, and these changes may also decrease biological diversity and compound existing environmental degradation
Global partnerships	• Global climate change is a global issue and response requires global cooperation, especially to help developing countries to adapt to the adverse impacts of climate change

Source: African Development Bank Asian Development Bank, Department of International Development, UK; Directorate-General for Development, European Commission; Federal Ministry of Economic Cooperation and Development, Germany; Ministry of Foreign Affairs, Development Cooperation, The Netherlands; OECD; UNDP; UNEP, World Bank. Poverty and Climate Change – Reducing the Vulnerability of the Poor through Adaptation, 2004.

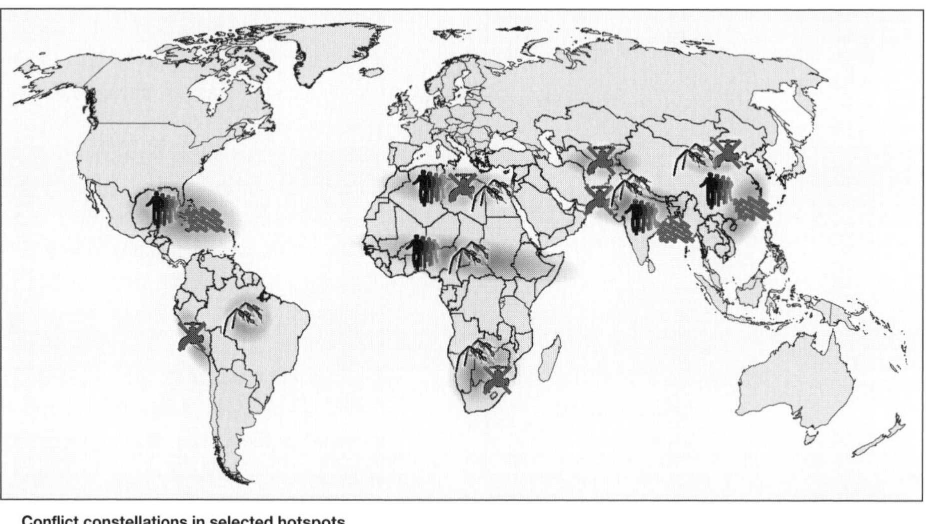

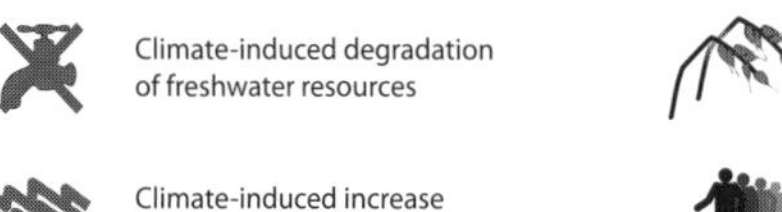

Figure 4.2 Potential areas where violent conflicts could emerge as a result of climate change.
Source: German Advisory Council on Global Change, World in Transition: Climate Change as a Security Risk. Summary for Policy-makers. Berlin, 2007. See Plate 9 for colour version.

Making development more sustainable

The answer to the threat of climate change and to many other threats, such as losing biodiversity, natural resource depletion, and extreme poverty, is to make development more sustainable. The notion of sustainable development was put on the political agenda by the Brundlandt Commission in its report 'Our Common Future' in 1987. They defined sustainable development as *'development that meets the needs of the present without compromising the ability of future generations to meet their own needs'*. But how can this be made operational? The most common interpretation is that there are three dimensions of development, i.e. economic, social, and environmental, that need to be in harmony. Often the institutional dimension (governance structures, democratic institutions, etc.) is added as 'cement' between the three pillars (see Figure 4.3).

This conceptual description of sustainable development does not answer the question of how much of each dimension needs to be there to be sustainable. Can progress on one dimension compensate lack of progress in another? Or are there minimum levels of progress for each of the dimensions to make development sustainable? And how do you measure sustainability? No unequivocal answers to these questions exist. There are different schools of thought: different definitions of 'sustainable' and different sets of sustainability indicators have been developed[4]. For the purposes of this book a practical approach is chosen that has three elements:

1. *Relation between the sustainability dimensions*: each dimension of sustainable development needs to be satisfied to such an extent that undermining the other dimensions is avoided. For example, climate change impacts need to be limited to a level that allows preservation of a healthy environment (look at the discussion in Chapter 3 and the section above on climate change impacts). It should also not endanger eradication of poverty and should guarantee a positive economic development (defined in a broad sense, including

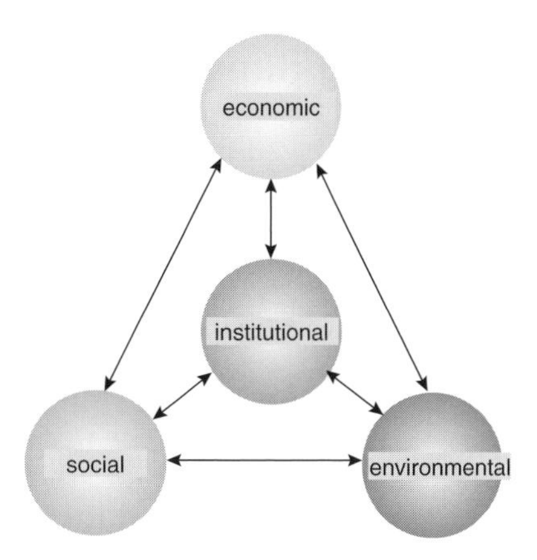

Figure 4.3 **The dimensions of sustainable development. The three main dimensions are shown as the corners of the triangle. Institutions form the 'cement' between the three pillars.**

Source: Munasinghe M. Making development more sustainable: sustainomics framework and practical applications, MIND Press, Munasinghe Institute for Development, Colombo, 2007.

the value of natural resources). This means there are minimum demands for each of the three main dimensions for a sustainable situation. Compensation can only happen over and above this minimum standard. The determination of the minimum standard for each dimension is a political, not a scientific issue.

2. *Indicators and metrics*: these are chosen so that they are appropriate for each particular sustainability dimension. For climate change impacts that could for instance be the number of people with access to sufficient water or the number of people affected by floods. No attempt is made to translate everything into monetary terms (so-called cost–benefit analysis, although this approach is discussed in Chapter 3).
3. *Operational approach*: the best practical way to make development more sustainable is through integration of all relevant elements of sustainability into development decisions. This is called 'mainstreaming'.

Sustainable development is of course not only connected to climate change. There are many other global problems that tend to undermine development: poverty, lack of social security, loss of biodiversity, overexploitation of oceans, loss of tropical forests, large scale air pollution, damage to the ozone layer, etc. In fact, many of these problems have causes similar to those of climate change. They are the result of an inherently unsustainable pattern of production and consumption. Taken together, they are a huge threat to our planet. Trying to make development more sustainable can therefore help address a whole series of problems if a broad and integrated policy approach is chosen.

An important notion is that meaningful discussions about the direction of development can only be held at a local level, i.e. at the scale of a city, region, or country. That is where decisions are taken and impacts are felt. And making such development decisions more sustainable brings climate change (and other global problems) to the heart of the political process. International progress towards more sustainable development is by definition the sum of local actions.

Mainstreaming climate change in development policies

How would mainstreaming work when it comes to specific development issues? Are there real synergies to be found between economic and social development goals and dealing with climate change? Or are there serious tradeoffs to be made? To find that out we will consider a number of development issues in more detail. In doing so we will look at the two aspects of climate change: possibilities for development towards a low carbon economy and possibilities for development of a society that is more resilient (i.e. less vulnerable) to climate change.

Modernizing industry to become competitive

For the manufacture of steel, aluminium, fertilizer, chemicals, paper, cement, glass, and ceramics and the refining of oil, large amounts of energy are needed. For example, an oil refinery uses about 15–20% of the crude oil in the process of producing oil products. Energy is responsible for about 15% of the costs of a tonne of steel[5]. So energy is an important component of production costs. With oil prices above US$100 per barrel and gas and coal prices sharply increased, energy costs have gone up enormously. Since these industries are generally competing on the world market, cost matters a lot. Energy efficiency improvement is thus a vital strategy for these industries to become or remain competitive (see Box 4.1).

Box 4.1

Energy efficiency in the fertilizer industry

Ammonia is the most important component of nitrogen fertilizers. Its production requires large amounts of fossil fuels. The majority of ammonia plants in the world use natural gas as the raw material. The production costs of ammonia are very sensitive to the costs of natural gas. An increase of the natural gas price by a factor of 2.5, which happened over the past few years, doubles the production costs. That makes energy efficiency a very important issue, and is why energy use per tonne of ammonia produced has gone down substantially over the years. A plant designed in 2005 uses less than half the energy per tonneof ammonia produced as one designed around 1960 (see figure). The newest plants therefore have the best efficiency and many of those are located in developing countries.

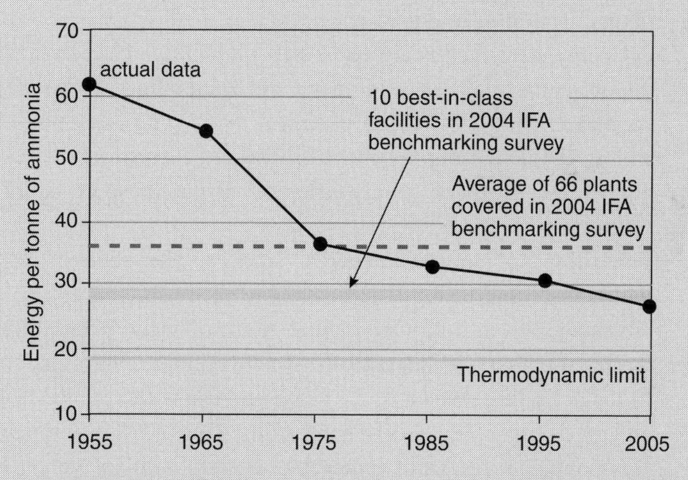

Energy consumption per tonne of ammonia for manufacturing plants around the world, designed in a particular year. The figure shows a continuous improvement of energy efficiency over time. The thermodynamic limit is the lowest theoretical energy consumption for this process.

(Source: IPCC Fourth Assessment Report, Working Group III, Chapter 7.4.3.2)

This is not just an issue for industries in industrialized countries, on the contrary. In 2003 42% of steel, 50% of aluminium, 57% of fertilizer production, and 78% of the cement industry was located in developing countries[6]. To be honest, these industries have moved there not only to serve the growing domestic markets in developing countries, but also because of closeness to raw materials and availability of cheap energy and labour. For example, after the 1980s oil crisis Japanese industries moved many of their energy intensive production processes overseas in order to escape high energy costs, supply security problems, and environmental legislation. In the current globalized economy relocation of plants is a common phenomenon. And energy costs are an important consideration. When greenhouse gas emission controls are added to these energy costs it could become attractive to move to countries where energy costs are lower and emission controls less stringent. This industrial relocation could thus lead to a deterioration of overall energy use and CO_2 emissions and act as a possible countermovement to modernization and energy efficiency improvement.

Improving energy security and reducing oil imports

Security of energy supply is a major concern of many countries these days. Oil imports are an increasing burden for the economy of many nations. Conflicts between Russia and neighbouring countries on contracts for gas supply have made importing countries realize their vulnerability when they are dependent on one country for a major part of their energy sources. There are in principle three different responses to this problem of energy security: more efficient use of energy, shifting to domestic energy resources, and diversification of imported energy.

Energy efficiency is always a win–win option. The lower the use of energy, the lower the dependence on energy imports and vulnerability to interruptions in supply and price increases. This is in fact in perfect synergy with reduction of CO_2 emissions, because energy efficiency improvement has low or even negative costs (i.e. is profitable). (See also the discussion on energy efficiency in the transport, buildings, and industry sector in Chapters 6, 7, and 8.)

Shifting to domestic energy sources, the second response strategy, becomes particularly interesting when these are renewable energy sources that have immediate advantages from a climate change point of view. Availability of renewable resources varies of course considerably between countries, but most countries have the potential to replace a large part of their fossil fuel use with renewable energy, given enough time and given cost reductions of some of the more advanced renewable energy technologies. Developing a domestic renewable energy industry can also create new jobs[7]. The Brazilian alcohol programme is a good example of a successful implementation of this strategy (see Box 4.2).

Box 4.2 **Fuel alcohol in Brazil**

The oil crisis of 1973 had a bad economic impact on Brazil: oil imports were rising to about half the value of all exports. In addition to launching a big exploration programme for oil and gas, the Brazilian government started an ambitious programme of producing alcohol from sugar cane. This was made relatively easy by low sugar prices on the world market at the time. Alcohol was sold in two forms: as a mandatory 20–25% blend in all regular gasoline and as pure alcohol. The blended product could be used by all cars with only minor modification. For pure alcohol use, engines needed more radical adjustment.

Through a combination of (gradually declining) subsidies to sugar cane growers, ethanol ready cars, and gas stations, a large scale ethanol supply system was built. Due to low gasoline prices after the oil crisis was over and fluctuations in the supply of alcohol, consumption of pure alcohol had its ups and downs. Only after the automobile manufacturers offered so-called Flex Fuel Vehicles (FFVs, with engines that automatically adjust to the gasoline–ethanol mixture available) in the late 1990s, did the market stabilize. In 2006 more than 80% of vehicles sold in Brazil were FFVs at an additional cost of about US$100 per vehicle. At the same time production costs of alcohol had gone down from about US $100/barrel of oil equivalent to about US$40 and in 2005 costs had further fallen to about US$30. Alcohol is now cheaper than gasoline.

The alcohol programme generated a large number of jobs in the sugar cane industry and the processing and distribution of alcohol. Oil import savings over the period 1975–2002 amounted to more than US$50 billion, while additional investments and subsidies are only a fraction of that. All of this was done without considering the reductions in CO_2 emissions.

(Source: IPCC Fourth Assessment Report, Working Group III, chapter 5.3.1.3; Goldemberg J. The ethanol programme in Brazil. Environmental Research Letters, vol 1 (October–December 2006) 014008; Kahn Ribeiro S, Andrade de Abreu A. Climate Policy, vol 8 (2008), pp 220–240)

In several countries, such as India, China, and the USA, coal resources are abundant and energy security concerns can easily lead (and actually have already led) to a shift to domestic coal, away from imported oil or gas. Coal is a cheap source for electricity production, but can also be used to produce gasoline based on the so-called Fisher Tropsch processes[8]. This coal-to-liquid process doubles the CO_2 emission per unit of fuel. The use of domestic coal thus provides a difficult trade-off problem with climate change and air pollution. The only way to reconcile such an energy security decision with controlling greenhouse gas emissions is to rely heavily on CO_2 capture and storage[9].

Diversification of energy imports has been a long-standing strategy of many countries to enhance energy security. Again, this often leads to considerable use of coal for reasons of costs (coal prices have gone up but are substantially lower than gas prices) and spreading the suppliers. A shift to nuclear power, also beneficial from a CO_2 point of view[10], would help to diversify energy sources, but there are only a few countries that have uranium resources. The more domestic or imported renewable energy can be developed the better the possibilities to implement diversification policies in synergy with CO_2 reduction strategies.

The response of China to deal with energy security, industrial competitiveness, and air pollution is shown in Box 4.3.

Box 4.3 China's sustainable development policies on energy

China has an extensive set of policies in place to deal with energy security, industrial competitiveness, air pollution, and greenhouse gas emissions:

- Shift to a less energy intensive economy: by promoting the growth of high tech manufacturing and services
- Energy efficiency improvement: in power supply (a 12% reduction in carbon intensity of electricity between 1990 and 2004; see note 1) by closing small inefficient coal fired power plants; in energy intensive industries (a 30% reduction in energy per tonne of steel and 20% per tonne of cement between 1990 and 2004; in buildings (energy efficiency standards for new buildings)
- Over the period 1991–2005 energy use per unit of GDP was reduced by 45%, double the improvement of the world on average*
- Expansion of energy efficiency programmes, such as technology modernization in the iron and steel, metals, oil and petrochemical, and building materials industries; stronger efficiency standards for buildings (Green lighting programmes, government buildings programme); introduction of fuel efficiency standards for new vehicles and removal of old inefficient ones
- Expansion of renewable energy capacity: by the end of 2005 there was 117GW hydropower capacity (23% of electric power capacity), 2GW biomass based power, 1.3GW wind power, 70MW solar photovoltaic (in remote regions), 1 million tonne per year in fuel alcohol, about 40 million solar water heaters, and 17 million households with biogas units (in addition to 1500 larger ones). Targets for expansion of renewable energy are: 10% of primary energy from renewable sources by 2010, 15% by 2020; 20GW wind power by 2010, and strong increase in number of biogas units and solar water heaters, solar PV, and biofuels
- Expansion of nuclear power and coal bed methane recovery
- Expansion of forest area: increase of forest cover from 14 to 18% between 1990 and 2005. Planned extension to 20% in 2010
- Banning of coal burning in a number of urban areas to address air pollution
- The overall target for improved energy efficiency of the economy is a 20% reduction of energy use per unit of GDP between 2005 and 2010.

The estimate of the Chinese government is that these programmes together will reduce CO_2 emissions by more than 1Gtonne per year (1000 million tonnes) in 2010, compared to the business as usual projection of around 7Gtonne per year. This is still a big increase compared to 2000, but also about a 15% reduction compared to business as usual.

* Energy and carbon intensity of electricity has increased between 2004 and 2007.

(Source: China National Climate Change Programme, National Development and Reform Commission, June 2007; 11th 5-year Plan, National Development and Reform Commission, Gao Guangsheng, Policies and measures of China under the framework of sustainable development, presentation at 2nd Dialogue Session, UNFCCC COP-12, Nairobi, 2006)

Table 4.2. Ranking of cities according to their liveability (December 2006)

Best			Worst		
Rank		Liveability (%)*	Rank		Liveability (%)*
1	Vancouver	1.3	132	Algiers	64.7
2	Melbourne	1.8	131	Dhaka	60.4
3	Vienna	2.3	130	Lagos	60.1
4	Perth	2.5	129	Karachi	58.6
5	Toronto	3.0	128	Kathmandu	54.7
6	Adelaide	3.0	127	Abidjan	53.9
7	Sydney	3.2	126	Dakar	53.2
8	Copenhagen	3.7	125	Phnom Penh	53.0
9	Geneva	3.9	124	Tehran	52.6
10	Zurich	3.9	123	Bogota	48.3

* Weighted index rating whereby 0% = exceptional quality of life and 100% = intolerable.
Source: Economist, August 22, 2007.

Providing efficient transport for people

Adequate transport is a basic need of modern societies and a prerequisite for development. Traffic congestion, health impacts from air pollution, and rising oil imports are however inherent to transport in many countries. Making transport more energy efficient, replacing fossil fuels, and shifting from private to public transport are effective ways to address these problems and at the same time reduce CO_2 emissions. Building more roads, a popular response to traffic congestion, is the only strategy that is not synergistic with CO_2 reduction, and even in terms of combating congestion it is not effective because it generally increases traffic and results in congestion in other places.

At the city level providing efficient public transport, careful urban planning, limiting car access, and creating safe walking and bicycling spaces and facilities can go a long way to create a 'liveable city', something that is highly appreciated by people. In the real world there are forces driving cities in the opposite direction. Think of the ever increasing tendency to locate large shopping centres in the outskirts of cities, generating a large transportation demand; freight transport and warehouse hubs for the distribution of industrial goods and retail products; and suburban (often uncontrolled) residential development in search of affordable housing, with inadequate public transport infrastructure. Controlling such developments requires strong municipal governments and adequate land use planning legislation. Huge differences exist in liveability of cities in developed and developing countries (see Table 4.2). With the fast growth in many cities in developing countries investments in urban planning and good public transport facilities will allow these cities to become more liveable and low carbon in the future.

Improving air quality to protect health

Air pollution in many countries and particularly in urban and industrialized areas is a growing problem in developing countries, with important health consequences and negative impacts on food production. More than 700000 people die prematurely every year as a result of urban air pollution (see Figure 4.4). Cities with small particle air pollution at least 50% above the World Health Organization guidelines are listed in Figure 4.5. Most of these cities are in developing countries. Successful abatement of air pollution in many industrialized countries was the result of eliminating pollution from coal fired power plants (sulphur oxides, nitrogen oxides, soot, and other small particles) and reducing traffic related emissions (nitrogen oxides, soot, and small particles from diesel fuel). So there are ample opportunities for achieving synergies between improving health by reducing air pollution and reducing CO_2 emissions from cleaner fuels and more energy efficiency.

A good example of a win–win strategy for reducing air pollution and CO_2 emissions is the Delhi natural gas-for-transport programme. The main cause of air pollution in Delhi is traffic. Since the early 1990s measures were taken to tighten emission standards for vehicles and improving fuel quality. But in 1998, as a result of a court case filed by an environmental activist, the Supreme Court ordered a complete replacement of diesel and gasoline by compressed natural gas for motorized rickshaws, taxis, and buses. This resulted in a significant improvement of air quality, in particular in terms of fine particles, one of the most dangerous components of polluted air. It also resulted, as a co-benefit, in reduced emissions of CO_2[11]. Other examples of win–win strategies in the transport sector are so-called Bus Rapid Transit Systems that have successfully changed the liveability of several South American and other cities (see Chapter 6).

Air pollution is affecting crop productivity (see Figure 4.6), which is also negatively influenced by climate change. So there is also a synergy between abatement of air pollution, food security, and addressing climate change.

Ensuring a strong agriculture and forestry sector

The forestry and agriculture sector are economically very important in many developing countries, in terms of food, fodder, and forest products as well as for the livelihoods of the people dependent on it. They also provide essential ecological functions through nature and biodiversity protection and water management. Policies aimed at these core development issues can have positive or negative impacts on greenhouse gas emissions and will determine vulnerability to future climate change.

Food security is one of the most important concerns of governments. Existing poverty in many countries means that many people are vulnerable to food scarcity and increasing food prices. Social unrest and even riots in times of rising food prices have happened frequently. Policies to enhance food security are manifold. Extending cropland and grassland, increasing crop productivity, efficient irrigation methods, erosion control, improved pest and fertilizer management, and creating incentives for farmers are all

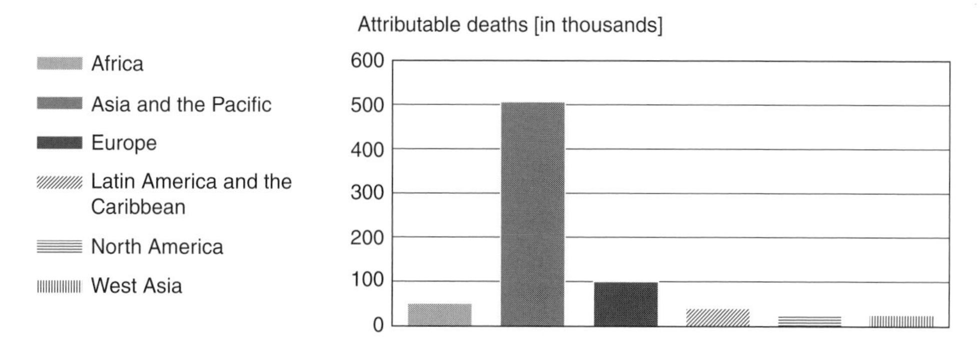

Figure 4.4 **Premature deaths due to outdoor urban exposure to small particle air pollution in the year 2000.**
Source: UNEP, Global Environmental Outlook, 2007, ch 2.

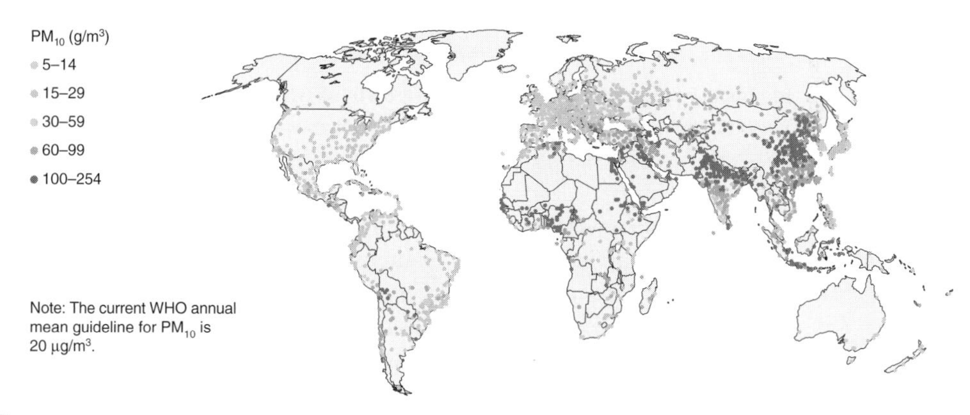

Figure 4.5 **Cities with annual mean concentrations of small particles (PM10) at least 50% above the current WHO air quality guideline.**
Source: Cohen, A. J., Anderson H. R., Ostro B. *et al.*, Mortality impacts of urban air pollution. In: Comparative Quantification of Health Risks: Global and Regional Burden of Disease Due to Selected Major Risk Factors, eds. M. Ezzati, AD Lopez, A. Rodgers, CJL Murray, vol. 2. World Health Organization, Geneva, 2004. p. 1374. See Plate 11 for colour version.

important policy objectives. Most but not all of these policies have synergies with reducing greenhouse gas emissions: higher crop productivity reduces the need for land and reduces deforestation, minimizing erosion through low tillage methods, retention of crop residues and agro forestry help to accumulate carbon in soils, and lower fertilizer use reduces emissions of nitrous oxide.

Several of these policies also help to reduce vulnerability to climate change, a very important issue for countries heavily dependent on agriculture. Increased soil carbon and agro forestry improve the water holding capacity of farmland, making it less vulnerable to droughts. Efficient irrigation makes it possible to continue farming if the climate gets dryer. Better erosion control helps to deal with extreme rainfall, something that is going to happen more often in a changed climate.

However, there are also conflicting policy objectives. Increasing populations and shifts to diets with more meat as incomes rise increase the need for cropland and grassland,

Figure 4.6 **The impact of local air pollution on the growth of wheat in suburban areas of Lahore, Pakistan.** *Source*: photo by A. Wahid. Published in *Global Environment Outlook-4: Environment for Development*, UNEP, 2007.

leading to land conversion and increased CO_2 emissions. Reducing subsidies for farmers (abundant in many industrialized countries) would reduce greenhouse gas emissions, but could also lead to a decrease in food exports, which might harm global food security. Creating new income for farmers through bioenergy crops is attractive, but could easily create tensions with food production and biodiversity protection[12].

Reducing deforestation and planting new forests to preserve ecological functions and water management have large positive impacts on retaining the stocks of CO_2 in trees and soil. They are also important in making forests resilient to climate change, by retaining groundwater and generating rain. A sustainable forestry industry also benefits from maintaining healthy forests. The pressure to convert forests to cropland or pastures is high however and the short term economic benefits of turning forests into cropland or grassland are big. In addition, there are often unexpected connections with other social and economic policies. An example is the devaluation of the Brazilian currency in 1999 by 50% against the US dollar. Together with the increase in soybean prices on the world market, it made it so attractive to convert forests into land for soybeans and meat production for export that this led to massive deforestation in the state of Mato Grosso (about one third of all deforestation in Brazil between 1999 and 2003[13]).

So potential synergies between development and sustainable development policies and climate protection in forestry and agriculture are abundant, but trade-offs between different policy objectives are inevitable[14].

General macro-economic policy

Taxes, subsidies, currency exchange rates, policies to stimulate economic growth, and trade policies seemingly have little to do with greenhouse gas emissions. This is not the case. On the contrary, taxes and subsidies determine energy, raw materials, and pollution costs and the cost of labour. 'Greening' taxes, i.e. shifting taxes to polluting activities and lowering them on labour and environmentally friendly activities, is being pursued in many countries now, maintaining the necessary tax base but helping to reduce greenhouse gas emissions. Eliminating energy subsidies, 95% of which are for fossil or nuclear energy (worth about US$250 billion per year globally) would lead to stronger economic growth of 0.1% on average per year and a reduction of CO_2 emissions of more than 6%[15]. Of course, caution is needed to avoid unwanted side effects and compensating measures for poor people may be needed when removing subsidies.

Trade policies have important implications for greenhouse gas emissions as well. In a globalizing economy production moves to the place where it is cheapest. Trade laws allow countries to ban harmful products as long as these rules also apply to domestically made products. Trade rules however make it difficult for importing countries to demand clean (i.e. low carbon emitting) production processes for the products they import. So-called border tax adjustments (for instance adding a tax to a product made with cheap energy because no CO_2 abatement is applied in the exporting country) are therefore currently not applied. This means that large scale shifts of export oriented production to developing countries, such as China, where production facilities are less energy efficient and the electricity fuel mix has a higher share of coal, increases global greenhouse gas emissions. For example, the current USA–China trade pattern (much greater import from China than export to China) has led to an increase of CO_2 emissions for the two countries together of about 100 million tonnes CO_2 per year[16] (equivalent to the total annual emissions of a country like the Philippines).

Existing import duties on climate friendly products are a barrier to wider use of low carbon technologies. A good example is the import duties on Brazilian sugar cane alcohol in the USA and Europe to protect domestic production of biofuels, even though Brazilian alcohol has a much greater benefit in terms of CO_2 emissions than the biofuels produced in the USA and Europe and is lower in price. Changes in trade laws can create incentives for low carbon products.

What about providing people with modern energy?

There are still 2.4 billion people that rely on wood, crop residues, charcoal, and animal dung for their energy needs and 1.6 billion people do not have access to electricity. Most of these people live in rural areas of developing countries. As a consequence they suffer from serious indoor air pollution caused by smoke, leading to acute respiratory infections in children and chronic lung disease in adults. These health problems are responsible for nearly all of the 1.6 million deaths each year from indoor air pollution[17], 98% of which happens in developing countries[18].

Providing these people with modern energy in the form of liquefied petroleum gas (LPG), kerosene, biogas, and electricity would reduce indoor air pollution by 95% and clearly has huge advantages. In doing so CO_2 emissions will inevitably go up (modestly), but that cannot be an argument not to improve the health conditions of course. If 2 billion people were provided with LPG for cooking, lighting, and heating global CO_2 emissions would go up by about 2%[19]. In practice the increase can be much lower. Biogas is an option in many rural areas to reduce CO_2 emissions considerably without adding to net CO_2 emissions (see Chapter 5). And since part of the wood and charcoal used currently is not produced sustainably, traditional biomass energy does have net CO_2 emissions. Experience from an LPG programme in Senegal showed a considerable drop in charcoal production that counters the CO_2 emissions from the LPG.

As far as electricity supply is concerned, it is possible to extend the grid so that it reaches many of the 1.6 billion people currently without electricity. But in many rural areas grid extension may take a very long time to materialize. Developing local mini grids, using a combination of renewable energy sources (small hydro, wind, biomass, solar), is a viable alternative in such areas. It is a better option than individual home solar systems, which limit the owners to small amounts of electricity and may make the village less attractive for grid extension (see Chapter 5 for more details).

Developing coastal regions while retaining natural coastal protection and ecologically valuable areas

Mangrove forests provide a natural protection against storm surges and hurricanes. In addition, they are hugely important as breeding grounds and shelter areas for fish and other sea life. So retaining them will make the system much more resilient against climate change and the resulting increase in sea level rise, hurricane intensities, and storm surges. They will also make fish populations less vulnerable and help ecosystems to survive under very different climatic conditions. Retaining or replanting mangrove forests will also contribute to maintaining or increasing reservoirs of carbon.

Building a good public health system

A good public health system is of course of prime importance for better living conditions and it is a prerequisite for sound social and economic development. Climate change will bring new threats from contagious diseases. A good health system, good sanitary systems, and controlling disease vectors (e.g. mosquitoes transmitting malaria) are extremely important to adapt to a new climate.

Nature and biodiversity protection

Nature and biodiversity protection has become an integral part of development. Ecological goods and services are an essential input in sound social and economic development. They are already under enormous stress due to the strong reduction in suitable areas for important ecosystems (see Figure 4.7). Species and ecosystems will be subject to increasing stresses in

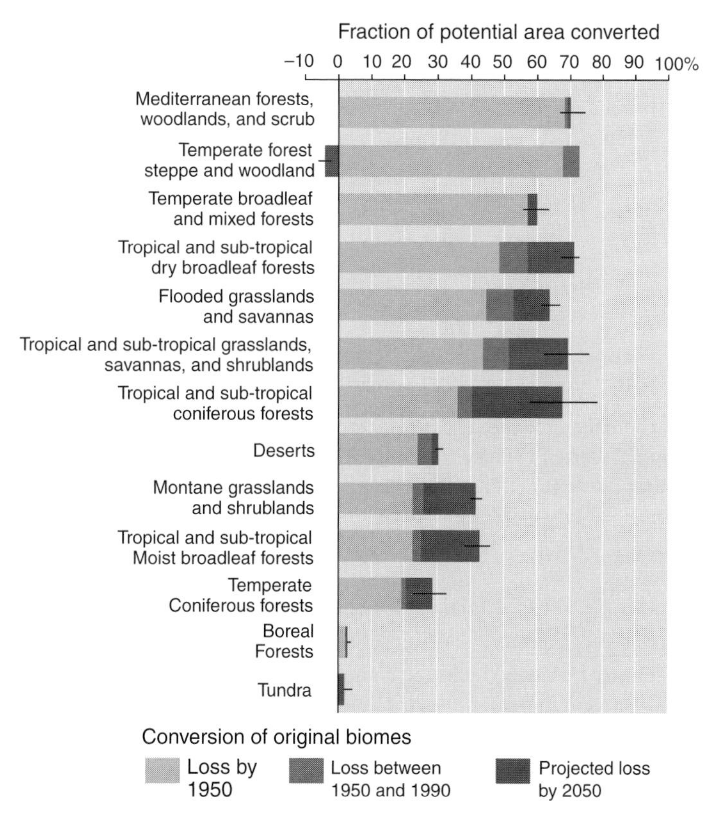

Figure 4.7 **Reduction in area covered by important ecosystems between 1950 and 1990 and projected further loss till 2050.**

Source: Millennium Ecosystem Assessment, 2005.

a changed climate and large numbers of species are threatened with extinction with further climate change. Conditions with respect to temperature and precipitation will no longer support species in a specific location. In practice many species are locked in certain locations by their required habitats and natural and man-made obstacles. Protected areas are often surrounded by cultivated and populated lands and cannot easily be moved. Different concepts and approaches are needed for nature and biodiversity protection in light of climate change. One of the most important is merging smaller ecologically valuable areas with larger ones and providing corridors between protected areas, so that species can more easily find appropriate conditions for survival. Protecting natural vegetation from decay will also help to retain carbon reservoirs.

Socio-economic development

The potential for sustainable development to deliver a significant contribution to controlling greenhouse gas emissions and controlling climate change can be illustrated

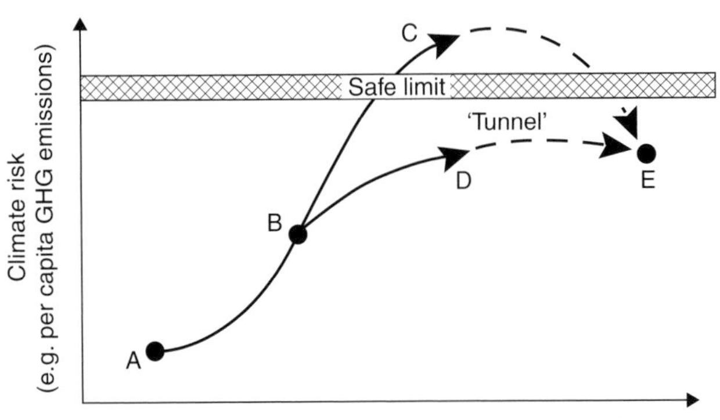

Figure 4.8 **Schematic drawing of development paths and resulting greenhouse gas emissions. The industrialized country trajectory A-B-C-E is not the model for currently developing countries that can follow trajectory A-B-D-E.**

Source: Munasinghe M. Making growth more sustainable, *Ecological Economics*, 1995, 15:121–124.

with the IPCC scenarios for socio-economic development and expected climate change over this century. As Figure 1.8 shows, the difference in temperature at the end of the century between the highest and the lowest scenario is about 2°C. And that is the difference between a sustainable development scenario (IPCC SRES B1) and a high growth fossil fuel based scenario (IPCC SRES A1FI). Both scenarios do not assume any specific climate policy. So the 2°C difference is purely the result of different development paths. It is as much or even more than what can be achieved with specific climate policy.

When you compare the way countries have developed in the past, a strong increase in emissions with increasing development is the dominant picture. However, that does not mean countries whose development has taken off only recently will have to follow the same path. Knowledge is now available of how to avoid serious social and environmental problems. Modern technologies are now widely available to produce goods and services with much lower greenhouse gas emissions. So newly developing countries should be able to skip the stage of high emissions and serious environmental problems by finding a 'short cut' or 'tunnel' to a modern low carbon society. Figure 4.8 shows this schematically.

Does this harm economic growth?

There is a generally held belief that sustainable development policies lower economic growth. This is often used as an argument to continue development in an unsustainable way. Short term economic growth means more jobs, more goods, improvement of living conditions for those participating in the market economy, and generally more money for governments to invest in education, health care, and infrastructure. The long term effects are often not taken into account.

Climate change damages as a result of unsustainable development have negative effects on economic growth, particularly in the long term. As discussed in Chapter 3 they can become so large (of the order of 5–20% of GDP) that they can really undermine development. So a somewhat slower economic growth in the short term caused by investments in a low carbon and climate resilient economy is more than compensated by the gains in the long term.

What do we know about the economic impacts of climate change mitigation action? The general findings are that average global economic growth rates, even allowing for an ambitious climate policy to control global average temperature change to something like 2°C above the pre-industrial period, are only marginally affected. It means lowering annual economic growth rates by no more than 0.1–0.2 percentage points, but often less, depending on the specific policies applied. We also know that new industries, such as those for producing wind turbines, solar panels, energy efficient machines, and equipment can bring many new jobs. In Germany about 250000 jobs in the renewable energy industry have been created in about 10 years. This number is already higher than that for all of the jobs in the coal mining sector. A study of the Confederation of European Trade Unions[20] shows that implementing the ambitious EU climate policies in the period to 2030 will lead to shifts in sector employment, but overall creates many opportunities for new jobs. There are many other co-benefits as well, not the least being the creation of a healthy environment for people to live in. (See more elaborate discussion in Chapter 11.)

Changing development paths is not so simple

Mainstreaming climate change into development decisions and making development more sustainable, the main thrust of this chapter, means changing development paths. So what do we know about these processes of change? What are the conditions to make it easier? And what are the obstacles to social and economic change?

The conditions that determine how well societies respond to the need for change are manifold. They can be grouped together under the term 'response capacity'. Drawing on studies of social change in general and adaptation to and mitigation of climate change in particular, a number of important factors related to the economy, institutions, resources, and governance can be identified. Before going into those in more detail, it must be emphasized that the response capacity of a society is both influenced by development paths (in other words it is path dependent) and helps shape development paths. This interaction is responsible for so-called 'lock-in' effects, i.e. creating such infrastructures, governance and institutions that are strongly geared towards the current development path, creating vested interest in the status quo, and making change more difficult.

Another important notion is that development paths do not emerge as a result of a set of conscious decisions by government[21]. They emerge as the result of interactions between governments, the private sector, civil society (citizens and non-governmental organizations – NGOs) and also to some extent due to international developments and pressures. So changing development paths is not just a matter of changing government policy,

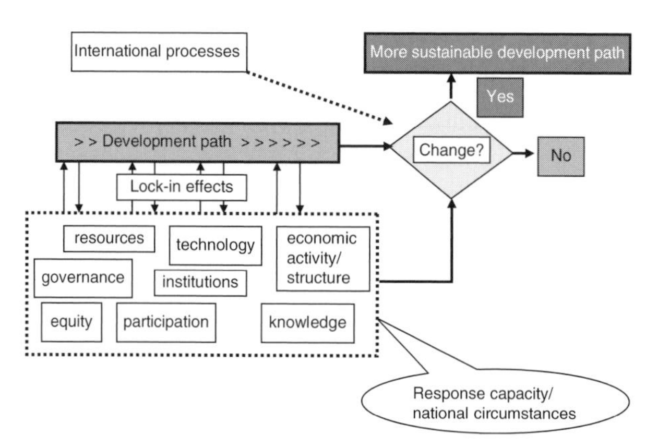

Figure 4.9 Schematic representation of the interaction of response capacity with development paths and the influence of response capacity and other factors on changing development paths.

although that is eventually a necessary condition to realize change. NGOs and business have shown to be able in many countries to initiate fundamental changes[22]. Figure 4.9 summarizes in a schematic way these interrelationships.

The main elements of response capacity are as follows.

Economic activity and structure

In a dynamic, growing economy change is easier to make. New activities can be taken up; older, less profitable activities can be stopped. If the labour market is flexible, i.e. skilled labour is available and employees can switch to new jobs easily, economic growth can be shifted to new activities. Moving to a service based economy with lower energy intensity requires the right skills in the labour force. Adapting agricultural practices to a new climate requires farmers to have the necessary skills. In a situation of economic stagnation, unfortunately the case in many developing countries, reorientation to new economic activities is much more difficult. Economic stability is a major determinant of attractiveness of a country for foreign investments that can help to make changes in the economy.

Resources

Natural resources vary from country to country. An abundance of natural resources allows a country to diversify the economy and to undertake new economic activities. On the other hand many countries are dependent on a limited set of natural resources. Abundant coal reserves and lack of natural gas will make it more difficult for a country to reduce the carbon intensity of the economy if it wants to preserve energy security at the same time. Strong winds will allow the development of wind energy. Abundant sunshine positions a country well for the development of solar energy. Financial resources are another crucial factor in determining response capacity. The possibility of attracting finance from abroad is strongly connected to political and monetary stability.

Technology

Availability of modern technology and capacity to use and maintain this technology is also important to respond to the need for change. This applies to efficient use of energy, renewable energy sources, efficient use of water, modern agricultural practices, health care systems, and many other practices that are needed to adapt to or prevent climate change. And this is strongly connected to the research, development, and innovation structure in a country as well as the ability to attract foreign investments in modern technologies. International arrangements and mechanisms can play an important role as well.

Knowledge

Availability of and access to state of the art information is another factor determining response capacity. This requires trained professionals in all sectors of the economy, up to date information management systems, and sharing knowledge amongst key institutions, not just government institutions. When it comes to adjusting existing practices to deal with climate change or to apply new technologies for renewable energy generation in specific locations, the use of local knowledge is crucial.

Institutions

Social and economic development depends on the availability of strong institutions in the field of security and justice, banking, insurance, research and development, education, power and water supply, business and professional associations, trade unions, and many others. They form the backbone of a stable society with predictable rules and enforcement of contracts and legislation. Without this network of institutions new economic and social initiatives are hard to establish. On the other hand some institutions may resist change, because it is perceived as going against their interests.

Equity

Widespread poverty and large differences in income or wealth create social unrest. Increases in food prices can create immediate problems for poor people and food riots have been shown to be a huge threat to political and economic stability.

Governance and participation

Government cannot realize social and economic change through legislation alone. It needs the cooperation and participation of the business community, civil society, and a wide range of institutions. In short: 'governance' instead of 'government'. It is a matter of getting the support of these groups, using their insights in workable solutions, and mobilizing power and motivating them to be instrumental in creating change. This allows for sharing information, bringing in better knowledge, improving the likelihood of successful changes in development paths, and motivating people to be part of the change. NGOs, including philanthropic foundations, have been particularly active on the issue of development and climate change[23].

How to make it happen?

As emphasized above, changing development paths is not just a matter of government decisions. When societies are facing important challenges, such as dealing with climate change, it matters how the necessary transitions are managed, i.e. how the coherence between different actions is maintained in overcoming barriers and getting the desired change.

There are many barriers to mainstreaming climate change into development policies. They range from lack of awareness about climate change risks, the complications and lack of 'media appeal' of integrated approaches, lack of cooperation between various ministries and institutions, lack of trained people, and last, but not least, an overload of issues for which integration with development is on the agenda.

Some experience in a number of countries has been gathered with strategies and approaches that have shown to be effective in overcoming these barriers.

Start at the top

One of the most critical things is to have all relevant ministries and government bodies share a common strategy and to have key ministries of planning, finance, and development take climate change as a serious and relevant issue. That will only happen if there is sufficient attention amongst the country's leaders to make integration of climate change into development policy a success. Only then will the government budget, the economic strategy, and the country's Poverty Reduction Strategy Plan – a precondition for getting international financial support – reflect mainstreaming of climate change. And only then will climate change get enough political attention to have a chance of changing the country's development path. This sounds self-evident. In practice however only very few countries have managed to achieve this so far. Tanzania (see Box 4.4) is a good example of successful application of this principle. In countries like Nepal or Bangladesh the main policy documents are silent about climate change, while there are serious threats of climate change undermining development[24].

Box 4.4 The example of Tanzania

Tanzania's vulnerability to the impacts of climate change is increasingly becoming a national concern. Extreme weather events impact negatively on agriculture. Infrastructure such as roads, railways, and bridges is destroyed by floods and cyclones. The economy, which grew by 6.9% in the year 2005, is expected to grow by 5.9% in 2006. The decline is mainly a result of drought, a climate related phenomenon.

The economy and the very survival of the majority of communities, like in many Least Developed Countries (LDCs), depend on such climate sensitive sectors. Tanzania's economy can aptly be described as a Climate-Sensitive Economy. It is because of this dependency and the current and projected impacts of climate change on such sectors that climate is a national priority and now, a national preoccupation.

Mainstreaming the environment and hence climate in the national development process is a prerequisite, with or without any international treaty. Mainstreaming entails integration of sustainability principles into a development strategy and, for most poor countries, building capacities at national and local levels for better identification of environmental concerns and opportunities.

This implies properly integrating actions into plans and budgets. Factoring environmental actions into the budgets of the key sector of the economy is an essential attribute of environmental mainstreaming. Tanzania's national budget for the fiscal year 2006/2007 has been dubbed as a 'green budget'. Environment now features prominently, with an increasing level of emphasis in the different national and sectoral policies and strategies. A number of initiatives have been undertaken, and policies, strategies, and programmes put in place to achieve environmental concerns. These include the National Environmental Policy; the Environment Management Act, 2004; Rural Development Policy; the Agricultural Sector Development Strategy (ASDS); the Tanzania Assistance Strategy (TAS); the National Strategy for Growth and Reduction of Poverty; and the Tanzania Development Vision 2025.

(Source: speech of Professor M. J. Mwandosya (MP), Minister of State (Environment), Vice President's Office, Tanzania at the Development and Climate Workshop, Paris Sept 2006, http://www.mnp.nl/en/publications/2006/IntegratedDevelopmentandClimatePolicies_howtorealizebenefitsatnationalandinternationallevel_.html)

Prepare a long term low carbon development plan

South Africa did a remarkable thing recently. It developed a Vision and Strategic Framework on long term low carbon development for the country[25]. It was based on a nationally coordinated long term scenario exercise that explored how the gap between current development plans and a sustainable long term society could be bridged[26]. The basic consideration for this strategic vision was the risk of serious climate change impacts in South Africa with respect to water and food security. It accepted the need for drastic emission reductions along the lines of the lowest IPCC stabilization scenarios that would lead to a limit of mean global mean temperatures to about 2°C above pre-industrial times.

It then looked at options in energy supply, industry, transport, buildings, agriculture, and other economic sectors on how development ambitions can be satisfied, while drastically reducing greenhouse gas emissions. It shows positive effects on poor people because of lower energy bills, increased employment, and overall a negligible or even slightly positive effect on GDP for the country. The strategic plan has been translated into specific policy directions aimed at building a strong renewable energy industry, drastically improving fuel efficiency standards for vehicles, and strengthening R&D, environmental education and policy coordination. It also contains plans to identify vulnerability and develop appropriate adaptation measures, which are to be included in the key performance standards for affected government departments.

Coordinated actions

Changing development paths is generally the result of a multitude of actions, as was explained above. Often these actions are not coordinated or even spontaneous as business and civil society take initiatives that are not in line with government policy. The more coherent individual actions are, the higher the probability that changes will lead to a more sustainable development path. One important element in such a transition management is coherence in government policy, something that is not at all self-evident. To be more effective, transition management should be extended to the role of business and civil society by creating dialogue, networks, and public-private partnerships and encouraging local action and experimentation to find promising approaches The capacity to manage transitions is thus an important condition for effective mainstreaming of climate change in development policy.

An example of such deliberate and coordinated transitions can be found in the way industrialized countries responded to the 1973 oil crisis. France, Germany, and Japan all faced the same problem. They chose different strategies to cope with it. France heavily invested in nuclear power and energy efficiency in buildings, reducing the dependence on fossil fuel. Germany built a strong export industry, compensating the trade balance deficit from the increasing cost of energy imports. Japan invested in making its industrial activities less energy intensive through energy efficiency and moving energy intensive production facilities overseas. All three managed to adjust to the new realities of expensive energy and maintained their economic growth, but in very different ways[27].

Climate proofing

This started as a way to check how well Danish development assistance projects were taking climate change risks into account. It was called 'climate proofing', i.e. systematically assessing how climate change was dealt with in these projects. It has now become a more widely applied practice by development assistance agencies and governments of developing countries to assess development policies, programs, and projects on their consistency with the goal of developing a low carbon economy (low emissions of greenhouse gases) and a society that is resilient to the impacts of climate change. DANIDA, the Danish Development Assistance Agency, has continued to champion this approach in collaboration with the governments of countries where they operate[28]. The approach is gaining ground now. It has been adopted by a variety of organizations such as the Asian Development Bank, United Nations Development Program (UNDP), and United Nations Environment Program (UNEP) (in the context of their climate change vulnerability management programmes), the UK Development Assistance Programme to assist African countries in dealing with climate change[29], and the Netherlands government to assess challenges to manage sea level rise and river flooding.

The key points from this chapter

The main message from this chapter is that a low carbon/high climate resilient society is the appropriate answer to the challenges of improving the living conditions of people around the world. It creates jobs, it improves energy security, it reduces health problems due to air pollution, and it avoids the most important damages from climate change. Even from a purely economic point of view it is the right thing to do. Changing development paths from the current fossil fuel/high greenhouse gas emission trajectory towards a low carbon, 'climate proof' one is a difficult process that requires close cooperation between governments, the private sector, NGOs, and civil society.

Notes

1. Agrawala S, van Aalst M. Adapting development cooperation to adapt to climate change, Climate Policy, vol 8, 2008, pp.183–193.
2. World Bank. Managing Climate Risk, 2006.
3. Baumert K et al. Navigating the numbers, WRI, 2005, http://pdf.wri.org/navigating_numbers_chapter6.pdf.
4. IPCC Fourth Assessment Report, Working Group III, 2007, ch 2.1, ch 12.1.
5. http://www.eia.doe.gov/emeu/mecs/iab/steel/page2d.html.
6. IPCC Fourth Assessment Report, Working Group III, ch 7.1.2; for a further discussion of industrial mitigation options see Chapter 8 of this book.
7. See section on co-benefits of climate policy in Chapter 10.
8. A Fisher Tropsch process is based on chemical reactions of gasified coal in the presence of water at high pressure, leading to the synthesis of gasoline or other chemicals.
9. See Chapter 5 for an in-depth discussion.
10. See the discussion about nuclear power and its problems in Chapter 5.
11. Shukla PR et al. Development and Climate: an assessment for India, Indian Institute of Management, Ahmedabad, India, 2003.
12. See also Chapter 9.

13. Fourth Assessment Report, Working Group III , ch 12.2.4.3.
14. See for instance IPCC Fourth Assessment Report, Working Group III, tables 8.10 and 8.11.
15. IPCC Fourth Assessment Report, Working Group III, ch 12.2.4.1.
16. IPCC Fourth Assessment Report, Working Group III, ch 12.2.4.4.
17. World Health Organization. Fuel of life: household energy and health, 2006.
18. IPCC Fourth Assessment Report, Working Group III, ch 6.6.2.
19. IPCC Fourth Assessment Report, Working Group III, ch 12.2.4.1.
20. Climate change and employment: impacts on employment in the European Union-25 of climate change and CO_2 emissions reduction measures by 2030, http://www.tradeunionpress.eu/Web/EN/Activities/Environment/Studyclimatechange/rapport.pdf.
21. IPCC Fourth Assessment Report, Working Group III, ch 12.2.3.
22. See also the section on Voluntary Actions in Chapter 11.
23. IPCC Fourth Assessment Report, Working Group III, ch 12.2.3.3.
24. See note 1.
25. South African Department of Environmental Affairs and Tourism. Government's vision, strategic direction and framework for climate policy, 2008.
26. South African Department of Environmental Affairs and Tourism. Long term mitigation scenarios: strategic options for South Africa, October 2007.
27. IPCC Fourth Assessment Report, Working Group III, box 12.5.
28. DANIDA, Danish Climate and Development Action Programme, http://amg.um.dk/NR/rdonlyres/C559F2DF-6D43–4646–80ED-C47024062FBD/0/ClimateAndDevelopmentActionProgramme.pdf.
29. UK Department for International Development. Climate proofing Africa: climate and Africa's development challenge, London, 2005.

5 Energy Supply

What is covered in this chapter?

Chapter 2 pointed out the fact that energy use is responsible for about two-thirds of greenhouse gas emissions, which is a good reason to explore in depth in this chapter what drives energy use, how it is supplied, why countries value energy security, where its main uses are, and what greenhouse gas emissions it produces. The chapter then focuses on electricity production. Improving the efficiency of power plants, shifting from coal to gas, nuclear power, renewable energy, and capture and storage of CO_2 from power plants can all help to reduce greenhouse gas emissions. The status of these technologies and their costs are discussed, as well as the competition between these technologies when reducing overall emissions from electricity supply. Economic, security, health, environment, and other considerations in choosing an optimal fuel mix for electricity generation are explored. Technology and economics present a fairly optimistic prospect for drastic emissions reductions. Implementing these opportunities however is hard. Selecting the right policies to provide incentives for implementation by business and individuals is crucial. On-the-ground experience is growing and lessons for effective policy choices can be drawn.

Energy and development

As outlined in Chapter 4, energy is an essential input for development. Historically there has been a very strong relation between income and energy use. Figure 2.10 shows that relationship for a number of countries for the period 1980–2004. Most countries show a steady increase in energy use per person when income per person goes up. Russia is an exception because of the economic recession after the collapse of the Soviet system. Canada and the USA show almost no increase in energy use per person over the period considered, despite an increase in incomes, because a kind of saturation has occurred. What is striking is the large difference in energy use per person between countries, for similar income levels. Some European countries and Japan use almost half the energy per person of that of the US and Canada. Differences in lifestyle, the structure of the economy

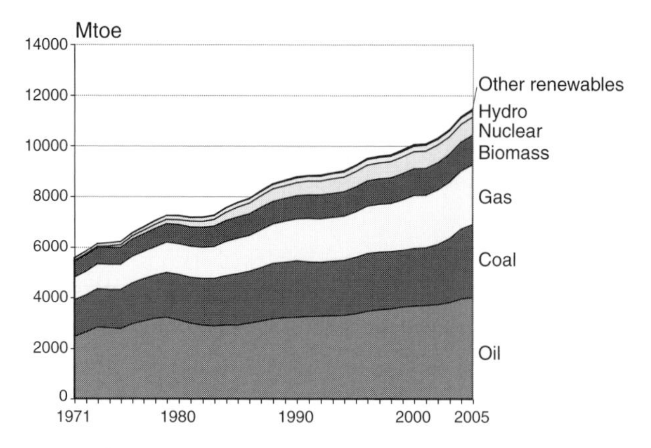

Figure 5.1 World primary energy supply, 1971–2005; for an explanation of 'primary energy supply' see Box 5.1.
Source: IPCC Fourth Assessment Report, Working Group III, figure TS.13.

(the role of energy intensive industries in the economy), and efficiency of energy use are the major reasons behind this.

Rapidly developing countries like China and India are still at an early stage of income growth. It is of great importance to the world's energy requirements how these countries develop. Will it be the American or the European/Japanese way? Or will they chart new territory by using less energy than other countries when going through a development transition?

It should be no surprise that overall energy demand has roughly doubled over the past 35 years and – since demand has to equal supply– so has energy supply. In 2005 fossil fuels (coal, oil, and gas) represented 80% of the total. Biomass, mostly traditional fuels like wood, agricultural waste, and cow dung, accounted for 10%. Nuclear energy was good for 6%, hydropower about 2%, and 'new' renewable energy (wind, solar, geothermal, modern biomass) less than 1% (see Figure 5.1).

Box 5.1 **Units for energy**

Amounts of energy are usually expressed in joule (J). Larger quantities can be expressed in kilojoule (kJ = 10^3J), megajoule (1MJ= 10^6J), gigajoule (1GJ = 10^9J) or exajoule (1EJ = 10^{12}J). Another unit for energy that is often used is Million tonnes oil equivalent (1Mtoe = 0.042EJ).

Capacity of power plants is expressed as the amount of energy that can be produced per second, or joule per second (J/s). 1J/s equals 1watt (W). Power plant capacities are therefore normally expressed in megawatt (1MW = 10^6W) or gigawatt (1GW = 10^9W).

Electricity produced is normally expressed in kilowatt hour (1kWh = 3.6MJ). Larger quantities as gigawatt hour (GWh = million kWh) or terawatt hour (1TWh = 10^9kWh).

To convert power plant capacity into electricity produced you need to factor in the so-called capacity factor (the proportion of the time the plant is operational). For fossil fuel and nuclear power plants the capacity factor is usually something like 80–90%. For wind turbines and solar plants it is much lower.

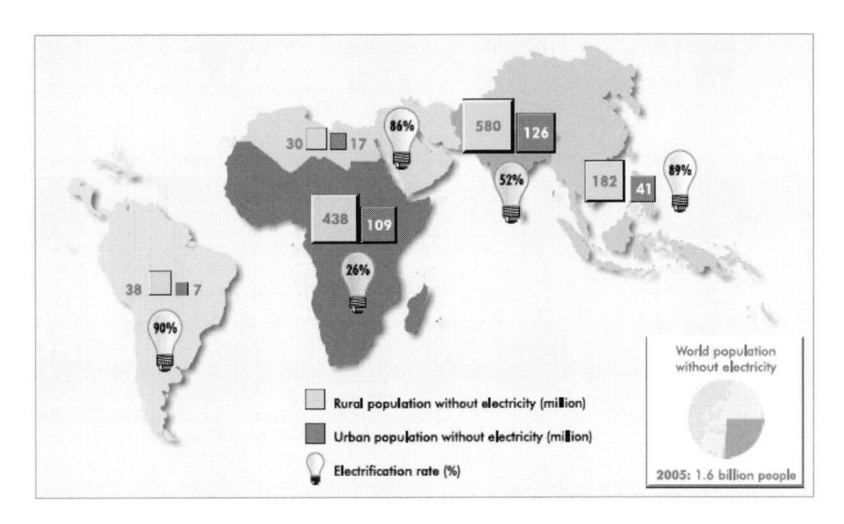

Figure 5.2 Population without access to electricity in 2005.
Source: IEA World Energy Outlook 2006.

Future energy demand

Access to modern energy, in particular electricity, is still a big issue for many developing countries[1]. Currently about 1.6 billion people in developing countries have no access to electricity and about 2.4 billion rely on traditional fuels (wood, agricultural waste, cow dung) for their cooking and heating needs. Most of these people are located in Africa and South Asia. China on the contrary has reached a 99% access rate (Figure 5.2). Giving all the people in the world access will require a strong growth in modern energy supply, which would no doubt increase CO_2 emissions. If all households that still rely on traditional biomass fuels were provided with LPG for cooking, global greenhouse gas emissions could increase by 2%. However the reduced deforestation as a result of that could then be subtracted. An LPG programme in Senegal that led to a 33-fold increase in LPG use resulted for instance in a 15% lower charcoal consumption (see also chapter 4).

On top of that, the need for improvement of the well-being of people in developing countries, the expected economic growth in industrialized countries, and the expected population growth will likely lead to a 50% increase in world energy demand by 2030[2].

How will that energy be supplied in the absence of policies to curb climate change? Basically the dominance of fossil fuel will continue. All projections for the period until 2030 show a substantial increase of hydropower and other renewable energy sources, but they remain a small fraction of the total. Fossil fuel use remains at about 80%. Opinions about the role of nuclear power vary widely. Given the risks of nuclear power (reactor accidents, radioactive waste, and nuclear weapon proliferation) you find both optimistic and pessimistic projections for the role of nuclear power. Figure 5.3 shows some recent estimates for the energy supply situation in 2030. What stands out is the large differences in total energy demand and the contribution of various energy sources between the individual

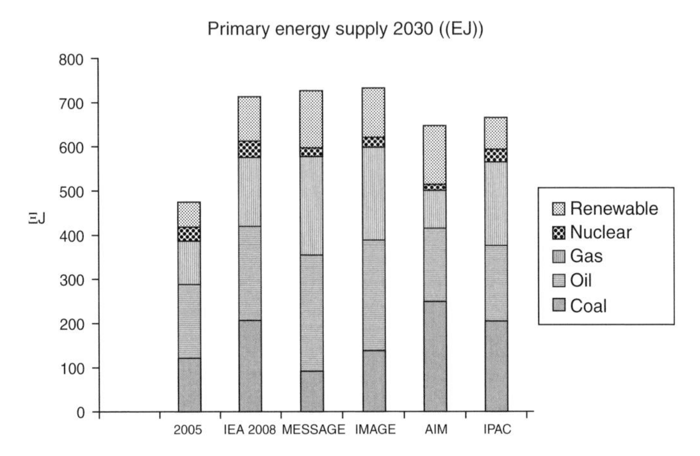

Figure 5.3 **Energy supply projections for 2030 without climate policy (baseline). Results from different models.**

Source: IPCC Fourth Assessment Report, Working Group III, chapter 3; IEA WEO 2008.

estimates. This is typical for projections over a 25 year period. It is more likely the energy demand will be on the high end of the range in light of the very strong growth in China and India over the past few years. As far as the projection for coal use is concerned, the IEA estimate is probably more in line with recent increases in coal use in China, India, and other parts of the world. This will make drastic reductions in CO_2 more difficult to achieve.

But fossil fuels are scarce, aren't they?

Contrary to the widespread belief that fossil fuels are scarce, there are in fact such big fossil fuel resources that there are no constraints to huge increases in the use of fossil fuels. To understand that it is important to make a distinction between 'reserves' and 'resources'. The fossil fuel industry defines 'reserves' as the quantities of oil, gas, or coal that have been proven to be available and economically attractive to extract. It is well known that other and larger quantities exist that are not economically attractive to exploit ('resources'). In other words, if the price of oil increases, the reserves of crude oil go up. Parts of the resources then become reserves.

There is another important distinction: between 'conventional' and 'unconventional' resources. For oil the unconventional resources are for example the so-called 'tar sands' and 'oil shales', basically oil containing soil or rock, from which oil can be released by heating it and extracting it from the raw material. This is a costly and energy intensive process, but at oil prices above US$60 per barrel it is economically attractive in many places and therefore adds to the oil reserves, if oil prices stay above US$60 per barrel. Another example is so-called 'natural gas hydrates' or 'clathrates', a kind of 'frozen' gas/water mixture that can be found in deep oceans. These hydrates are currently not economically attractive to use, but the quantities are so big (20 times all conventional gas

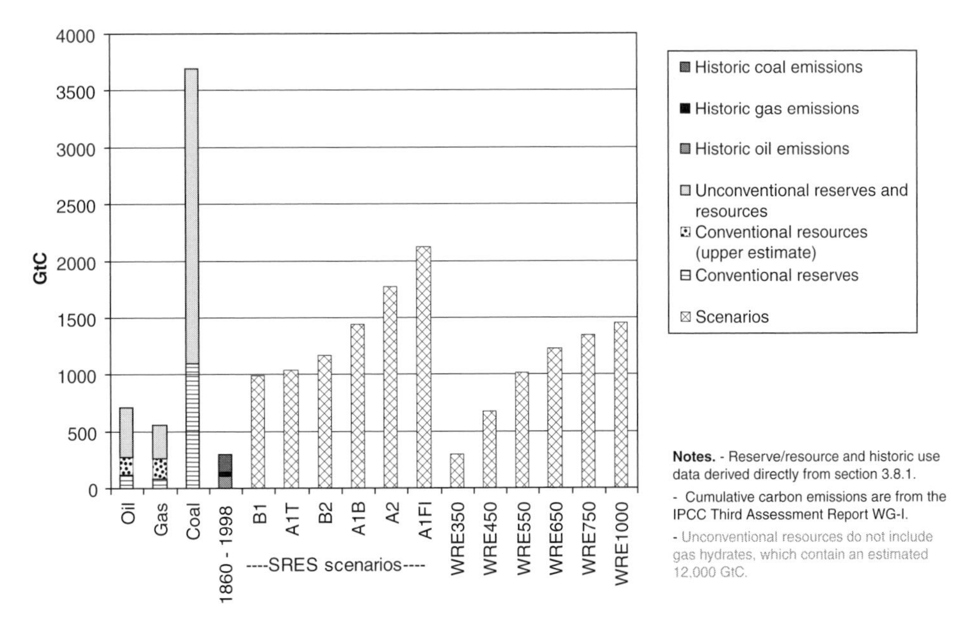

Figure 5.4 **Fossil fuel reserves and resources, compared with fossil fuel use in various scenarios for future energy use and carbon released in the atmosphere for various atmospheric CO_2 concentration stabilization scenarios.**

Source: IPCC, Third Assessment report, figure SPM.2.

resources) that they may well be exploited in the future. Conventional and unconventional fossil fuel resources together add up to enormous amounts.

Taking only the proven conventional reserves and resources, today's oil, gas, and coal use can be continued for about 60, 130, and 800 years, respectively[3]. Add to that the unconventional and 'yet-to-find' resources (note that nobody is really looking for coal these days), current use could be maintained for more than a thousand years. Or, to use a different perspective, the total amount of fossil fuel that would be needed during the 21st century if we assume strong economic growth and heavy fossil fuel use, is only a fraction of the fossil fuel resources. Figure 5.4 (left and middle part) shows how these quantities compare.

You might argue that the situation for oil is different. It is and it isn't. Oil resources are more limited and geopolitical tensions can easily lead to scarcity and price increases. There are also questions on how fast oil can be produced, even if there were large amounts available. This is the so-called 'peak-oil' issue. There are claims that geological formations would not allow production rates to be increased and that productions rates would start to fall in many oil producing regions. It is more likely however that the real reason for limits to production rates may be the national oil companies in the Middle East, China, Brazil, and elsewhere, the importance of which has grown enormously over the past 10 years. These national oil companies are behaving very differently from international oil companies, such as Shell, BP, Exxon, etc[4]. In any case, the technologies exist and are commercially viable to turn gas and coal into liquid transport fuels. This technology was used extensively in South

Africa when is was hit by an oil boycott during the apartheid regime. Today there are several such plants operating in the Middle East and China. Oil therefore is not the limiting factor for increased fossil fuel use.

Figure 5.4 shows on the right hand side of the diagram how much carbon would be released into the atmosphere for various scenarios of stabilizing CO_2 concentrations in the atmosphere. Keeping CO_2 concentrations in the atmosphere below a level of 1000ppm (which is way above what most people would consider acceptable) would limit fossil fuel use even more than the high growth scenario described above. In other words, as the Stone Age did not end due to a lack of stone, so will the fossil fuel age end long before fossil fuels are exhausted.

Energy security

Since energy plays such a crucial role in development of countries, it matters politically how secure the supply of energy is. Is the country self-sufficient in energy or does it need to import? And if it imports, are the foreign suppliers reliable or is there a risk of political instability? Are the imports spread over many different suppliers or are there only a few? All those factors contribute to what is called energy security.

Oil, the fuel on which the world's transport runs, is well known to be a heavily traded energy source. There is a limited number of suppliers, particularly the Middle East, Russia, and some Latin American and African countries, who export large quantities. The main importers are North America, Europe, Japan, but also China and India and other developing countries (see Figure 5.5). This means there is a strong dependence on oil imports for many countries and this dependence tends to increase over time.

Energy security of course is not limited to oil. It equally applies to coal and gas. Coal resources are more widespread than oil. Some of the biggest energy using countries

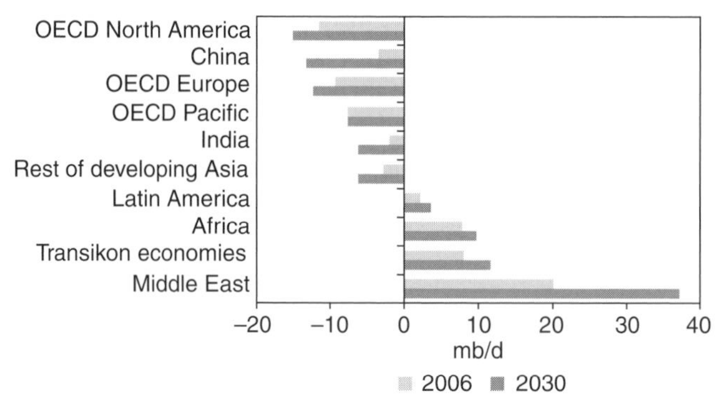

Figure 5.5 **Net oil trade (in million barrels/day (mb/d)) between regions for 2006 and projections for 2030. Transition economies are countries of Eastern Europe and the former Soviet Union.**
Source: IEA, WEO, 2007.

Table 5.1. Proven coal reserves by country

Country	Per cent of world coal reserves
USA	27
Russia	17
China	13
India	10
Australia	9
South Africa	5
Ukraine	4
EU	4
Kazakhstan	3
Rest of world	8

Source: IEA WEO, 2006.

(USA, China, India, Russia, and Australia) have abundant coal reserves (see Table 5.1). Relying on a big domestic energy source is of course very good from an energy security point of view. The share of coal in the energy for electricity production in China and India for example is 89% and 82%, respectively. But many countries also import coal. The world average share of coal in electricity production is about 45%. By relying on a mixture of energy sources countries improve their energy security.

For natural gas there is a big mismatch between use and production. North America currently imports about 2% of its gas, but that is projected to increase to 16% by 2030. Most of this gas will be imported from Venezuela and the Middle East as LNG (liquefied natural gas). Europe already imports 40% (mostly from Russia and North Africa by pipelines), and this is expected to increase to almost 70% by 2030. For Japan the numbers are even more staggering: 97% is imported now, going up to 98%, most of it as LNG from Indonesia, the Middle East, and Australia. China and India are expected to import about 50% and 60%, respectively, of their gas by 2030[5].

Where is energy used?

Energy is used in all sectors of the economy. About 45% goes into electricity generation and (to a small extent) centralized heat production for district heating purposes. Close to 20% each goes into transport (as fuel), industry (fuel and raw materials), and residential and commercial buildings (as heating and cooking fuel) and agriculture (see Figure 5.6).

In this chapter the supply of power and heat will be discussed. Energy used in the transportation, building, industry, and agriculture/forestry sectors (both the direct energy as well as the power and heat coming from the energy supply sector) is covered in Chapters 6, 7, 8, and 9.

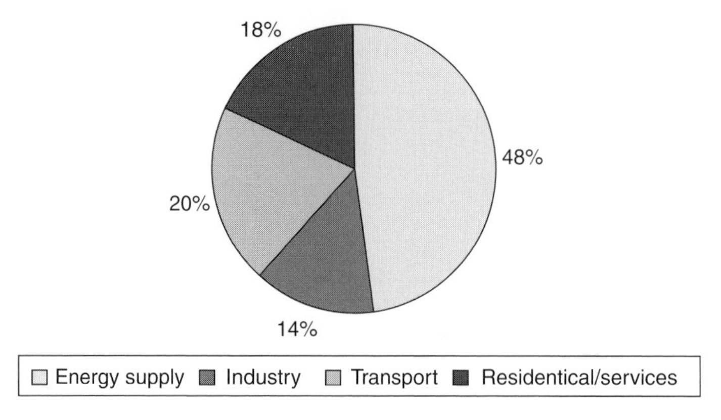

Figure 5.6 **Share of energy going into the economic sectors.**
Source: based on IEA, WEO 2008.

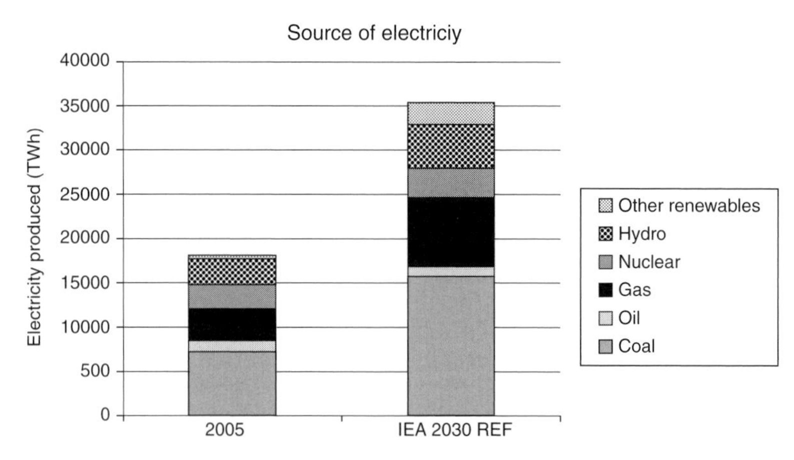

Figure 5.7 **Energy sources used for electricity production now and for the 2030 baseline situation. TWh, terawatthour.**
Source: IEA, WEO, 2007.

Electricity generation

Coal and gas are the dominant energy sources for power generation. Coal alone has a share of 40%. Together with a little bit of oil they cover about two-thirds of the energy sources. Nuclear and hydro both have about a 15% share. By 2030 the role of coal and gas is expected to be even stronger. Figure 5.7 shows the contribution from the various energy sources, based on the share of the electricity produced. It also shows electricity demand is expected to almost double by 2030.

You can also look at the generating capacity installed in the form of power plants, wind turbines, solar power, etc. Because wind and sun are not always available, wind

Table 5.2 **Comparison of share in installed capacity and in electricity produced**

Energy source	Installed capacity 2006 (GW)	Contribution in 2005 as % of installed electric power capacity	Electricity produced 2006 (TWh)	Contribution in 2005 as % of electricity produced
Coal	1382	32	7756	41
Oil	415	10	1096	6
Gas	1124	26	3807	20
Nuclear	368	8	2793	15
Hydro	919	21	3035	16
Other renewables	135	3	433	2

Coal fired and nuclear power plants are typically baseload installations, meaning they are operating almost continuously.
Source: IEA, WEO 2008.

turbines and solar power plants cannot operate continuously, unlike nuclear and fossil fuel plants, so for the same capacity their contribution to actual electricity production is less. Table 5.2 shows the difference in contribution when looking at the installed capacity of the various power sources. For an explanation of the units used see Box 5.1.

It is also useful to make a distinction between 'primary energy' (the raw energy sources) and 'secondary' or 'final' energy (the energy carriers that are actually used), because there is a significant energy loss when converting primary energy sources such as coal or biomass into energy carriers, such as electricity. Box 5.2 gives an explanation.

Box 5.2 **Energy supply and energy end-use**

So-called 'primary' energy sources (coal, oil, gas, uranium, water (hydro), wind, solar radiation, geothermal energy, ocean energy) are converted to energy carriers (called 'secondary energy' or 'final energy') such as electricity, heat, or solid, liquid or gaseous fuels. During the conversion process, such as in electricity production, a significant part of the primary energy can be lost. The conversion efficiency (or, in the case of electricity, the efficiency of an electric power plant) is therefore a crucial element of the energy system. When energy carriers are used to deliver certain services (light, transport, heat), another conversion process happens, where energy losses are happening. For instance, the amount of energy obtained from a traditional light bulb in the form of light is only 2% of the electric energy used (and that electricity contained only 35% of the energy that was used to produce it; see figure). The overall efficiency of the energy system is therefore determined both by the supply side efficiency and the so-called end-use efficiency.

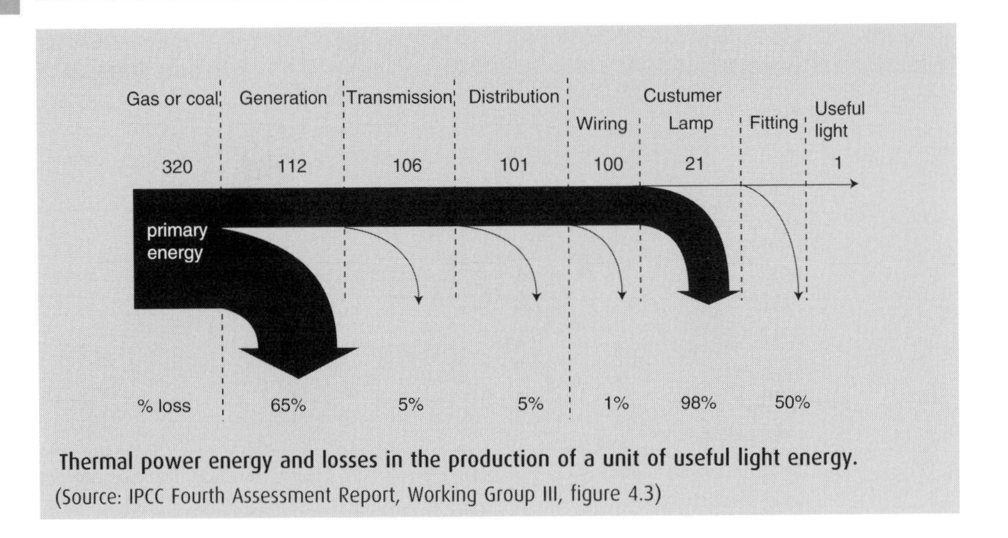

Thermal power energy and losses in the production of a unit of useful light energy.
(Source: IPCC Fourth Assessment Report, Working Group III, figure 4.3)

Greenhouse gas emissions

Energy supply and use in 2005 was responsible for about 64% of all greenhouse gas emissions. CO_2 alone accounted for about 60%, the rest came primarily from methane. The electricity supply sector is the biggest emitter, followed by industry, transport, and buildings (see Figure 5.8). Note that there is a small amount of emissions of CO_2 from cement manufacture (coming from the raw materials) and industrial nitrous oxide and fluorinated gases that are not energy related (shown separately).

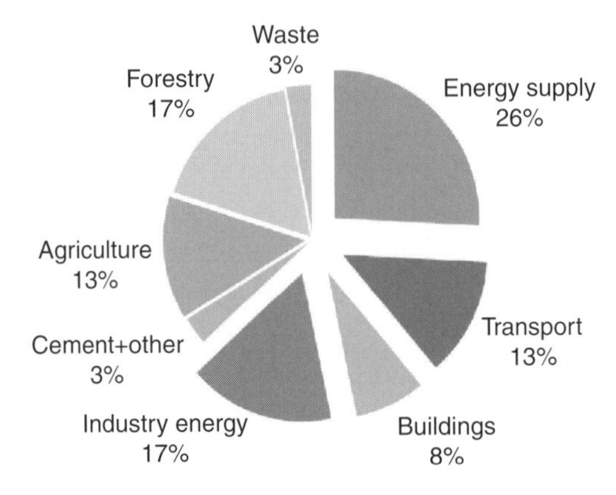

Figure 5.8 **Greenhouse gas emissions from energy supply and use in 2004 as percentage of total emission. Only energy related emissions are covered in the energy supply, transport, buildings, and industry shares.**
Source: IPCC Fourth Assessment Report, Working Group III, ch 1.

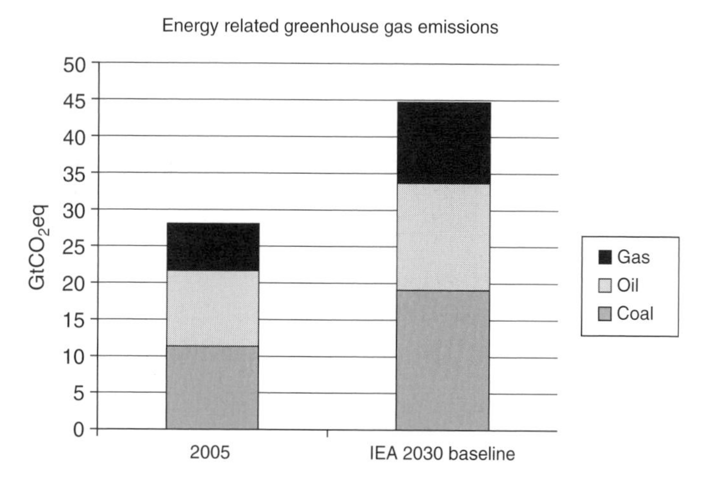

Figure 5.9 **Increase in energy related greenhouse gas emissions 2005–2030.**
Source: IEA, WEO 2007.

With the 50% increase of overall energy use expected by 2030, greenhouse gas emissions from energy supply and use will also rise strongly. The expected increase in emissions is about 45%, based on the latest International Energy Agency's scenarios[6] (see Figure 5.9).

The electricity sector and the emissions reduction challenge

Greenhouse gas emissions from electricity generation are dominated by coal. Gas is responsible for about 30% and oil is becoming a negligible factor with something like 4% expected in 2030 (see Figure 5.10). Total emissions from the power supply sector are projected to grow by about 70% until 2030. So the challenge in reducing emissions from the power supply sector lies in finding alternatives to the use of coal and gas. In the next section these alternatives will be explored.

Emission reduction options in the electricity sector

Improving the efficiency of power plants, shifting from coal to gas, nuclear power, various renewable energy sources, and capture and storage of CO_2 from power plants can all help to reduce CO_2 emissions. They will be described briefly in terms of the status of their technology, their costs, and availability.

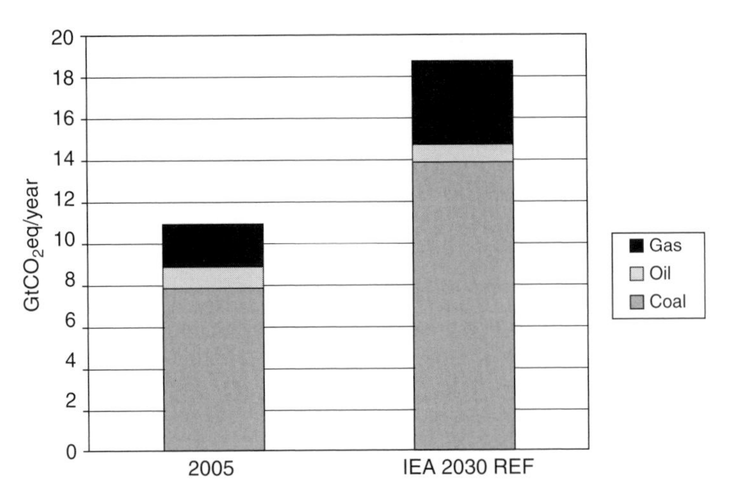

Figure 5.10 **Greenhouse gas emissions from electricity generation without climate policy 2005–2030.**
Source: IEA WEO 2007.

Often the discussions about the contribution of renewable energy are far too optimistic. Proponents tend to overestimate the speed at which renewables can penetrate the power supply market, costs tend to be too low, and problems in scaling up these technologies tend to be underestimated. This chapter wants to avoid those pitfalls and present a realistic picture.

On the other hand many discussions are too pessimistic about the potential of renewable energy by stressing the low share, the many obstacles to its introduction, and the resistance to promoting renewables. They forget that, even from a low starting point, annual growth rates of more than 10% can lead to enormous growth over a 20 year period. A 2% annual growth rate means a 50% overall increase over a 20 year period. A 20% annual growth rate means a 30-fold (!) increase over such a period. Several renewable energy systems have annual growth rates even beyond that. In addition, when investment in renewable energy systems really catches on, these barriers will be much less likely to play a serious role. This chapter aims to be realistic in this respect too.

Power plant efficiency and fuel switching

Electricity generation in thermal power plants is a wasteful operation. Most power plants operating today lose 50–70% of the energy that is put in (i.e. their efficiency is only 30–50%). Gas fired plants normally have a better efficiency than coal plants. Many coal fired plants operating today run at 30% efficiency. Newly built coal fired plants (so-called supercritical plants) reach an efficiency of about 42%, with some running at close to 50% efficiency. The most advanced coal fired plants, so-called 'integrated gasification combined cycle plants (IGCCs)', which first gasify the coal

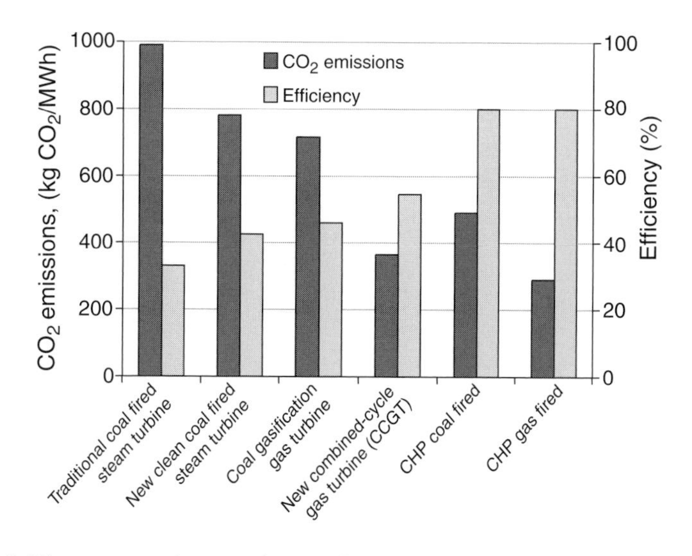

Figure 5.11 **Efficiencies of different types of power plants and the CO_2 emissions per unit of electricity produced.** *Source*: IPCC Fourth Assessment report, Working Group III, figure 4.21.

before burning it, can reach comparable efficiencies, while at the same time allowing better cleaning of exhaust gases to minimize air pollution. Modern gas fired plants, so-called 'combined cycle gas turbines (CCGTs)', can reach 55% efficiency. These efficiencies are expected to increase due to further technological development (see Figure 5.11).

Because of the advantages of gas in terms of efficiency, the lower emission of CO_2 per unit of energy (see Box 2.2, Chapter 2), and also the much shorter construction time for gas fired plants, it is attractive to shift from coal to gas for electricity production. Recently, however, gas prices have gone up, mostly as a result of increasing oil prices, making this option less attractive.

A big step forward in terms of efficiency is the so-called Combined Heat and Power (CHP) plant. By using the heat that traditional power plants waste, CHP plants can get 80% useful energy from primary energy input (see Figure 5.11), so wherever heat is needed, CHP plants can make a strong contribution to reducing CO_2. CHP plants are used predominantly in industrial complexes that need a lot of process heat (with electricity partly exported to the grid) or in areas where district heating is used (providing heating to residential and commercial buildings through a network). The CHP principle can also be used at small scale (down to household size) and for a variety of fuels, such as biomass (see section on Bioenergy).

Power plant efficiency improvement happens autonomously when a new fossil fuel plant is built. Any new coal fired power plant basically has the best efficiency available. What is not happening autonomously is shifting to IGCC type plants, shifting to gas, or turning the power plant into a CHP plant. Those choices bring additional costs that will only be incurred if climate policy or other incentives make it attractive. The potential for emission reduction therefore is limited.

Table 5.3. Top 10 countries with nuclear power plants in operation.

Country	Number of nuclear power stations in operation by end 2006	Number of nuclear plants under construction by end 2006	Number of nuclear reactors planned	% of electricity from nuclear power (2006)
USA	103	0		19
France	59	1		78
Japan	55	1	13	30
Russia	31	5	50% increase	16
Korea	20	1	60% increase	39
UK	19	0		18
Germany[a]	17	0		31
India	16	7	16	3
Ukraine	15	2		48
China	10	4	28–40	2
Sweden[a]	10	0		48

Nuclear power plants under construction as well as percentage of electricity produced by nuclear power are shown.

[a] Pledged nuclear phase-out.

Source: IAEA. Energy, Electricity and Nuclear Power Estimates for the Period up to 2030, 2007 edition.

Nuclear power

By the end of 2006, 442 nuclear reactors, with a total capacity of 370GW, were producing electricity, accounting for about 16% of world production. They are spread over 31 countries, with 10 countries responsible for 75% [7](see Table 5.3). Since the Chernobyl accident in Russia in 1987, no new nuclear plants have been built in North America and Europe and only about 50 plants have been built elsewhere.

The biggest reason for the stagnation of nuclear power is the discussion about its risks: (1) radioactive materials could escape from nuclear reactors or nuclear fuel processing and transport; (2) finding safe storage of radioactive waste with extremely long lifetimes from used nuclear reactor fuel is still problematic; and (3) possibly spreading the production of nuclear weapons by giving more countries access to nuclear technology. High investment costs, liability issues in case of accidents, and long regulatory procedures in light of the risks are an additional factor.

Safety of nuclear reactors has been an issue of concern, particularly after the Chernobyl accident in 1987. Reactor designs have been improved over time and development of safer designs is ongoing (see Box 5.3)[8].

When uranium fuel from a nuclear reactor needs to be replaced, the material is highly radioactive. The biggest problem is the long-lived highly radioactive material that takes thousands of years to decompose. More than 95% of the total radioactivity of all waste generated from the nuclear fuel cycle (uranium processing, reactor waste, waste processing) is in this high level waste, but this represents only 5% of the volume. One 1000 MW nuclear plant produces about 10m^3 of high level waste per year. If this waste is reprocessed, i.e. when usable uranium and plutonium and other highly radioactive materials are separated, this goes down to 2.5m^3. Reprocessing facilities are operating in France, Russia, UK, and Japan, while the USA has so far refrained from building one to reduce the risk of diversion of plutonium for nuclear weapons production.

Deep geological storage of this waste material is generally seen as the safest way to deal with it. The radioactive waste is then embedded in glass and packed in containers to make leakage very difficult. There is widespread consensus amongst experts that this is a safe way to store the waste. However, all of the proposed storage projects are facing serious resistance from the general public or citizen groups. In Finland and the USA deep geological storage sites have been chosen, but controversy still remains, and

Box 5.3

Nuclear power reactors and safety

Nuclear power reactors produce heat from the fission of uranium atoms as well as from plutonium formed during operation. They use uranium oxide in which the concentration of uranium-235 (the fissionable isotope) is increased from 0.7% to usually 4–5%. The core of the reactor (where the rods of uranium fuel are) is cooled with water or gas. The hot water or gas is then used to generate electricity via steam or gas turbines. About 80% of operating nuclear reactors use water (Boiling Water Reactors or Pressurized Water Reactors); most of the others use gas (helium or carbon dioxide).

Cooling of the reactor core is the most critical issue in terms of reactor safety. If cooling fails, the reactor core can melt and the molten reactor fuel can melt through the reactor vessel and get dispersed outside. Reactor safety is therefore strongly dependent on maintaining cooling at all times. Water based cooling systems have been improved over time by reducing the number of pumps and pipes (lowering the risk of leakage), adding several additional emergency cooling systems, and using gravity and natural circulation rather than electricity to operate the cooling system. The other safety element that has been strengthened in reactor design is the containment: modern reactors have double or triple containments, protecting against attacks from outside, and able to keep even a melted reactor core inside the building. These modern designs have considerably reduced the chances of releasing radioactive materials, but have not reduced that risk to zero.

Advanced high temperature gas cooled reactor designs use special fuel 'balls' that can resist very high temperatures (so-called 'pebble bed' reactors). No melting of reactor fuel would occur in these reactors, even when cooling completely fails. These reactors are under development in South Africa and China. A disadvantage of this design is that the capacity is 5 times as small as that of water cooled reactors, requiring multiple units that increase costs.

Most modern reactor types are able to use recycled uranium and plutonium (from used fuel rods) in so-called mixed-oxide (MOX) fuel. This reduces the need for new uranium, but of course requires the processing of the used fuel in special processing plants and therefore increases the risk of plutonium being diverted to nuclear weapons production. The gas cooled 'pebble bed' reactor fuel cannot be reprocessed, which would be an advantage from a proliferation point of view.

Special so-called 'breeder' reactors are designed to produce more nuclear fuel (in the form of plutonium) than is put in, thereby reducing the need for uranium imports. Such reactors are not yet commercially operating but the subject of active development in several countries, including China and India. India is putting a lot of effort in developing thorium based breeder reactors, because it has only small uranium reserves. The advantage of thorium is that is does not produce plutonium, which would reduce proliferation risks.

(Source: IPCC Fourth Assessment report, Working Group III, ch 4.3.2; Richter B. Nuclear Power: A Status Report, Stanford University Programme on Energy and Sustainable Development, Working Paper #58, September 2006)

detailed design studies are continuing. Actual operation is not expected to start before 2020. In Sweden, Germany, and France procedures for choosing sites are ongoing.

Nuclear weapons grade uranium can be obtained from extreme enrichment of uranium (much more than needed for a nuclear power reactor) and nuclear weapons grade plutonium from processing spent reactor fuel. Acquiring these technologies allows countries in principle to develop nuclear weapons[9]. In addition to the USA, UK, France, Russia and China, India, Pakistan and Israel now also possess nuclear weapons. The Treaty on Non-proliferation of Nuclear Weapons tries to limit that risk by a series of information and inspection obligations, overseen by the International Atomic Energy Agency (IAEA). Not all countries are a member of this treaty however. India, Pakistan, Israel, and North Korea have so far refused to sign. There have been several instances (Iraq, North Korea, Iran) where suspicions arose about possible intentions of countries to develop a nuclear weapon.

Uranium, the energy source for nuclear power reactors, is produced from uranium ore that is mined in a limited number of countries. Canada, Australia, Kazakhstan, and Russia account for almost 70%; Niger, Namibia, Uzbekistan, and the USA for another 25%. Reserves (identified amounts and those economical to produce) are good for 85 years at current consumption. Including all conventional resources brings this figure up to several hundreds of years[10]. When plutonium recycling from so-called 'breeder reactors' (see above) is included, resources would last several thousands of years. These kinds of reactors and the necessary fuel processing would however bring additional risks.

With climate change becoming an important political issue, nuclear power is seeing something of a revival in the USA and Europe. In the USA new legislation was passed in 2005 that simplifies licensing procedures, extends the limitation of liability of companies in case of accidents, and provides a subsidy (in the form of a tax deduction) of almost 2USc/kWh. In Finland and France a decision was made to build a new nuclear power plant.

So what are the prospects of nuclear power as a greenhouse gas emission reduction option? Projections for nuclear power in the future are very uncertain. On the one hand countries like Japan, China, Korea, and India are planning significant expansions of nuclear power. In many other countries plans for additional nuclear plants have been shelved and some countries have pledged a nuclear phase out. The International Atomic Energy Agency estimates for nuclear power capacity in 2030 therefore show a great uncertainty: between 280 and 740GW. The IEA projects 415MW nuclear capacity by 2030 without additional policy, but 30% more if climate policy is assumed[11]. In terms of greenhouse gas emissions nuclear power is attractive, although it does not have zero emissions. Because of the energy needed for uranium mining, waste processing, and eventual decommissioning of nuclear reactors, emissions are estimated at about 40g CO_2/kWh of electricity produced. There are however some estimates that give much higher total emissions of 80–120g CO_2/kWh due to uranium ore processing, construction, and decommissioning (see also Table 5.5 below for a comparison with other supply options). The cost of nuclear power from existing plants is between 1 and 12ct/kWh, reflecting the different local circumstances. The low end of this range makes nuclear power competitive with coal. By 2030 the cost of electricity from newly built plants is estimated to be between 2.5 and 7USc/kWh.

The role that nuclear power will play in a world with limitations on CO_2 emissions will depend on two main issues. The first is how the risks of nuclear power are going to be perceived. A meaningful contribution of nuclear power to reduction of CO_2 emissions would require a substantial expansion of countries with access to nuclear technology. In a world where international terrorism is likely to remain a fact of life, it could be a matter of time before nuclear weapons are made by terrorist groups from diverted plutonium. While the safety and waste disposal risks may be technically manageable, this risk may not be. The second important factor is the costs and availability of other low carbon alternatives with which nuclear power would compete in a carbon constrained world. This issue will be revisited after other alternatives have been discussed (see below).

Hydropower

Hydropower is already supplying 16% of all electricity and more than 90% of all renewable energy[12,13]. Most of this is coming from large scale hydropower stations with a capacity of more than 10MW to more than 10GW. The biggest hydropower project, the Three Gorges Dam in China, will have a capacity of more than 22GW when it is finally fully operational (that is 15% of the whole electric power capacity of India in 2005). Large scale hydro projects have become controversial because of the displacement of large numbers of people. A small percentage (0.1–9%, data are very uncertain) comes from mini (<10MW) and micro (<1MW) hydropower systems, mostly without reservoirs, but using river flows. These systems generally operate in rural areas (see Figures 5.12 and 5.13). The global installed hydropower capacity is about 850GW.

Hydropower installations normally have a capacity factor (the percentage of the time they are operational) of 80% or more, so they are usually operated to provide so-called base load power. Some hydropower installations however are operated for peak supply in

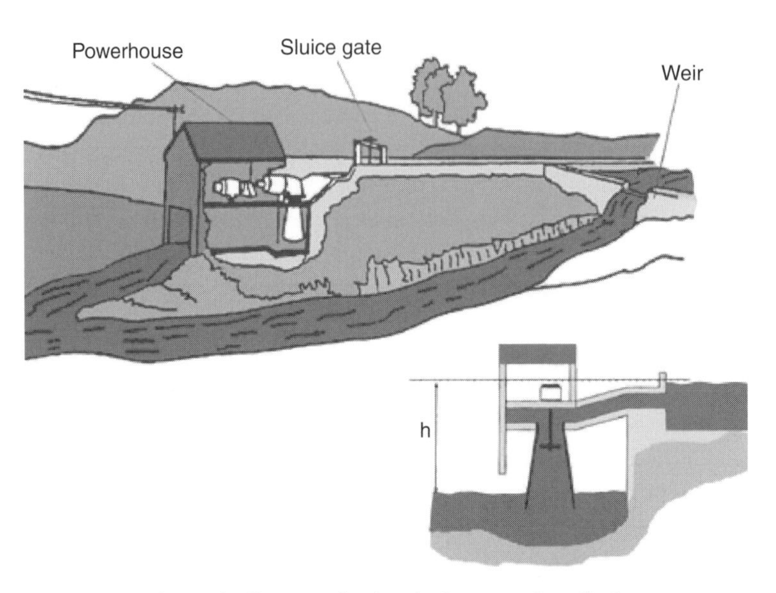

Figure 5.12 **Schematic diagram of micro hydropower installation.**
Source: Fraenkel P et al. Micro-hydro power: a guide for development workers, Practical Action, London, 1991.

Figure 5.13 **Picture of a floating turbine.**
Source: http://www.hydro-turbines.com/id72.html.

combination with pumped storage, meaning that at times of low demand water is being pumped up to a reservoir and at times of peak demand this water is flowing down again to generate electricity.

Hydropower is attractive as a low carbon energy source, although methane emissions from reservoirs due to rotting vegetation can be significant in some places. A study of Brazilian hydro reservoirs showed that some deep reservoirs emitted about as much per kWh electricity as a modern gas fired power plant (i.e. about 400g CO_2-eq/kWh). On average hydropower emissions are estimated at 10–80g CO_2-eq/kWh of electricity produced.

The cost of hydropower is currently 2–10USc/kWh and is estimated to be 3–7USc/kWh by 2030. The increase of the low end of the cost range indicates that the best hydro power sites already have been occupied.

Technically and economically there is room for more than a threefold increase of hydro power capacity. Climate policy will provide strong incentives. The future contribution of hydropower will however depend on managing the social problems created by large new reservoirs. For small scale hydro these problems don't exist. This could add 50% to the current capacity (more than 400GW), which would create excellent opportunities for providing electricity to rural areas, where many people still lack access. Without climate policy a 40–60% increase of hydropower is expected by 2030. Ambitious climate policy scenarios assume this can go up to more than 100%, although the relative costs of other renewables may become so attractive that this figure could be much lower.

Wind

Wind power capacity increased from 2.3GW in 1991 to about 94GW at the end of 2007[14,15]. Capacity has grown by about 25–30% per year since 2000. However, it only produced 0.5% of global electricity. More than 50 countries are using wind power as part of the commercial electricity supply. The biggest capacity can be found in Germany (22GW), USA (17GW), Spain (15GW), India (8GW), China (6GW), and Denmark (3GW). Italy, France, and the UK have a capacity of more than 2GW. New wind power development is aiming more and more at offshore locations, where higher wind speeds and absence of land use restrictions allow for significant expansion, albeit at higher costs. The average wind turbine sold in 2006 was about 2MW, but the largest that are commercially available now are 5MW. These windmills have a rotor diameter of about 120m and a height of more than 100m (see Box 5.4). Small wind turbines with capacities below 100kW are also widely used in many places.

Box 5.4 The influence of wind turbine scale

Wind turbines have been scaled up enormously since the 1980s. Currently about 5600 turbines deliver 20% of Danish electricity. In 1980 about 100 000 turbines would have been needed to produce 10% and by 2025 less than 2000 turbines could produce 50%.

Large windmills benefit from the fact that the wind speed increases with height, the power produced is proportional to the cube of the wind speed (a 2x higher wind speed gives a 2x2x2 = 8x higher power), and the power is proportional to the square of the rotor diameter (a 2x larger rotor gives a 2x2 = 4x higher power).

Figure 5.14 **Pictures of (a) large wind turbine and (b) small domestic wind turbine in rural New South Wales, Australia.**

Source: (a) European Wind Energy Association (b) Shutterstock.com, © Phillip Minnis, image #31168516.

In some countries electricity from wind is reaching a significant percentage of total supply: in Denmark it is about 20%, in Northern Germany 35%, and Spain 8%. An important issue with wind power is the fact that the wind is not always blowing. In 2005 the average capacity factor (percentage of the time the turbines were delivering electricity) was 23%. That means supply needs to come from other sources at times and that backup capacity should be available. Managing the stability of supply requires good forecasting of wind speeds. When wind is integrated in networks that extend over large areas this becomes less of a problem, since 'the wind will blow somewhere' at any point in time.

Another issue with wind (and other dispersed and fluctuating renewable energy sources) is network access. Electricity grids are typically designed to take power from a limited number of big power plants. With an increasing number of small electricity suppliers, access to the network is becoming more difficult. In a number of cases it has already led to delay or cancellation of wind power projects.

The fluctuating character of wind adds to the cost of it, in as far as backup capacity needs to be built if wind power contributes a large percentage to total electricity production (typically at 20% or more). In the worst case scenario it would add something of the order of 1USc/kWh to the cost of wind power[16] (wind power cost is currently

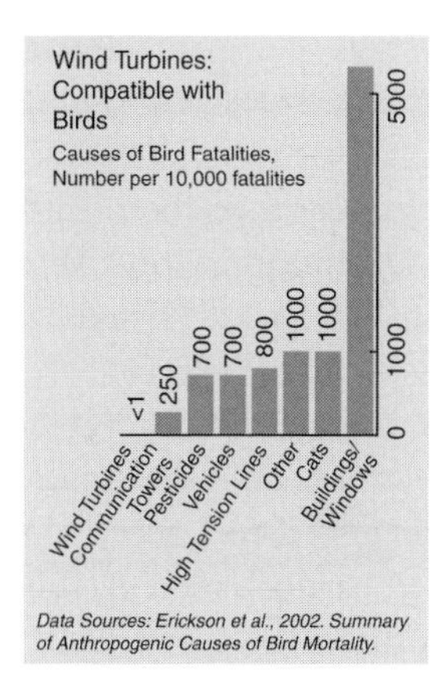

Figure 5.15 **Causes of bird fatalities.**
Source: American Wind Energy Association, Wind Energy Outlook 2007; Erickson et al. Summary of Anthropogenic Causes of Bird Mortality, 2002.

4–9USc/kWh), but very likely much less. Offshore costs are about 1USc/kWh higher than land-based wind power costs. Costs also go up when sites with lower wind speeds are used, but costs will come down as a result of further development and the influence of large scale production of wind turbines. Projections for 2030 on average indicate a cost of 3–8USc/kWh, which will be competitive with fossil fuel based electricity. The advantage of wind power is that it requires relatively low investment per unit of capacity and that it can be built relatively fast.

Prospects for the contribution of wind power to reduce CO_2 emissions are good. The technical potential, not influenced by costs or acceptability, is at least 500 times the current capacity. Taking into account limitations of acceptability and costs however would most likely keep capacity below a 25-fold increase by 2030 under a stringent climate policy scenario, i.e. a contribution of about 8% of total electricity supply. Even without climate policy a 10-fold increase is likely. Much of this new capacity would be offshore. To give an idea of the numbers of wind turbines needed, a 10-fold increase (with on average 2MW turbines) means about 200 000 new turbines. A 25-fold increase means 500 000 new turbines till 2030. The number of wind turbines produced in 2004 was about 6000. The wind turbine industry would have to expand considerably to meet those numbers, but given the historic growth rates that is certainly feasible.

Acceptability of wind turbines is an issue. Many people, even in a country like the Netherlands that has a long history with windmills, object to them because they spoil the landscape. In densely populated areas this limits the siting of new wind capacity seriously. This is an important factor in the move towards offshore locations. Mortality of birds, as a result of being hit by wind turbines, is also a much debated issue. Siting of wind parks away from bird migration routes can reduce those problems and the contribution of wind turbines to bird fatalities should not be overestimated as Figure 5.15 shows.

Bioenergy[17]

Biomass is a major source of food, animal feed, fibre for products like paper, cotton, etc., and last but not least of energy. About 2 billion people in developing countries still rely on traditional fuel such as wood, charcoal, or animal dung for their cooking and heating[18]. That is by far the biggest part of the energy use of biomass. There are many other forms of biomass used for energy: forestry and wood based industry residues, crop residues or whole crops, solid municipal and industrial waste and waste water. Outside the traditional biomass sector, these bioenergy products are usually transformed into different bioenergy carriers: modern solid biomass (as pellets, woodchips, etc.), liquid biofuels (alcohol, diesel fuel), or biogas. These carriers are used either directly as fuels or turned into electricity and heat. Figure 5.16 gives a schematic diagram of this bioenergy system.

In total, biomass supplies about 10% of current primary energy supply. Traditional biomass represents three-quarters of that; modern biomass one-quarter (i.e. 2.5% of total energy supply). About one-third of the modern biomass is used for electricity and heat production, industrial use also accounts for a third. Liquid biofuel only covers 10% of modern bioenergy.

Growth of biomass electricity and heat production is high (50–100% per year) in some OECD countries, like Germany, Hungary, the Netherlands, Poland, and Spain. Small projects in rural areas are also growing fast in some developing countries, such as

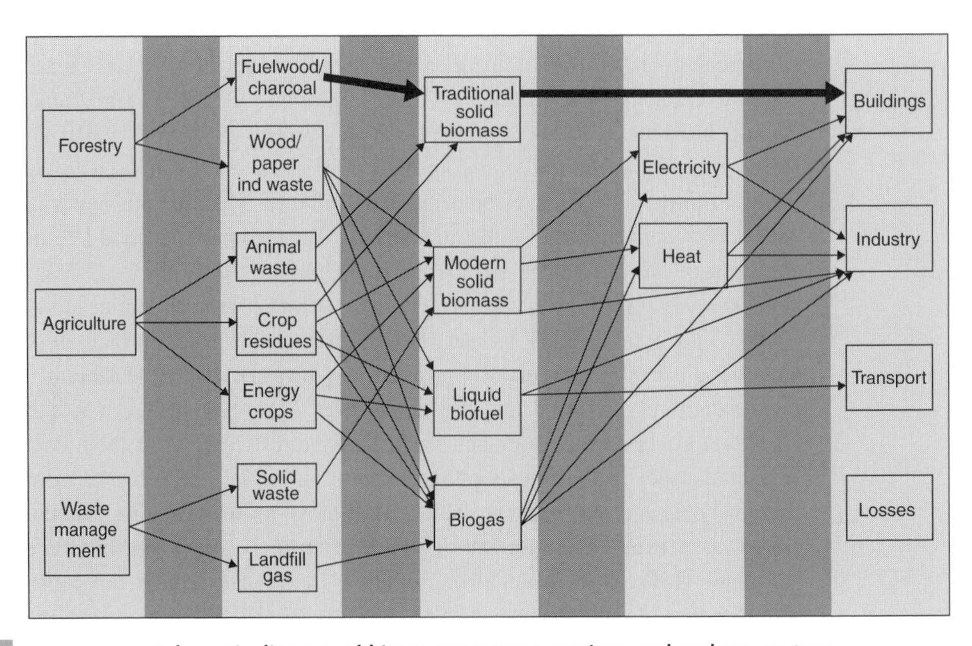

Figure 5.16 Schematic diagram of bioenergy sources, carriers, and end use sectors.
Source: based on IPCC Fourth Assessment report, Working Group III, figure 4.14.

Thailand. All growth however is from a very small base. Total installed capacity of biomass based power generation was about 45GW by the end of 2006[19].

Technology

The dominant technology for modern solid biomass use in electricity and heat production is so-called co-generation (Combined Heat and Power (CHP); see section on Power plant efficiency and fuel switching) with direct firing of the biomass. Different technologies are being used, depending on the type of biomass available. These biomass systems are relatively small: they typically have a capacity of 50MW or less, compared to coal fired plants that have a range of 100–1000MW. Heat is normally used for district heating, industrial processes, or greenhouses[20].

Recently co-firing of biomass in coal fired power plants and co-firing biogas (from landfills and biogas plants) has gained interest. More than 150 coal fired power plants currently have operational experience with co-firing, using a large variety of biomass materials, including wood chips, wet and dry agricultural residues, and energy crops. It is a relatively simple and low cost method of using bioenergy. However, the supply of large quantities of biomass to big power plants may be a problem (see below). Municipal solid waste incineration, one of the widely used waste management technologies, is a form of co-firing: organic material combined with plastic and paper is used to generate heat and electricity.

Biomass can also be gasified by heating it in an environment where the solid material breaks down to form flammable gas. After cleaning the gas can be burned in a gas turbine or more simple combustion engine. This technology is just beginning to become commercially available.

Gasification of biomass, in the form of animal waste and waste water, can also be done biologically in biological digesters in the absence of oxygen (called anaerobic digestion). This technology is being used in many places. In Europe alone, more than 4500 installations were operating in 2002. In several developing countries rural application of biogas has developed strongly (see Box 5.5 on biogas in China). Given the small scale of biogas digesters, the overall contribution is limited on a global scale, but can be significant in rural areas.

Box 5.5 **Biogas digester programme in China**

China has the biggest programme of installing biogas digesters in rural areas. By the end of 2005 about 18 million household digesters were installed and plans for reaching 84 million by 2020 are in place. It is now one of the priority areas in the central government's policy of improving the condition of people in rural areas. Subsidies are now available from the central and local governments for installing biogas units. Digesters are constructed from bricks and concrete and increasingly mass produced from fibreglass reinforced plastic. In recent times the emphasis is put on integrating the biogas plant in the farm: improving sanitation and crop productivity, and providing cooking gas.

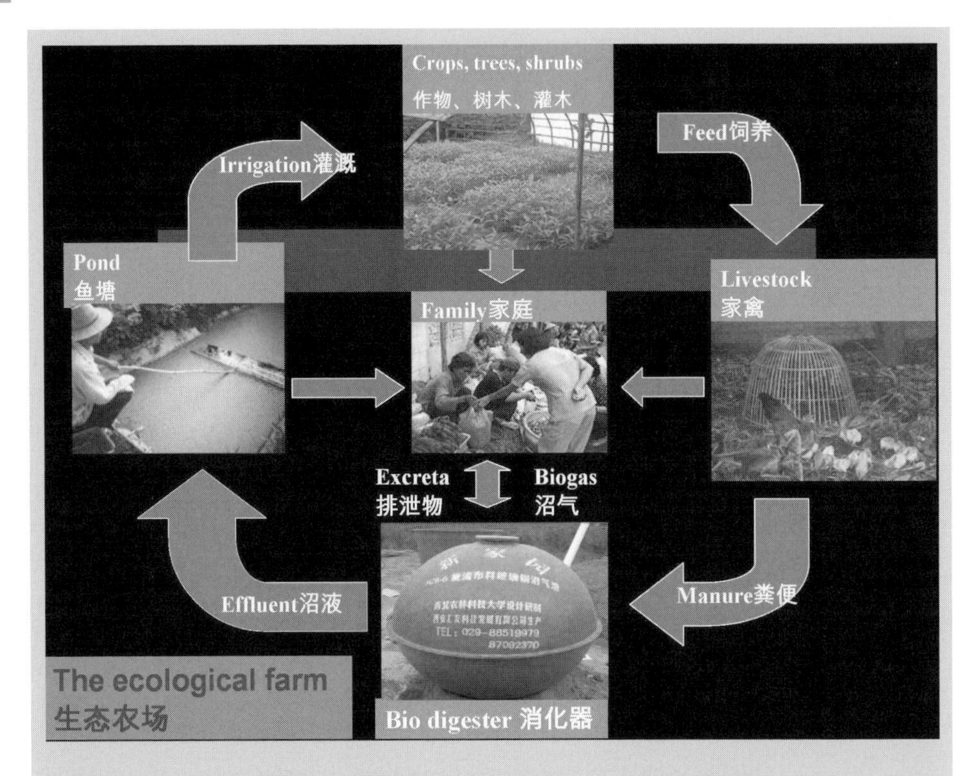

(Source: Zhang Mi. Chengdu Energy-Environment International Corporation. Presentation at Biogas meeting, Dhaka, March 2007)

Biomass sources

Figure 5.16 above already showed the various sources of biomass that are relevant to energy production. The bulk of the biomass becomes available in rural areas and a smaller fraction in urban areas (landfill and waste water treatment gas, municipal solid waste and industrial waste from processing of wood and agricultural products). One particular problem is that the energy density of biomass resources is low and it has to be collected from a wide area to supply the amounts needed at a CHP bioenergy plant. This explains the relatively small size of these units as outlined above.

Another issue is the sustainability of the biomass supply. For true waste materials, such as municipal solid waste, waste water and industrial waste materials, no problem exists. In many cases however there is competition. Crop residues for instance are often ploughed back into fields to keep up the organic content of the soils, animal manure is used as fertilizer, and land dedicated to energy crops would not longer be available for food or cash crop production. In rural areas there may also be competition for labour. Need for additional land for energy crops could lead to additional conversion of forests or biologically rich natural vegetation into agricultural land and thereby destroy ecosystems and lead to loss of species and biodiversity. This displacement effect is often neglected when discussing bioenergy.

Impacts of energy crops on food production and on biological systems need to be carefully considered in decisions to develop bioenergy potential. There are indications that the currently used energy crops, such as maize, soybeans, sugar beet, and oilseeds, may have some negative impacts on food prices and biological systems, but the increased demand for food, animal feed, and industrial use is likely to be the real reason for price increases. This issue will be discussed in more detail in Chapter 9 on Agriculture and Forestry.

Costs

Costs of biomass based electricity vary between 5 and 12USc/kWh. They can be lower if the biomass used has a negative value, i.e. in cases where waste otherwise would have to be disposed of. The relative small scale of biomass based CHP units and the relatively costly collection and preparation of the biomass explain this high cost. By 2030 further technological development and economies of scale should lead to lowering these costs to about 3–10USc/kWh. In Sweden, where there is a long experience with biomass electricity, each doubling of the installed capacity of CHP plants led to a 7–10% reduction in costs per kWh[21].

Reduction potential[22]

The first issue, when discussing the CO_2 reduction potential from bioenergy, is the net gain in terms of CO_2 emissions. Bioenergy in principle has an advantage over fossil fuel in the sense that it captures CO_2 from the atmosphere when the biomass grows. However, it takes energy to grow and harvest the crops and to transport and process the biomass, and the efficiency of the electricity generation may be less than that for coal or natural gas. In case of liquid biofuel or biogas production there is also the energy to run the process and refine the product. There may also be a net loss of carbon, when forest or land with natural vegetation is converted to land for energy crops. Or, in the case of crop residues, carbon lost from the agricultural soil. The CO_2 generated from this additional energy use and land use change needs to be subtracted from the gains made by using bioenergy. Unfortunately there is still a large controversy about the right numbers.

The other issue is the demand for bioenergy. This demand depends on the relative costs of bioenergy in terms of CO_2 avoided, compared to other reduction options, the level of ambition of climate policy, and the question of whether there is enough supply of biomass available. As far as biomass for electricity and heat is concerned, supply is not the limiting factor. Without having to rely heavily on energy crops, ample amounts are available to supply the demand for electricity and heat in a number of scenarios until 2030. Modern biomass could increase its share from about 2% of total electricity generation in 2030 without climate policy to something like 6% under a moderate climate policy. This would mean electric power capacities of 200–400GW by 2030. For liquid biofuels the situation is less clear. This issue will be discussed in more detail in Chapter 6 on transportation.

Geothermal energy[23]

Hot water and steam from deep underground in volcanic areas of the world are being used to generate electricity and to provide heat for warming buildings. In addition, heat from shallow soils and ambient air can be captured with heat pumps for warming of individual buildings, which will be discussed in Chapter 7. There are more than 20 countries where geothermal energy makes a significant contribution to electricity supply (see Table 5.4). Iceland, a country with high volcanic activity, gets more than 25% of its electricity and 87% of its home heating from geothermal energy. El Salvador (20% of electricity), Philippines (18%), Costa Rica, and Kenya (both 14%) are also forerunners. On a global scale geothermal electricity covers less than 0.3% of the electricity supply.

More than 40 countries use geothermal heat for purposes other than hot baths[24]. About half is being supplied to industries, greenhouses, and buildings from centralized systems, the other half by individual heat pumps.

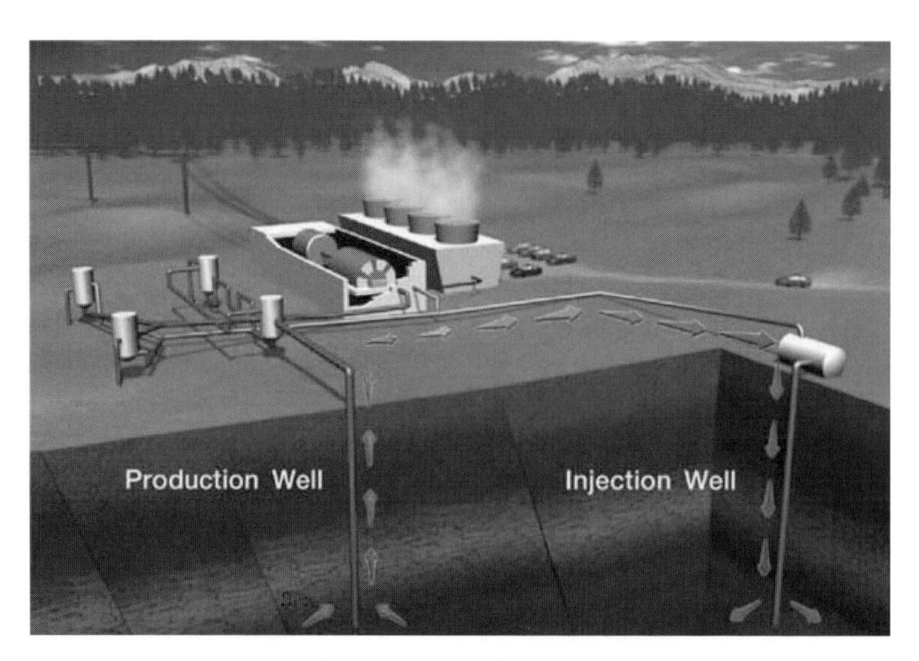

Figure 5.17 **Schematic drawing of geothermal power plant.**

Source: geothermal education office http://geothermal.marin.org/GEOpresentation/sld002.htm.

Table 5.4. Countries with highest geothermal electricity generating capacity by end 2005

Country	Geothermal electricity capacity (MW)
USA	2540
Philippines	1930
Mexico	950
Indonesia	800
Italy	790
Japan	535
New Zealand	435
Iceland	320

Source: International Geothermal Association.

Most geothermal power plants get a mixture of steam and water from drilled wells at depths of less than 2km, where temperatures are above 250°C (see Figure 5.17). In some cases water is pumped into hot dry rock formations. Water is often pumped back into the geological formations to keep pressure up and avoid water pollution from contamination. Some geothermal wells also produce CO_2 from volcanic origin, which comes up with the steam from deep wells and may annihilate the gains of geothermal energy in terms of CO_2 reduction.

Costs of geothermal electricity are currently 4–10USc/kWh, depending strongly on the local circumstances. This is well above the cost of coal fired power. Costs are projected to come down somewhat due to improvement of the technology by 2030 to 3–8USc/kWh. Emissions of CO_2 are often not equal to zero, because of volcanic CO_2 coming up with the water and steam. No reliable data are available however.

The technical potential of geothermal energy is very large, around 10 times current total primary energy use. Only a fraction of this can be tapped however by 2030, even with an ambitious climate policy in place. A share of about 1%[25] of total electricity supply by 2030 would probably be an upper limit, which is roughly equivalent to a fourfold increase of geothermal electricity and is consistent with the 7.5% per year growth that geothermal energy has shown over the last 35 years.

Solar

The solar radiation reaching the earth surface is more than 10 000 times the current annual energy consumption. The intensity varies, with the best areas in the subtropics (see Figure 5.18). Solar radiation can however be captured anywhere in the world, albeit with lower efficiency. There are three ways of capturing this energy:

Figure 5.18 **Areas with strong solar radiation (>400 GW/km^2).**
Source: Shine WB, Geyer M. Power from the sun, http://www.powerfromthesun.net.

- *Concentrating solar power*: concentrating the solar radiation with mirrors, heating a fluid, and using that heated fluid to generate electricity.
- *Solar photovoltaic*: generating electricity directly in a light sensitive device made out of silicon semiconductors (a photovoltaic (PV) cell)
- *Solar heating and cooling*: collecting direct heat of the sun in a system to heat water for domestic or other use or use solar radiation to drive a cooling system.

Concentrating solar power[26]

Concentrating solar radiation can be done in several ways (see Figure 5.19).

The most mature form is a set of mirrors in the form of a 'parabolic through' that concentrates solar radiation on a tube containing the working fluid from which electricity is produced. These systems have reached an overall efficiency of about 20% (i.e. 20% of the incoming radiation is converted to electricity). The biggest commercial plant is a 150MW facility in California (see Figure 5.20).

The other system operating at scale is a so-called 'solar tower': a set of flat mirrors that follow the sun ('heliostats') and concentrate radiation onto a tower where the working fluid is heated (see Figure 5.21). A few tower systems are operating in the USA and the EU (Spain) at a scale of about 10MW. The Spanish system is planned to be expanded to 300MW by 2013.

The total installed capacity of CSP is currently about 400MW, with most of it dating from the early 1990s, when tax credits in California led to construction of 350MW capacity plants. The recent addition of the Seville plant, as a result of a new feed-in tariff law in Spain, and plans for another 1400MW plant in 11 countries, indicate a more favourable situation. CSP systems are best placed in areas receiving high levels of solar radiation. They also have the advantage of a fairly high energy density, i.e. the land required for delivering significant amounts of energy from CSP installations is smaller

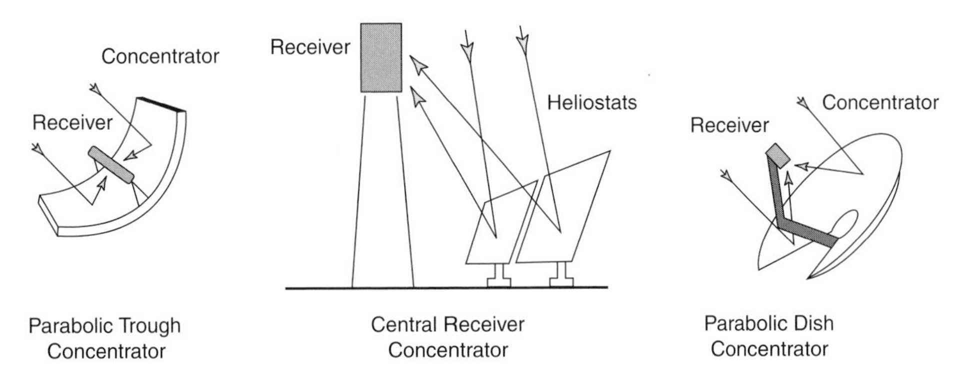

Figure 5.19 Different concentrating solar devices.
Source: Shine WB, Geyer M. Power from the sun, http://www.powerfromthesun.net.

Figure 5.20 Picture of the world's largest concentrating solar facility in California. It consists of a 150 megawatt concentrating solar power system that utilizes parabolic trough collectors.
Source: Desertec-UK, http://www.trec-uk.org/images/.

Figure 5.21 Picture of a 11MW concentrating tower system, Seville, Spain.
Source: Abengoa Solar, Spain.

than for most other renewables. Output is about 125 GWh/year for 1 square km (when a 10% conversion efficiency is assumed). That means only 1% of the world's deserts (240 000km^2, half the area of France) would be sufficient to produce all the electricity needed in the world by 2030. Of course, this is only in theory, because so far no practical solutions are available to transport that power to where the users are. Research is ongoing on developing high voltage direct current grids to make that easier.

Costs of electricity from CSP systems are currently between 10 and 45USc/kWh. Costs for new installations by 2030 are estimated to be much lower at 5–18USc/kWh. The potential for contributing to CO_2 reductions will depend strongly on the cost reduction achieved over the next 20 years. For the time being costs of CSP are higher than many other low carbon options. Once the 5USc/kWh level is reached, CSP would become competitive. For the price to come down that far many installations need to be built in order to gain experience. The current rate of cost reduction is about 8% for each doubling of capacity. Driving the cost down to the 5c/kWh level will require considerable subsidies.

Estimates of the contribution of CSP to low-carbon electricity are modest and very uncertain. For all solar power together no more than a 1.5% contribution to electricity supply is expected by 2030 under strong climate policy[27].

Solar photovoltaic[28]

Silicon semiconductor based PV cells are currently the dominant technology. The cells come in different varieties: monocrystalline silicon with about 18% efficiency (33% of the market), cheaper polycrystalline cells with15% efficiency (56% of the market), and even cheaper thin-film cells with 8% efficiency (9% of the market). There is a clear trade-off between costs and efficiency. They are applied on a wide variety of scales: from miniature cells powering a watch or a few PV panels on a roof, up to large arrays of PV panels generating more than 10MW[29] of electricity. There are even serious plans to build a 100MW PV plant in China[30].

About 70% of the total installed capacity by the end of 2007 was connected to the grid (about 8GW). In the case of home systems that means electricity generated that is not needed for the building is delivered back to the grid (and the grid supplies when the PV cells do not generate enough electricity). This grid connection has for a long time been discouraged by electricity companies by offering very low pay-back rates. With feed-in tariffs becoming popular in many countries this is rapidly changing. A typical grid connected home system is shown in Figure 5.22.

In areas where no grid connection exists, particularly rural areas in developing countries, many individual solar home systems have been installed. This is part of the roughly 3 GW solar PV capacity that is not connected to the grid. These systems normally have a limited capacity, enough for a few light bulbs and a TV set. Increasingly there has been resistance by individual people to invest in these systems, because they fear that having a solar home system will make it unlikely that the government will invest in a grid system for their particular village. Mini-grid systems at village level may be the solution for this problem.

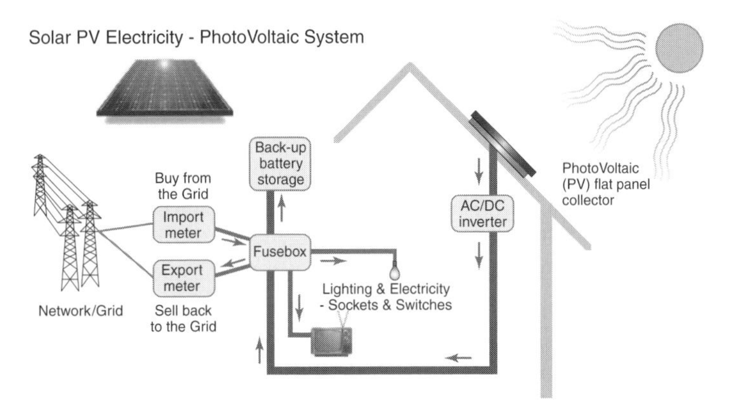

Figure 5.22 **Grid connected solar home system.**
Source: http://www.saveenergyuk.com/solar_lighting_electricity.htm.

The prospects for solar PV are very good. Growth rates of installed capacity of grid connected systems have been about 50–60% per year, albeit from a very small base (currently 0.004% of global electricity). Annual production of PV panels is now about 2GW. Costs of electricity from solar PV are however still high: from 25USc/kWh in very sunny areas to about 1.6$/kWh in less attractive areas and somewhat older systems. Costs are coming down rapidly however (about 18% for every doubling of installed capacity) and by 2030 the cost could be 6–25USc/kWh, which would bring the cheapest systems into the range where solar PV can compete with other low carbon options. Much effort is currently put into developing PV integrated building materials, such as wall and roof panels (more about that in Chapter 7 on Buildings).

Estimates for the contribution that solar PV can make by 2030 towards CO_2 reductions vary widely, in light of the costs. As indicated above, total solar electricity by 2030 is likely to be less than 1.5% of total supply. In the longer term, beyond 2030, the potential for solar PV could become significant though, as costs continue to fall rapidly.

Solar heating and cooling[31]

Solar hot water heaters for domestic housing are the most common form of solar heating that is found today. Other applications are for space heating, swimming pool heating, and industrial processes. It is discussed in Chapter 7 for buildings and Chapter 8 for industrial processes.

Ocean energy[32]

In principle a lot of energy could be obtained from waves, tidal flows, ocean currents, and from temperature differences between the ocean surface and the deep ocean.

The economically exploitable potential for the period until 2030 is however small. Currently there are only a few tidal flow installations with a capacity of not more than 260MW.

Wave energy contributes even less: there are only two commercial projects with a total capacity of 750kW. Most wave energy technologies operate at the surface, either through using the up or down movement of waves or the breaking waves at the shore, to operate a generator to produce electricity. There are many different types under development. One system, the Archimedes Wave Swing system[33], operates on the basis of a submerged buoy, 6m below the surface, filled with air and attached to the sea floor, that moves up and down with passing waves; the up and down movement is then converted into electricity.

Ocean thermal energy conversion (OTEC) systems, which aim to obtain electricity from temperature differences in the ocean, are currently only at the research and development stage, as are turbine systems positioned in areas of strong ocean currents and systems designed to obtain energy from salinity gradients.

It is very hard to predict when ocean energy systems could become commercially attractive, given the absence of large scale experience and realistic cost estimates. Theoretical calculations of the potential for wave power along the world's coasts show that 2% of the 800 000km of coast has a high enough wave energy density to make wave power systems attractive. Assuming a 40% efficiency in converting wave energy to electricity this would mean a 500GW electrical capacity. At this stage however these are purely theoretical calculations.

CO_2 capture and storage and hydrogen

The last option for reducing CO_2 emissions from the electricity sector is not to move away from fossil fuel, but to make fossil fuel use sustainable by capturing CO_2 before it is emitted, to transport it, and then either use it in some industrial process or to store it safely[34]. This is called CO_2 Capture and Storage (CCS).

CO_2 capture

The technology of capturing CO_2 from gas streams has been applied at commercial scale for a long time in refineries and fertilizer manufacturing plants (to separate CO_2 from other gases) and natural gas cleaning operations (to get rid of high natural CO_2 levels in some gas fields). Application at large scale coal or gas fired power plants has not yet happened. There are three different systems for CO_2 capture at power plants[35]:

- *Post combustion capture:* removing CO_2 from the flue gas that comes from the power plant, before it enters the smoke stack. The most common method for CO_2 removal is to let the flue gas bubble through a liquid that dissolves CO_2 and then to heat that fluid again and drive the pure CO_2 out. This is in fact an add-on technology that could be used at any coal or gas fired power plant. The technology is basically the same as that used in natural gas treatment facilities, where CO_2 is removed from the gas stream before transporting it to users via pipelines.

- *Pre combustion capture:* in this system the fuel, mostly coal, but also applicable in principle to biomass, is gasified and converted in a chemical process (so-called Fischer Tropsch process) to hydrogen and CO_2. The CO_2 is then separated from the hydrogen with a liquid absorption as described above or a different process. The hydrogen is used in gas turbines to generate power. For coal fired plants this system is called Integrated Gasification Combined Cycle (IGCC). This technology is very similar to the one used in hydrogen production in refineries and in fertilizer manufacture.
- *Oxyfuel combustion and capture:* CO_2 in flue gases from a traditional coal or gas fired power plant is mixed with a lot of nitrogen and oxygen from the air that was used in the combustion. CO_2 in an IGCC is mixed with hydrogen. That means large quantities of gas have to be pumped through a CO_2 separation unit, which is costly. Therefore a third system was developed in which coal or gas is not burned with air, but with pure oxygen. This produces a flue gas stream with high CO_2 content, making the CO_2 removal simpler and cheaper. This technology has so far only been demonstrated at relatively small scale.

Figure 5.23 gives a schematic diagram of these three systems and also shows for comparison the systems used in natural gas treatment and industrial processes. Only about

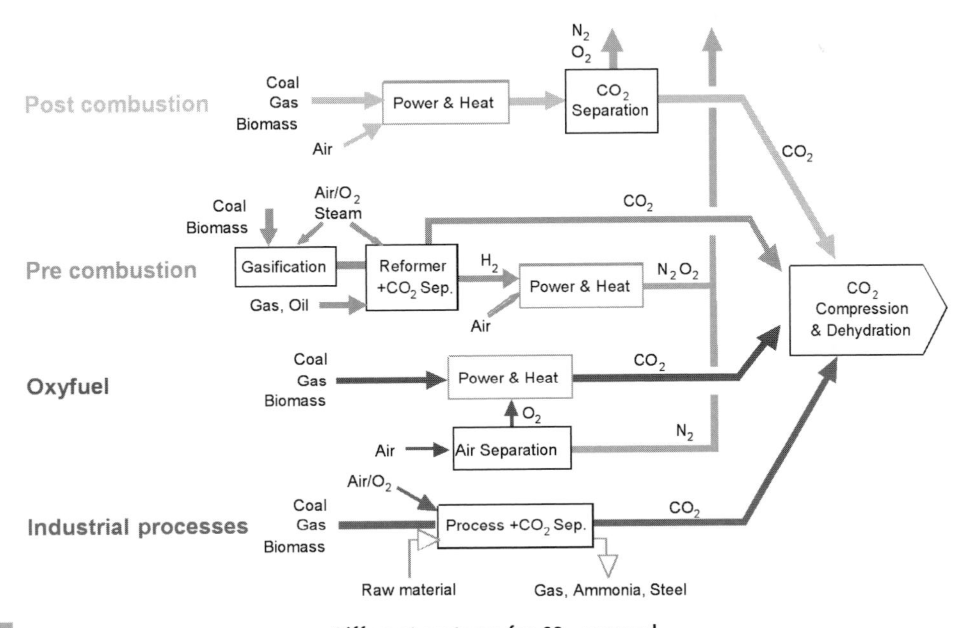

Figure 5.23 **Different systems for CO_2 removal.**
Source: IPCC Special Report on CO_2 Capture and Storage.

90% of CO_2 is captured, because going to higher percentages would require very costly installations. The capture process requires large amounts of energy, which means that the electricity output of a power plant goes down significantly when CO_2 capture is added. The amount of energy used for capture is 10–40% of the total. In other words, the net efficiency of the plant with CCS is reduced compared to the same plant without CCS. Technological development is aiming therefore at reducing this energy requirement by using more effective fluids or membrane systems. Expectations are that the energy loss can be reduced by 20–30% over the next 10 years.

Storage or usage of CO_2

There are some industrial processes that use CO_2, such as production of urea fertilizer, carbonated drinks, refrigeration, and food packaging. There is also some usage of CO_2 (normally through burning of gas) in agricultural greenhouses to enhance plant growth. Captured CO_2 can be used in such cases, but quantities are very small compared to the amounts power plants produce[36]. Processes to convert CO_2 in chemicals or biological material such as algae are under development, but it remains to be seen if they can reach a positive energy balance. So far these processes require more energy than they produce.

So the only meaningful way is to store CO_2. Geological formations (depleted oil and gas fields, unusable coal seams, and water bearing formations that have no use for drinking or industrial water supply) are the preferred storage medium. Storage of CO_2 by dissolving it into oceans is still very much at the research stage (see below). In principle there is yet another storage method: letting CO_2 react with minerals to form a solid carbonate and to dispose of this solid waste. Costs and waste management problems are such however that there are very poor prospects for this method.

'Depleted oil and gas fields' (these fields still contain sizeable amounts of oil or gas, but are no longer economical) are prime candidates for CO_2 storage because their geology is well known and they have often contained gas for millions of years. That is very important because a good CO_2 storage site would have to retain CO_2 for a very long time (thousands of years). In addition, a well known technique for enhanced oil recovery (getting more oil out of a field than through traditional pumping) is to pump CO_2 into a 'depleted' oil field to 'sweep' additional oil out. Thus CO_2 storage in oil fields can be combined with getting additional oil out, provided that it is ensured that the CO_2 does not escape, which occurs in traditional CO_2 enhanced oil production. This approach is currently being used at large scale at the Wayburn oil field in Canada. The same principle can be applied to depleted gas fields. BP's natural gas cleaning plant in In-Salah, Algeria, where about 1 million tonne per year of CO_2 is captured and stored, uses this approach.

There are large water bearing geological formations, so-called aquifers, that are not used for other purposes. If these aquifers had a structure that would prevent CO_2 from escaping to the surface, then they could be used for CO_2 storage. The Sleipner CO_2 capture plant in Norway (removing CO_2 from natural gas) pumps about 1 million tonnes of CO_2 annually into a nearby aquifer (see Figure 5.24).

Figure 5.24 **Picture of Statoil's Sleipner CO_2 separation and injection platform.**
Source: StatoilHydro, Image courtesy of Marcel Fox, image at http://www.mfox.nl/experiences4.html.

There are many coal seams that are uneconomical to exploit. These coal seams can in principle be used to store CO_2. When pumping CO_2 into the coal, it is adsorbed and so-called coal-bed methane is driven out, producing a useful gas stream. This technology is still in the development stage and so far there are still problems with getting the CO_2 to penetrate the coal seam in an even manner. Figure 5.25 shows schematically the different geological storage methods.

Hydrogen

CCS is the key to hydrogen as a future clean energy carrier. As outlined above, coal gasification combined with a Fischer Tropsch chemical conversion and CO_2 capture produces hydrogen. Natural gas can be converted in a similar way to hydrogen. Currently this is the cheapest way to do it, and its use is widespread in refineries, chemical plants, and fertilizer manufacture. When the CO_2 produced during hydrogen manufacture is properly stored, the hydrogen then is a low carbon fuel.

Hydrogen can in principle be produced by electrical decomposition of water. Low carbon electricity (nuclear, renewable) could therefore also produce a low carbon hydrogen. Costs are however much higher.

Hydrogen is a very clean fuel for heat and electricity production; it only produces water as a combustion product. Hydrogen also has the potential to be a clean transportation fuel, if hydrogen fuel cells are used in vehicles (more about that in Chapter 6 on transportation). For it to become a significant energy carrier, a hydrogen infrastructure needs to be developed in the form of a pipeline network. Currently there are only a few regional hydrogen pipelines in heavily industrialized areas like North-Western Europe. With a hydrogen pipeline network low carbon electricity and heat could be produced in multiple locations. Use of stationary fuel cells, having a higher electrical efficiency than gas or steam turbines, would then be possible.

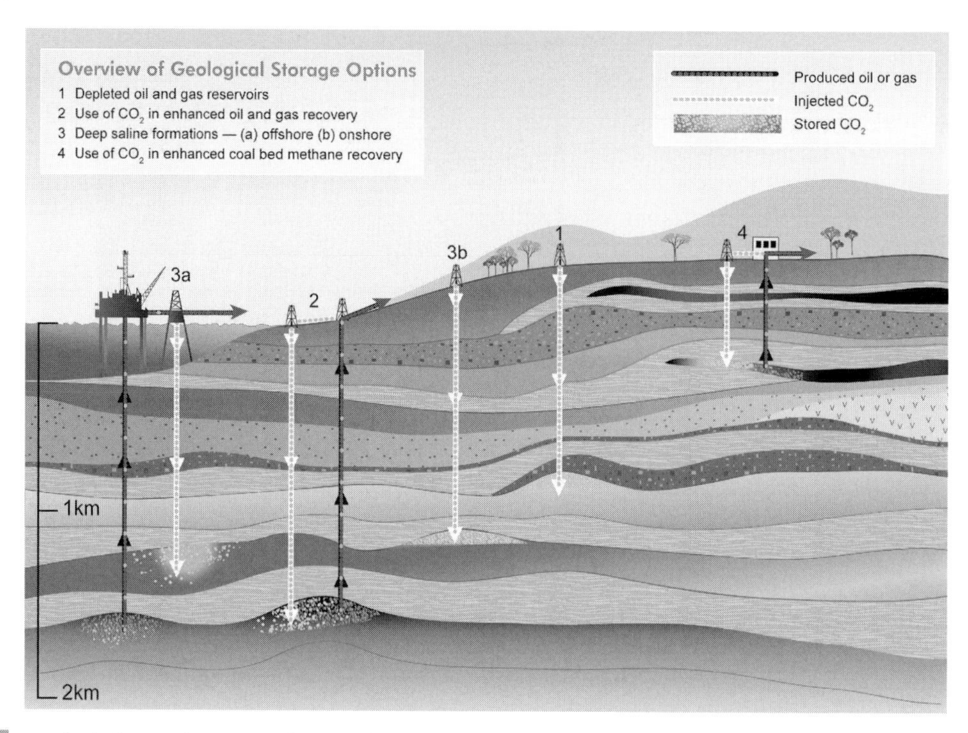

Figure 5.25 **Methods for storing CO_2 in deep underground geological formations. Two methods may be combined with the recovery of hydrocarbons: EOR (2) and ECBM (4).**

Source: IPCC Special Report on CO_2 Capture and Storage, figure TS.7. See Plate 10 for colour version.

Transport of CO_2

CO_2 needs to be transported from the place where it is captured to a storage site. The preferred method is by pipeline, after compressing the CO_2 to become a fluid-like substance. Since costs of pipeline transport are basically proportional to the length of the pipe, distances for this type of transport need to be limited to about 1000–1500 km. Beyond that, in cases where that would be economical, transport by ships, comparable to current LNG tankers, is better. Pipeline transport of CO_2 is a known technology. In the USA alone there is more than 2500 km of CO_2 pipeline. CO_2 shipping is not practiced yet.

Safety

There are risks involved in CO_2 capture, transport, and storage. Concentrated CO_2 is dangerous because it is colourless and odourless and at levels of more than 7–10% in air, it can kill after exposure of less than 1 hour. Handling concentrated CO_2 therefore requires stringent safety measures, comparable to those for handling toxic and flammable products from the oil and chemical industry. Pipelines need to be constructed from special corrosion resistant materials.

For storage similar considerations apply, since leaking of CO_2 to the surface from a geological storage site could concentrate CO_2 in the basements of houses. It can also cause harm to animals and plants. In addition, the biggest concern is that CO_2 could leak back to the atmosphere, which would make the whole operation of capture and storage a pointless exercise. Geological formations for storage therefore need to be carefully characterized in terms of their ability to retain gas. Monitoring of the distribution of the CO_2 underground needs to be performed and emergency measures to close a possible leak need to be prepared. If properly handled in that way, it is unlikely that CO_2 storage sites would leak more than 1% in 1000 years.

Costs

CCS is not cheap and that explains why it has only been applied in two large scale installations in the gas treatment industry (Sleipner and In-Salah). For use at power plants the costs of adding CCS are currently 1–3 USc/kWh for gas plants and 2–5 USc/kWh for coal fired plants. Or, expressed in $/t$CO_2$ avoided: 20–70 US$/t. This is higher than many other reduction measures available today. The potential for cost reduction is however significant, so that by 2030 costs could go down significantly.

Potential for CO_2 reduction

The reduction potential of CCS is very large. The total storage space available is more than 2000 $GtCO_2$[37], which would be sufficient to store 80 times total current global CO_2 emissions and about double the amount that would be required this century, even under very ambitious climate policy assumptions. So it is basically competition with other reduction options that will determine the role of CCS in the period till 2030. Depending on climate policy, expectations are that CCS could become commercially applicable around 2020 and by 2030 could be applied at about 10% of all coal fired power plants in the world.

The prospects for hydrogen produced from natural gas or coal with CCS is very uncertain, because it depends on a hydrogen infrastructure. That infrastructure would probably only make sense if there were a significant demand from transportation. And as hydrogen fuel cell vehicles are not expected to become commercially available in significant numbers before 2030, these prospects are very uncertain at this moment.

Comparing CO_2 emissions

Table 5.5 gives an overview of the CO_2 emissions per kWh of the various power supply options that were discussed above, as well as the contribution these options can make to electricity supply.

Table 5.5. CO_2 emissions per kWh for different electricity supply options

Option	CO_2 emissions (gCO_2-eq/kWh)	2006 electricity supply (TWh)[a]	2030 BAU electricity supply (TWh)	2030 ambitious climate policy (TWh)[b]
Coal	680–1350	7760	14600	4230
Gas	350–520	3810	6720	4190
Coal CCS	65–150	0	0	1740
Gas CCS	40–70	0	0	670
Nuclear	40–120	2790	3460	5430
Hydro	10–80	3040	4810	6640
Modern biomass	20–80	240	860	1730
Wind	0–30	130	1490	2750
Geothermal	n/a	60	180	220
Solar	10–100	4	350	720
Ocean	n/a	1	14	50

[a] From IEA WEO 2008 reference scenario.
[b] From IEA WEO 2008 450ppm CO_2-eq stabilization scenario; new renewable shares estimated based on relative share in 550ppm scenario.
Source: IPCC Fourth Assessment Report, Working Group III, fig 4.19, IPCC Special Report on CCS, fig TS.3 and IEA WEO 2008.

Comparing costs

Figure 5.26 summarizes the cost per unit of electricity for the various options. Low carbon options become competitive with coal and gas fired power plants, as costs for fossil fuel go up and costs of low carbon options come down. When climate policy leads to a price on carbon, more low carbon options become attractive. For instance, if it became a requirement that coal and gas fired plants were equipped with CO_2 capture and storage (CCS), many low carbon electricity generation options would become competitive by 2030, including the lowest cost solar power (CSP and PV).

These expected cost developments explain why the contribution of low carbon electricity is growing even without policy intervention. Policy intervention will further enhance these contributions.

Does that also work for climate policy?

For climate policy the cost of electricity is not the primary issue to look at. The main concern is to reduce emissions and therefore it makes sense to look at the cost per tonne of CO_2-eq avoided and then to take the cheaper measures first. These avoidance costs can only be calculated when two alternatives are compared. For example,

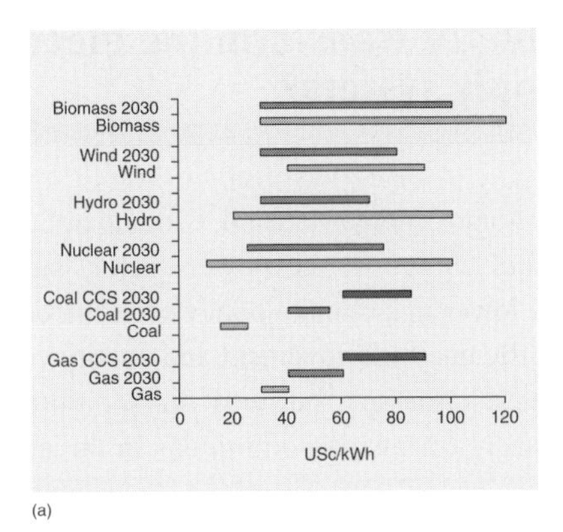

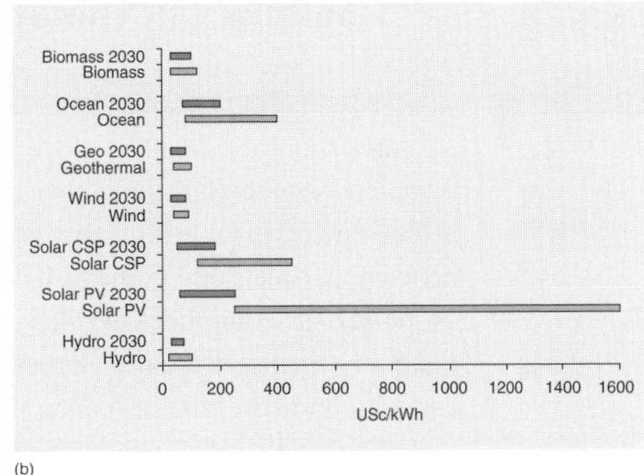

Figure 5.26 **(a) Cost of different electricity supply options, current and for 2030; for coal and gas the 2030 cost including CCS is also given. (b) The same for renewable electricity options, indicating the strong expected cost reduction between now and 2030.**
Source: IPCC Fourth Assessment Report, Working group III, table 4.7.

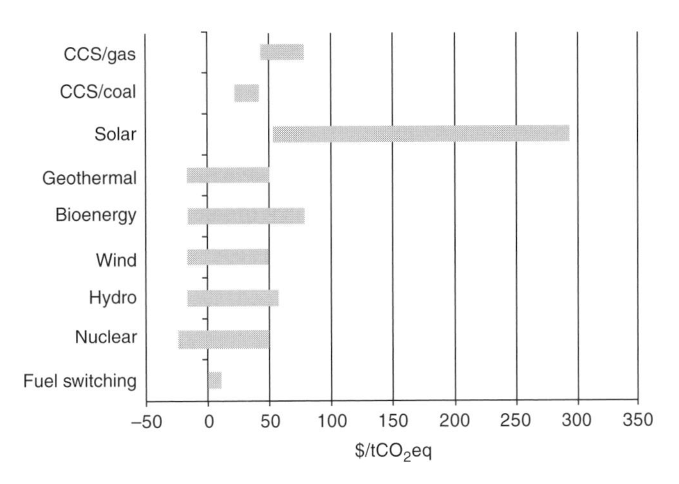

Figure 5.27 **Relative cost in US$/tonne CO_2-eq avoided for different electricity supply options, compared with a coal fired power plant.**
Source: IPCC Fourth Assessment Report, Working Group III, ch 4.4.3.

the avoided CO_2 from wind power, compared with coal fired power, can be calculated, as well as the additional costs of using wind instead of coal. That would give a cost per tonne of CO_2 avoided. Comparing wind with gas fired power gives a different answer. Figure 5.27 summarizes the cost per tonne of CO_2-eq avoided for the various low carbon options, compared with coal fired power. It shows that many renewable energy options and nuclear power have a low-end cost per tonne of CO_2-eq avoided that is negative.

So how can climate policy transform the electricity supply system?

The first issue to consider is the reduction of energy demand. Climate policy will not only affect power plant efficiency and fuel choice in the power sector, but also end use efficiency. Electricity demand will be lower in a climate policy situation compared with a no policy situation, because energy efficiency improvement and energy conservation are generally cheap. Reduced demand will influence the need for additional power plant capacity and that will influence the share of low carbon options in the electricity mix.

The second issue concerns the choice of low carbon options. An economically rational approach for climate policy would pick the cheapest options first. That means options with the lowest costs per tonne of CO_2-eq avoided will be prioritized. In practice this ideal is not met. There are preferences and interests that will lead to less than optimal cost outcomes. In some countries nuclear power is not considered, in others there may be resistance against the building of many windmills, while in yet others considerations of energy security point in the direction of maintaining coal use. Such less than optimum choices can still deliver a low carbon electricity system, but at higher cost.

The third issue is the choice of policy instruments. When policy intervention is done in the form of a so-called renewable portfolio standard (requiring a minimum share of renewable energy), this normally does not lead to the least cost outcome for the electricity system as a whole. The same happens with feed-in tariffs (guaranteed prices paid for renewable electricity by the electricity distributors) or subsidies that are used to stimulate the penetration of renewable energy in many countries, because the specific tariffs may not be set in an optimal fashion.

Last but not least the stringency of the policy will of course have a dominant influence.

Figure 5.28 shows the share of the various electricity options for a number of policy scenarios for replacement of coal and gas. The scenarios have a different ambition level:

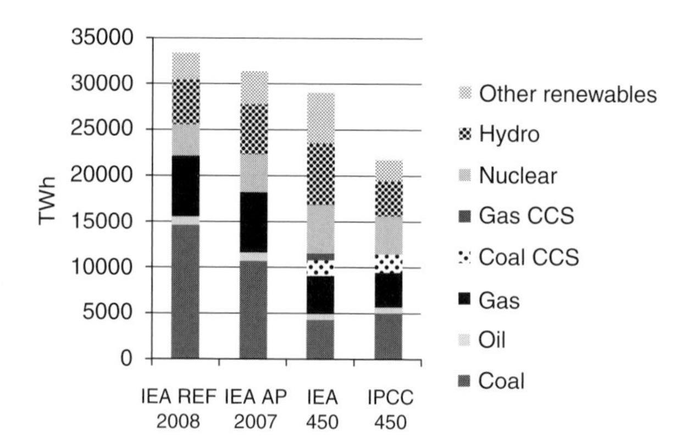

Figure 5.28 Share of low carbon options in the electricity sector in 2030 for a reference and two policy scenarios. CCS is not included in the IEA AP 2007 policy scenario, but it is in the IEA and IPCC 450 scenarios.
Source: IEA WEO 2007 and 2008.

the IEA 2007 Alternative Policy scenario roughly corresponds to a 550ppm CO_2 equivalent stabilization and the IEA 2008 450 and IPCC 450 scenarios[38] correspond to stabilization at 450ppm CO_2 equivalent. Only the IEA 450 scenario assumes early retirement of existing fossil fuel plants (i.e. before they are 50 years old). The scenarios generally use an approach in which the costs of the options are used to allocate the shares to the new plants to be built, but with restrictions on nuclear power in light of its acceptance problems.

Total electricity consumption for the policy scenarios varies significantly as a result of energy efficiency improvement assumptions in the end use sectors (transportation, buildings, industry). These will be discussed in more detail in Chapters 6, 7, and 8. Also the assumptions about the share of nuclear power vary between the scenarios. All this is well within the technical potential for the various low carbon options.

What does this tell us about what needs to be done?

Ambitious climate policy, aiming at stabilization at 450ppm CO_2-eq, corresponds to carbon prices of about \$100/tCO_2-eq avoided or more. It requires massive changes in the electricity sector. There is enough economic potential in the various low carbon options to make that happen. However this means a drastic change from business as usual. In the first place, it means building very few new gas or coal fired power plants without CCS worldwide after 2012. In some scenarios it is even necessary to close some of the older coal fired plants prematurely. It also means building more than 400 coal or gas fuelled plants with CCS, investing in 200–500 nuclear power stations, increasing hydropower capacity by up to 50%, doubling or tripling the number of biomass fired CHP plants, increasing the number of wind turbines from 50 000 to more than 500 000 by 2030, and a 100-fold increase in the capacity of solar CSP and PV.

Is that possible? It is good to keep in mind that wind power has seen a growth of 25–30% per year. Maintaining 25% growth means a 70-fold increase over 20 years. Solar PV capacity grew by 50–60% per year. Maintaining a 30% growth per year means a 140-fold increase over 20 years, and producing more than 400 coal and gas fired power plants with CCS in 20 years is a modest figure compared with the 100 coal fired power plants that China built in 2006 alone.

Investments also change drastically. In the first place it means that investments will shift: more will be invested in low carbon electricity supply and more in demand reduction and less in fossil fuel plants without CCS. The US\$22 trillion that will be needed for expanding and upgrading the world energy system between now and 2030 will thus be spent in a different way to that under a business as usual scenario. The 450ppm CO_2-eq scenario will also require additional investments. Newly built plants will on average have about double the investment costs per unit of capacity. The much better efficiency of energy use lowers the need for supply capacity and thus lowers the investment needs. For the ambitious policy cases considered however this does not compensate the higher power generation investments. The highest estimates of additional investment needs are about US\$9 trillion (partly as a result of early closing of coal plants)[39]. Reduced expenditure on fuels will however save about US$6 trillion. Still, these large investment flows are less than 1% of global GDP.

What policy intervention is needed?[40]

Low carbon power supply options are going to be used to some extent without climate policy, because they are cheaper than fossil fuels and energy security and air pollution considerations make them attractive. But they are not going to be implemented at the scale needed for drastic greenhouse gas emission reductions without additional policy intervention. However, in most countries governments no longer have direct control over electricity generation. In the past power supply was in the hands of government owned monopolies, but the sector has been thoroughly liberalized in many countries. So what are the policies that governments have available and what are the policies that would work?

A price for CO_2

The most important policy intervention is to create a price for CO_2, i.e. charging a fee when CO_2 is emitted. Traditionally, CO_2 emissions were free, which means that the actual costs of CO_2 emissions in terms of damaging the environment are not included in the price of the energy used. This is surprising, because charging a fee for emission of air pollutants or requiring abatement of such emissions at some cost is quite common practice.

How do you give CO_2 a price? The simplest way is to put a tax or a fee on every tonne of CO_2 (or other greenhouse gas) when emitted. Norway, Sweden, Denmark, and the UK have introduced such a direct tax for large companies. Many other countries have indirect energy and carbon taxes, levied on energy use of smaller consumers. Taxes are not very popular however. Attempts to introduce an enery/carbon tax in the EU for power companies and large energy using industries failed because of massive resistance. This eventually led to the introduction of the EU Emissions Trading System (see Box 11.5, Chapter 11). In the Netherlands the carbon tax and its exemption for renewable energy resulted in a booming market for 'green electricity' in households. Because prices of green electricity were the same as for regular electricity, about 20% of households shifted to green electricity.

Cap and trade[41]

Another way to create a price for CO_2 is to limit the amount of CO_2 that can be emitted by a company and allow trading in these allowances, a so-called 'cap and trade system'. Companies that want to emit more than their allowances permit can buy additional emission allowances from others. Companies that find ways of reducing emissions can sell spare allowances. Just as in any other market a price for CO_2 emerges. If there are few allowances available for sale and demand for additional allowances is large, the price is high. Conversely, if there is not much demand for additional allowances (for instance because the companies were given ample allowances to begin with), the price will be low. The amount of allowances given out to companies therefore determines the price. Cap and trade systems have been used for other air pollutants in the past and have been used in very different sectors, such as the milk quota in the EU to control surplus milk production by farmers. In 2005 the EU introduced the European Emission Trading Scheme for CO_2. It applies to large electricity producers and large energy users. It covers about 40% of all CO_2 emissions emitted across the EU. Since the electricity sector is not very sensitive to international trade, the cap and trade systems are quite effective in this sector.

Subsidies

A more indirect way of establishing a price for CO_2 is to change the relative cost of low carbon versus fossil based electricity. There are still subsidies on fossil fuel electricity production, in the form of subsidized fuel (e.g. on domestically produced coal) or subsidized electricity. In total about US$250–300 billion are spent annually on such fossil fuel subsidies. Doing away with those subsidies (which is good for the economy, but politically difficult to achieve) would narrow the gap between low carbon and fossil fuel based electricity.

Changing the relative cost can also be done by a giving a subsidy on low-carbon electricity. This is done in many countries, but in different forms. The most successful method is the so-called 'feed-in tariff' system. Suppliers are given a guaranteed price for renewable electricity and electricity distribution companies are required to buy the renewable electricity at that price. The additional costs are then shared between all consumers. These feed-in tariffs can be adjusted over time, to reflect decreasing costs of low carbon electricity. More than 35 countries have introduced such a system.

The third method is to give direct subsidies to producers of low carbon electricity. Again, this can take several forms. Competitive bidding is used in several countries. In this system a low carbon power supplier can offer a certain amount for a specific price. The lowest bidders get the contract and the government pays the difference with the regular wholesale price. In the UK this system was abandoned in 2002 because it attracted only limited interest. It was replaced by a system of renewable portfolio standards (see below).

Another form is a subsidy on the initial investment, either as a rebate or a tax reduction. The idea behind that is to overcome the resistance against the high initial investment required for putting up solar PV panels, solar water heaters, windmills, or biomass fired CHP plants. China has been using this system in providing more than 700 rural villages with combined PV, wind, and hydropower systems. Japan managed to become the world leader in solar PV systems by providing for a long time a 50% subsidy on the initial

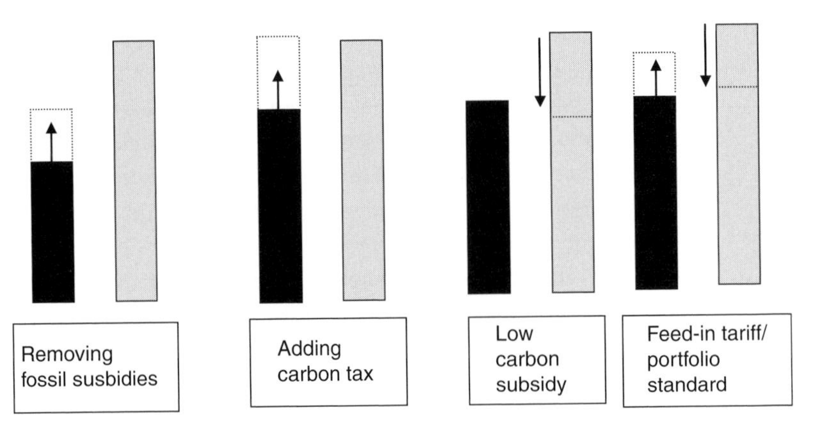

Figure 5.29 Schematic drawing of various subsidy schemes for making renewables penetrate the market. Dark bars represents fossil fuels, light grey bars renewable energy products.

investment of homeowners and project developers. After reducing the subsidy in 2002 from 50% to 12%, investments declined strongly.

Figure 5.29 gives a comparison of these various subsidy schemes.

Regulation

A regulatory system that is used widely, is the so-called 'renewable portfolio standard' approach. Electricity suppliers are required to have a certain minimum percentage of the electricity they sell from renewable energy sources. This percentage can be increased over time. Suppliers absorb the additional costs into their general prices. More than seven countries and many US States are using such a system. Since this system does not directly stimulate generators of low carbon electricity, effectiveness depends on the way the obligation is enforced.

New Zealand recently used legislation to put a 10 year moratorium on the building of new coal fired power plants in order to avoid lock-in while climate policy was put in place[42].

Risk reduction

The policy instruments described above are particularly relevant for promotion of renewable energy by narrowing the cost difference for fossil fuel based electricity and heat. For nuclear power however cost is not the most important barrier. As mentioned above, new legislation in the USA has simplified licensing procedures and has extended limitations of liability in addition to providing a 2$c/kWh subsidy in the form of tax reductions. International efforts to control the risk of proliferation of nuclear weapons also forms part of the policy package to stimulate nuclear power.

Technology policy

For CO_2 capture and storage yet another issue is the main problem: making the technology commercially viable. Particularly for power plants, CCS has not been applied yet at large scale. The current carbon price is too low to make investment in a large CCS facility attractive. For that reason technology stimulation policies are applied. Apart from creating information exchange mechanisms between researchers and commercial companies (the EU CCS platform, the international Carbon Sequestration Leadership Forum, the IEA Implementing Agreement on greenhouse gas R&D), government financed demonstration programmes will be established. The EU decided to create 10–12 large scale demonstration plants, subsidized by member state governments. These plants should be operational by 2015. With the experience gathered, CCS costs coming down, and carbon prices going up, it is expected that by 2020 CCS in coal and gas fired power plants will become commercially attractive.

So what does this mean?

Technical possibilities for large reductions of greenhouse gas emissions from the energy supply sector are available. By 2030 many of these options will be cost competitive with fossil fuel based power, particularly when fossil fuel based plants have to be equipped with CCS. But the only way to get there is by strong policy action to make it attractive to invest in low carbon technologies and to avoid building many more fossil fuel based power plants that would lock-in the electricity infrastructure further into a fossil fuel future.

Notes

1. IEA, World Energy Outlook 2006.
2. IEA, World Energy Outlook 2008.
3. IPCC Fourth Assessment Report, Working Group III, table 4.2.
4. Stanford University, Program on Energy and Sustainable Development, http://pesd.stanford.edu/research/oil/.
5. IEA, World Energy Outlook 2008.
6. See note 5.
7. IAEA, Energy, Electricity and Nuclear Power Estimates for the Period up to 2030, 2007 edition.
8. Richter B. Nuclear Power: A Status Report, Stanford University Program on Energy and Sustainable Development, Working Paper #58, September 2006.
9. Richter B. Nuclear Power and Proliferation of Nuclear Weapons, Stanford University Program on Energy and Sustainable Development, 22 February 2008.
10. http://www.iaea.org/NewsCenter/News/2006/uranium_resources.html.
11. IEA, WEO 2007.
12. IPCC Fourth Assessment Report, Working Group III, ch 4.3.3.1.

13. See note 11.
14. REN21. 2008. "Renewables 2007 Global Status Report" (Paris: REN21 Secretariat and Washington, DC:Worldwatch Institute)."
15. Global Wind Energy Council, Global Wind Energy Outlook 2008.
16. If a coal fired power plant were needed as a full back-up for wind (which is a worst case assumption), the costs of coal fired power minus the fuel costs, i.e. 70% of about 2$c/kWh, would be the additional cost.
17. IPCC Fourth Assessment Report, Working Group III, ch 4.3.3.3.
18. Karekezi S et al. Traditional Biomass Energy: Improving its Use and Moving to Modern Energy Use, Thematic Background Paper, Renewable Energy Conference, Bonn, January 2004, http://www.renewables2004.de/pdf/tbp/TBP11-biomass.pdf.
19. See note 14.
20. See EPA, Biomass Combined Heat and Power: catalog of technologies, September 2007, http://www.epa.gov/CHP/basic/catalog.html.
21. Junginger M et al. Energy Policy, Volume 34, Issue 18, December 2006, pp 4024–4041.
22. IPCC Fourth Assessment Report, Working Group III, ch 11.3.1.4.
23. IPCC Fourth Assessment Report, Working Group III, ch 4.3.3.4.
24. See http://geothermal.marin.org/GEOpresentation/sld073.htm.
25. IEA WEO 2007.
26. IPCC Fourth Assessment Report, Working Group III, ch 4.3.3.5.
27. IEA, WEO 2007, 450ppm stabilization case.
28. IPCC Fourth Assessment Report, Working Group III, ch 4.3.3.6.
29. Largest PV plants are operating in Germany (10MW), Portugal, and Spain; source http://www.pvresources.com/en/top50pv.php#top50table; and REN21. 2006. Renewables Global Status Report 2006 Update (Paris: REN21 Secretariat and Washington, DC:Worldwatch Institute, 2006).
30. http://www.redherring.com/Home/19866.
31. IPCC Fourth Assessment Report, Working Group III, ch 4.3.3.7.
32. IPCC Fourth Assessment Report, Working Group III, ch 4.3.3.8.
33. The Engineer, 29 October–11 November 2007.
34. IPCC Special Report on CO_2 Capture and Storage, 2005.
35. IPCC Special Report on CO_2 Capture and Storage, 2005, ch 3.
36. IPCC Special Report on CO_2 Capture and Storage, 2005, ch 7.
37. IPCC Special Report on CO_2 Capture and Storage, 2005, ch 5 and IPCC Fourth Assessment Report, ch 4.
38. IPCC Fourth Assessment Report, Working Group III, ch 11, appendix.
39. See note 2.
40. IPCC Fourth Assessment Report, Working Group III, ch 4.5.
41. See also Chapter 11.
42. http://www.treehugger.com/files/2007/10/new_zealand_dec.php.

6 Transportation

What is covered in this chapter?

Managing CO_2 emissions from transportation cannot be separated from managing congestion, air pollution, and oil imports. All of these problems emerge from the ever increasing transportation needs. The solutions overlap to a great extent. Understanding the drivers and the trends is a must. In terms of strategies to address the issues there is a hierarchy: reduce demand, shift transport modes, improve efficiency, and change the fuel. Policy intervention is needed to realize the many good options to drastically reduce congestion, air pollution, oil consumption, and CO_2 emissions.

Need for transportation

Mobility is an essential human need. Social relations and earning an income require transportation. Industrialization and specialization have created the need for shipments of large amounts of goods over short and long distances. Globalization of the economy has strongly accelerated this. Transportation of people and goods therefore is crucial for economic and social development.

In the year 2000 average transport per person ranged from 1700 km in Africa to 21 500km in North America, reflecting the strong influence of income. About half the total passenger-kilometres was covered by cars[1], of course with huge differences between countries. The share of the different transport modes (so-called modal split) in cities varies a lot. Figure 6.1 shows the large share of public transport and walking/cycling in cities in areas other than North America and Oceania.

Historically there has been a strong correlation between income and car ownership (see Figure 6.2). At the same time there are marked differences. For similar income levels vehicle ownership in the USA is almost twice that in Denmark and 50% higher than in Switzerland, and in New Zealand it was more than twice that in South Korea despite having the same income. The most important consideration is of course how countries like China and India, still at the bottom of the curve, are going to develop. Will they go the American way or the Japanese/European way, or can they manage to keep car ownership at a lower level?

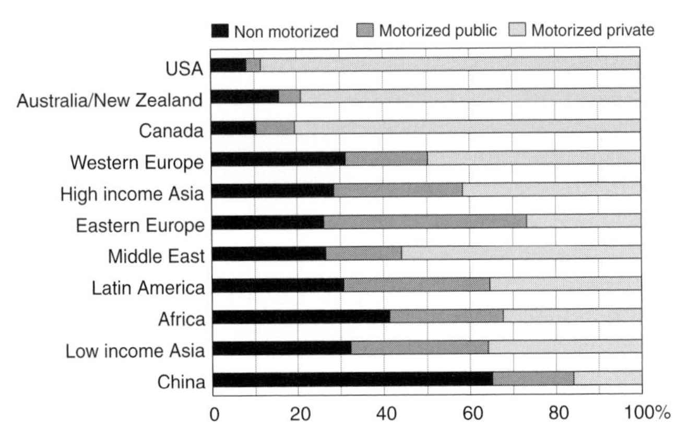

Figure 6.1 **Share of different transport modes in selected cities in 1995. Indicated is the percentage of trips taken with the respective transport mode.**

Source: IPCC Fourth Assessment Report, Working Group III, figure 5.17; original data from Millennium Cities Database for Sustainable Transport.

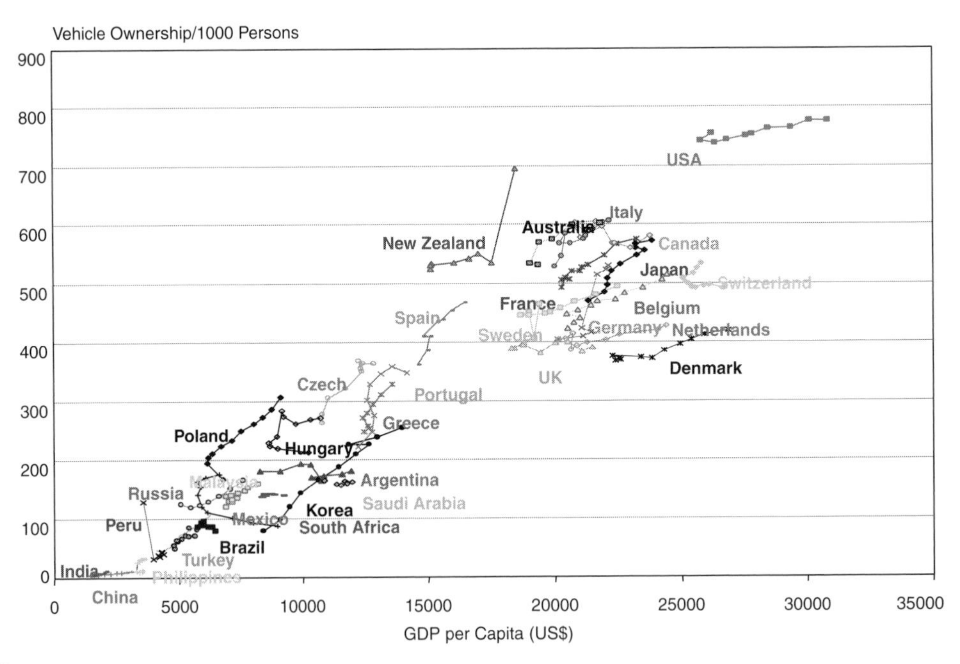

Figure 6.2 **Comparison of vehicle ownership between countries as a function of per capita income. Data for the period 1990–2000 (with some differences for specific countries).**

Source: IPCC Fourth Assessment Report, Working Group III, figure 5.2. See Plate 12 for colour version.

Freight transport has roughly doubled over the past 30 years. Globally, in the year 2000, 70% of freight (measured as tonnes x kilometers) was transported by sea going ships[2], 15% by rail, and 15% by road. The rail/road ratio varies strongly between countries. In Western Europe it is now 16/84, in Eastern Europe 35/65 (in 1990 it was still 65/35!)[3], and in the USA it is 42/58[4]. The trend towards more road transport and away from rail has been stimulated by specialization in industry: production of parts and half-products, assembly, and processing are happening in different places; large retail firms have centralized their distribution centres; and companies have downsized their warehouses, resulting in 'just-in-time delivery', which means their supplies are on the road in a truck rather than in their warehouses.

Development and climate implications

Energy security implications

The transport sector used about 20% of total primary energy in 2006[5], almost all of it in the form of oil products. It consumes about half of all oil. Road vehicles represent more than three quarters of this, with passenger vehicles alone accounting for 45% of the total energy, trucks 25%, and buses 8%. Aviation, shipping, and rail transport together cover 20% of energy use[6]. Energy use in the transport sector has almost doubled between 1970 and 2000 and is still growing strongly at a little less than 2% per year. Oil imports in many countries are high (see Table 6.1), causing energy security concerns and putting pressure on foreign currency reserves.

Traffic congestion and health impacts

Traffic congestion has become an almost universal problem in urbanized areas of the world. On weekdays downtown traffic speeds in Bangkok, Manilla, and Mexico City are 10 km/hour or less and in Sao Paolo and Kuala Lumpur 15 km/hour or less. Bicycles are faster. Public transport costs in Rio de Janeiro and Sao Paolo have increased by 10% and 16% respectively due to congestion. In Jakarta, Lagos, Manilla, and Kinshasa city trips on average last more than 1 hour. Economic losses due to congestion in many developing country cities are between 2% and 6% of local GDP, even though vehicle ownership in many of these cities is still relatively low[7].

In Chapter 4 the growing problem of air pollution in cities is discussed. More than 700 000 people die prematurely every year due to exposure to small particles in air. Traffic is a main source of that air pollution.

Greenhouse gas emissions

Greenhouse gas emissions from the transport sector were about 13% of the global total in 2004 (see Figure 6.3). In most countries more than 95% of all emissions is in the form of CO_2, with small contributions from N_2O (from vehicles with catalytic converters) and fluorinated gases (from air conditioners)[8].

Table 6.1. Oil import dependency of selected countries

Country	Oil import as % of consumption 2007	Expected oil import as % of consumption 2030
USA	65	62
EU-27	82	92
Australia/New Zealand	92	89
China	51	74
India	72	92

Source: IEA, WEO 2008 reference scenario.

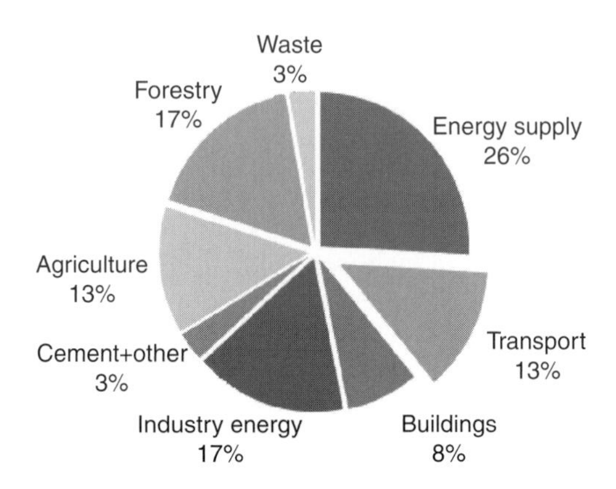

Figure 6.3 **Transport sector contribution to global greenhouse gas emissions in 2004.**
Source: IPCC Fourth Assessment Report, Working Group III, ch 1.

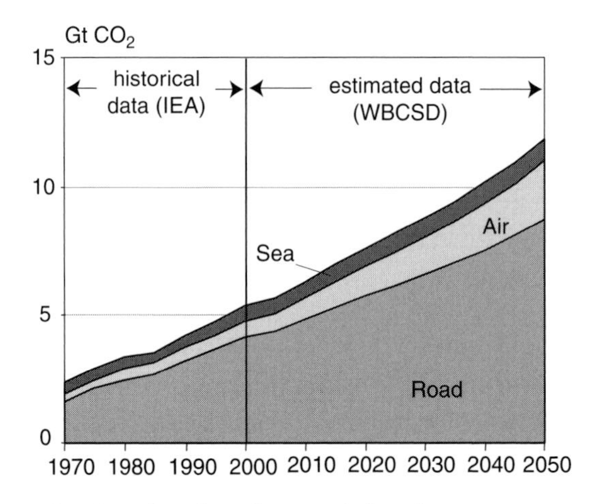

Figure 6.4 **Historic and projected CO_2 emissions from transport.**
Source: IPCC Fourth Assessment Report, Working Group III, figure 5.4.

The number of vehicles in the world today is about 900 million. By 2030 it is expected to be around 2.1 billion[9]. Freight transport (tonnes x kilometres), total energy use for transportation, and transport emissions in 2030 are projected to be roughly twice current levels. Developing countries' share of emissions, now about one-third, is expected to reach close to half of the total by 2030 (see also Box 6.1). Figure 6.4 shows the projected growth in CO_2 emissions from global transport.

Box 6.1 Transport in India

Transport in India is going through a rapid transition. Cars still provide less than 10% of all passenger kilometres, and public transport (rail and bus) about two-thirds. Car sales however are now growing by about 30% a year, showing that incomes in India are at the tipping point of car ownership. The introduction in 2007 of the TATA Nano, a UA$2500 car for the Indian market, and plans of other automakers to launch similar cheap models, reflects that as well. Car ownership is projected to grow from less than 10 million now to about 100 million in 2030. In addition there is a strong growth in motorbike ownership: 17% annual growth between 1990 and 2000 and an estimated increase to about 300 million motorbikes in 2030. Public transport is loosing ground because of lack of quality and traffic congestion (current average bus speed in cities is 6–10 km/hour). Serious air pollution from traffic, increasing oil imports, and one of the highest rates of road traffic accidents complete the picture.

Several policy initiatives have been developed to counter this trend: the National Urban Transport Policy, a combination of city planning, building metro and modern bus systems, providing dedicated space for public transport and introducing parking fees; the National Auto Fuel Policy, promoting cleaner fuels such as Compressed Natural Gas; and the Jawaharial Nehru Urban Renewal Mission, to provide cities with funding for structural changes in city planning. Analysis of what these policies can mean in moving towards a more sustainable transport system shows that transport energy use (and oil imports and CO_2 emissions) could be 30% lower than in a scenario without policy, a reduction equivalent to all of the transport energy used in India today. It also would lead to retaining a share of public transport of about 60% of all passenger travel (provided by modern and clean buses and trains), which would be an enormous contribution to creating liveable cities.

(Source: Schipper et al. CO_2 emissions from land transport in India, Transportation Research Record, 2009, in press)

Whether there is sufficient oil to fuel all this new transportation is addressed in Chapter 5. The conclusion drawn there is that conventional and unconventional oil resources, and liquid fuel production from coal and gas, can easily supply the required amounts, albeit with high CO_2 emissions.

Aviation and shipping[10]

The contribution from aviation is growing rapidly. Expected annual growth rates of 4–5% for passenger traffic and around 6% for freight traffic in the period till 2030 will mean that aviation will have a share of about 15% of the transport emissions by the end of that period, despite better fuel efficiency of aircraft. And that is 15% of a total that is much bigger than today. In addition, emissions from aviation are more harmful than other transport emissions, because of the altitude where they take place. Water vapour and NOx emissions alter the concentrations of the greenhouse gases ozone and methane and form condensation trails. Together this leads to a warming effect that is 2–4 times as big as that from CO_2 alone[11].

World shipping (expressed as tonnes freight x distance) is growing by about 5% per year. CO_2 emissions are now 0.8 $GtCO_2$/year, double the amount that was mentioned in the IPCCs 4th Assessment[12]. For 2020 the increase is about 30%; for 2050 it could be 300%.

How can transport emissions be reduced?

Asking about reduction of emissions is actually the wrong question to start with. The first question should be how transport itself can be controlled to get rid of congestion, air pollution, and rising oil imports, issues that are high on the political agenda in most countries. CO_2 emissions will follow. Except for building more roads to ease congestion, a strategy that has generally failed because it just attracts more traffic, solving congestion, air pollution, and growing oil imports goes hand-in-hand with reducing CO_2 emissions. An integrated transport policy would consist of the following elements:

- *reduce demand*: lower need for transport
- *shift means of transport*: shifting to other space saving and less polluting and CO_2 emitting transport modes
- *improve efficiency*: reduce the fuel consumption of vehicles, ships, and aircraft
- *change the fuel*: shift from oil products to less polluting and CO_2 emitting fuels

Reducing demand

The place where people live, work, go to school, or take part in recreational activities drives transport. About half of all people in the world now live in cities and that percentage is expected to increase to 70% by 2050[13]. The way cities are designed therefore determines transport needs. But so does income. With more money to spend there is the option of moving to suburban areas with nicer homes, requiring a commute to work. People with a higher disposable income also want to travel more for leisure. In many cities people are forced to live further and further away from the city centre, because they cannot afford to live closer, and to commute to work in the city centre. That puts pressure on governments to keep travel cheap, although making it more expensive would provide an incentive to choose a place to live closer to work. Then there is the 'law of constant travel time'. With faster means of transport people are travelling (to work) longer distances, keeping the time spent on commuting constant. Finally there is the possibility of so-called telecommuting (working at home made possible by the internet). These are the factors that influence transport demand.

So let us look at how much reducing transport demand can deliver in terms of reduced congestion, air pollution, oil imports, and emissions. Building compact cities helps, with workplaces, shops, schools, recreational, and residential areas not too far apart. This would require conscious decisions to go against the trend, but it is possible. What also helps is to make it affordable and attractive to stay in city centres, or creating suburbs with a good mixture of work and living places. Experience makes it very clear: the higher the density of people and jobs, the lower the demand for transport. Cities like Vancouver, Melbourne, Vienna, Perth, and Toronto, the top 5 in the Economist's Most Liveable Cities list[14], have managed to achieve this, in addition to providing enough space and green areas (see also Box 6.2 on Copenhagen). The building of cities however is a slow process and changing their design even more so. Only in rapidly growing cities in developing countries can this make a difference for transport emissions in the near term.

Information on the effect of the price on transport demand usually covers both demand reduction (such as car-pooling) and shifting away from car travel to other transport modes. So it is very hard to give an accurate picture. Parking charges proved to be quite effective in promoting car-pooling[15]. For poor people the effect is the clearest: an increase in bus or bush-taxi fares can prevent them from travelling. But that is of course not what should be aimed for. Basic services should be preserved. For air travel over longer distances no real alternative exists. Price increases do have more of an effect here. The relative cost of different transport modes and the possibility of people choosing other means of transport will be discussed in the next section.

Experience with telecommuting (working from home via internet connections) shows rather modest results in terms of reducing travel. In the USA, the reduction in vehicle kilometres travelled is estimated to be not more than 2%[16]. Some of the gains by not commuting to work are lost via additional trips for other reasons during the day.

Freight transport demand reduction would have to come from a change in the current 'just-in-time' delivery structure and reduced national and international specialization. This would only happen in the longer term and needs a strong price signal. If it becomes cheaper to build warehouses and have deliveries done in big volumes than keeping stock on the road, then something may change; likewise if it becomes cheaper to carry out processing of materials in one place, rather than in multiple locations. Fuel price increases are probably not going to be sufficient. Road taxes or tolls would also have to increase substantially.

For the period till 2030, the prospects for significant changes in transport demand compared to the trend (not counting modal shift) is poor, except for rapidly developing cities in developing countries. Over longer periods of 50 years or more however the strategy of reducing demand through structural changes in urban planning and industrial manufacturing can have a much bigger impact.

Shifting transport modes

Moving people from A to B in private cars is very inefficient. Average occupancy of cars in Melbourne, Australia is about 1.2 people per car. In the UK it is 1.6. In developing countries it is usually much higher. Figure 6.5 shows how much more space is required to move 100 people with cars, compared to that for bus or bike transport. Car travel produces more pollution, more CO_2, and requires more fuel.

The carbon intensity of passenger transport modes varies greatly. Walking and cycling produce zero CO_2 emissions. Emissions from buses, trams, metro, and trains vary with the fuel or electricity used and the occupancy, but are generally lower than for private vehicles or motorbikes. Occupancy has a large influence. For developing country circumstances (with relatively high occupancy per vehicle) Table 6.2 gives a comparison of emissions per passenger kilometre. Note that even when bus occupancy is down to 10 people, it is still lower in CO_2 emissions than a car with 2.5 people. Single occupancy cars (heavily used for commuting to work everywhere) cannot compete with anything.

What drives the choice of transport mode? Income and costs are very important. Currently, more than 3 billion people in the world cannot afford a car. They rely on walking, bicycles, motorbikes, and buses. With rising incomes, more people will one day realize their dream of buying a car (see Figure 6.2). This tells us three things: (1) the majority of people are better served by good public transport and safe cycling and walking spaces than by building roads (note: the car industry is politically very influential); (2) if public transport is good, people may postpone or refrain from buying a car; and (3) even if people have a car, it is still possible to encourage them to use public transport. Increasing the cost of parking, congestion charges as applied in London,

(a) (b) (c)

Figure 6.5 Space required for transporting the same number of people by (a) car, (b) bus, or (c) bicycle.
Source: UNEP GEO4; original picture from the city of Muenster in Germany.

Table 6.2. Greenhouse gas emissions from different transport modes in developing countries

	Load factor (average occupancy)	CO_2-eq emissions per passenger-km (full energy cycle)
Car (gasoline)	2.5	130–170
Car (diesel)	2.5	85–120
Car (natural gas)	2.5	100–135
Car (electric)[a]	2.0	30–100
Scooter (two-stroke)	1.5	60–90
Scooter (four-stroke)	1.5	40–60
Minibus (gasoline)	12.0	50–70
Minibus (diesel)	12.0	40–60
Bus (diesel)	40.0	20–30
Bus (natural gas)	40.0	25–35
Bus (hydrogen fuel cell)[b]	40.0	15–25
Rail Transit[c]	75% full	20–50

Note: All numbers in this table are estimates and approximations and are best treated as illustrative.
[a] Ranges are due largely to varying mixes of carbon and non-carbon energy sources (ranging from about 20–80% coal), and also the assumption that the battery electric vehicle will tend to be somewhat smaller than conventional cars.
[b] Hydrogen is assumed to be made from natural gas.
[c] Assumes heavy urban rail technology ('Metro') powered by electricity generated from a mix of coal, natural gas, and hydropower, with high passenger use (75% of seats filled on average).
Source: IPCC Fourth Assessment Report, Working Group III, table 5.4.

Singapore and a few other cities[17], and lowering the cost of public transport are important measures.

Time spent on commuting is another important driver for deciding on the means of transport. If the car can get you to work quicker (and more comfortably), then taking public transport does not look very attractive, even if it is cheaper. Of course this only applies if income is not a constraint. So developing an efficient and streamlined public transport system on the one hand, and making car travel slower through speed limits and reducing capacity on the other, should go hand-in-hand. Many cities have understood this and are now reserving specific lanes for buses, taking the capacity away from private vehicles and increasing the speed of public transport.

Figure 6.1 shows the large differences in the contribution of private cars to transport in cities around the world. Developing countries can go in two possible directions: the North American/Australian way with a dominance of private cars or the high income Asia/ Western European way with a much larger share of non-motorized and public transport.

City planning is key to shifting transport modes. Ensuring new housing developments have excellent access to public transport, reducing parking space, limiting access and speed of vehicles, pedestrian streets, good walking and cylcing facilities, and clean and reliable public transportation can make a big difference. Asian cities like Singapore, Hong Kong, and Shanghai are applying these principles. Maintaining urban density is a key condition for making such an approach cost effective (see Figure 6.6 and Box 6.2 on Copenhagen).

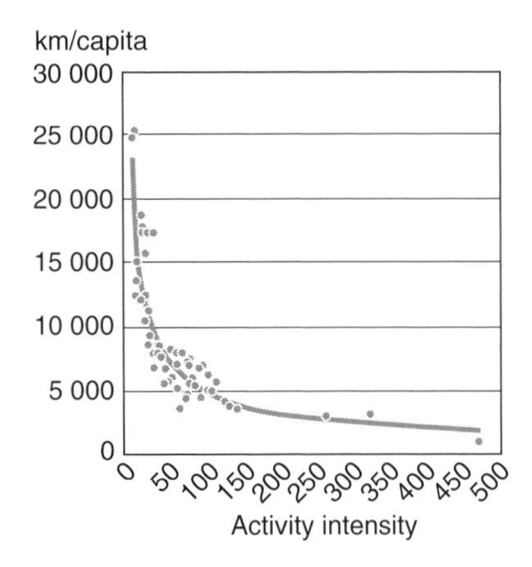

Figure 6.6 Personal car use as a function of the density of people and jobs in 58 higher income cities.
Source: UNEP GEO4.

Box 6.2 Copenhagen's 10-step programme towards a low-car/high-bike liveable city

1. Convert streets into pedestrian thoroughfares

The city turned its traditional main street, Strøget, into a pedestrian thoroughfare in 1962. In succeeding decades they gradually added more pedestrian-only streets, linking them to pedestrian-priority streets, where walkers and cyclists have right of way but cars are allowed at low speeds.

2. Reduce traffic and parking gradually

To keep traffic volume stable, the city reduced the number of cars in the city centre by eliminating parking spaces at a rate of 2–3% per year. Between 1986 and 1996 the city eliminated about 600 spaces.

3. Turn parking lots into public squares

The act of creating pedestrian streets freed up parking lots, enabling the city to transform them into public squares.

4. Keep scale dense and low

Low-slung, densely spaced buildings allow breezes to pass over them, making the city centre milder and less windy than the rest of Copenhagen.

5. Honour the human scale

The city's modest scale and street grid make walking a pleasant experience; its historic buildings, with their stoops, awnings, and doorways, provide people with impromptu places to stand and sit.

6. Populate the core

More than 6800 residents now live in the city centre. They've eliminated their dependence on cars, and at night their lighted windows give visiting pedestrians a feeling of safety.

7. Encourage student living

Students who commute to school on bicycles don't add to traffic congestion; on the contrary, their active presence, day and night, animates the city.

8. Adapt the cityscape to changing seasons

Outdoor cafés, public squares, and street performers attract thousands in the summer; skating rinks, heated benches, and gaslit heaters on street corners make winters in the city centre enjoyable.

9. Promote cycling as a major mode of transportation

The city established new bike lanes and extended existing ones. They placed bike crossings – using space freed up by the elimination of parking – near intersections. Currently 34% of Copenhageners who work in the city cycle to their jobs.

10. Make bicycles available

People can borrow city bikes for about US$2.50; when finished, they simply leave them at any one of the 110 bike stands located around the city centre and their money is refunded.

(Source: http://www.metropolismag.com/html/content_0802/ped/)

Latin America led the way with development of modern, clean, and fast bus systems (Bus Rapid Transit systems; BRT). Curitiba in Brazil was for a long time an isolated example, but now BRT systems are found in more than 40 cities around the world and many more big cities are planning BRT systems (see Box 6.3).

Box 6.3 Bus Rapid Transit Systems in Latin American cities

Bus Rapid Transit is a bus-based public transport system for big cities, characterized by: safe, clean, comfortable, and modern buses; high speed by using dedicated bus lanes, preferences at traffic lights and intersections, and high frequencies; and integration with follow-up transport to residential areas. They started in Latin America and now altogether more than 40 cities on six continents have a BRT system, including in developed countries (see table below). BRT systems have proven to be relatively low cost (US$1–8 million per kilometre, compared to light rail US$10–30 million and metro US$50–300 million), allowing systems to operate without subsidies. They can be installed relatively quickly (1–3 years from inception). They can have capacities of 13 000 to 45 000 passengers per hour in each direction of a BRT line. They reach speeds of 23–30 km/hour on average. Key success factors for BRT systems are:

- Careful analysis of transport demand and selection of bus corridors
- Including provision of good 'follow-up' transport in the form of safe ways to walk, cycle, or use smaller buses
- Easy and tamper free fare collection systems (coin machines, magnetic cards, smart cards)
- Friendly staff and security personnel
- City planning to concentrate residential and commercial buildings around the bus corridors
- Public participation in the design stage and an active marketing strategy during operation

BRT systems are revolutionizing public transport – see table below (from Bus Rapid Transit Planning Guide).

Cities with BRT systems, as of March 2007

Continent	Country	Cities with BRT systems
Asia	China	Beijing Hangzhou, Kunming
	India	Pune
	Indonesia	Jakarta (TransJakarta)
	Japan	Nagoya (Yutorito Line)
	South Korea	Seoul
	Taiwan	Taipei
Europe	France	Caen (Twisto), Clermont Ferrand (Léo 2000), Lyon, Nancy (TVR line 1), Nantes (Line 4), Nice (Busway), Paris (RN305 busway, Mobilien, and Val de Marne busway), Rouen (TEOR), Toulouse (RN88)
	Netherlands	Amsterdarn (Zuidtangent), Eindhoven, Utrecht
	UK	Bradford (Quality Bus), Crawley (Fastway), Edinburgh (Fazdink), Leeds (Superbus and Elite)
	Germany	Essen (O-Bahn)
Latin America and Caribbean	Brazil	Curitiba (Rede Integrada), Goiania (METROBUS), Porto Alegre (EPTC), Sâo Paulo (Interligado)
	Chile	Santiago (Transantiago)
	Colombia	Bogotá (TransMilenio), Pereira (Megabus)
	Ecuador	Quito (Trolé, Ecovía, Central Norte), Guayaquil (Metrovía)
	Guatemala	Guatemala City (Transmetro)
	Mexico	León (Optibus SIT), Mexico City (Metrobus)
North America	Canada	Ottawa (Transitway)
	United States	Boston (Silver Line Waterfront), Eugene (EmX), Los Angeles (Orange Line), Miami (South Miami-Dade Busway), Orlando (Lynx Lymmo), Pittsburgh (Busway)
Oceania	Australia	Adelaide (O-Bahn), Brisbane (Busway), Sydney (T-Ways)

Cities with BRT systems under construction as of March 2007

Continent	Country	Cities with systems under construction
Africa	Tanzania	Dar es Salaam
Asia	China	Jinan, Xi'an
Europe	France	Evry-Sénart, Douai, Clermont-Ferrand
	Italy	Bologna
Latin America and Caribbean	Colombia	Bucaramanga, Cali, Cartagena, Medellin
	Venezuela	Barquisimento, Mérida (Trolmérida)
North America	United States	Cleveland
Oceania	Australia	Canberra
	New Zealand	Auckland (Northern Busway)

(Source: Bus Rapid Transit Planning Guide, http://www.itdp.org/documents/BRTPG2007%202007%2009.pdf and Fulton L. Emissions and Transport: a global perspective, ADB Conference on Climate Change Mitigation in the Transport Sector, Manilla, 2006)

Information is very scarce on how much emissions of CO_2 can be reduced through land use planning and policies on modal shift. Studies for Delhi (India), Shanghai (China), and Santiago (Chile) suggest that a strong policy package can halve emissions in cities compared to business as usual by 2020[18]. However, caution is needed, because in many instances new and cheap public transport facilities have drawn their users from other forms of public transport or from those who walked or cycled[19].

Reliable cost estimates are even scarcer and, if available, usually allocate all costs of the package to CO_2 reduction. That is of course very unfair, because the benefits of reduced congestion, reduced air pollution, reduced oil imports, and, maybe even more importantly, a more liveable city also need to be taken into account. Some calculations for Bus Rapid Transit Systems in some Latin American cities cite a cost of about US$30 per tonne of avoided CO_2 for a package of measures that would reduce emissions by 25% compared to business as usual[20]. Bringing in the other benefits would drastically reduce the costs, which makes these policies very cost effective.

Freight transport and modal shift

For freight transport, rail and shipping have about 5 times lower CO_2 emissions per tonne kilometre than road transport (see Table 6.3). Nevertheless the trend has been away from rail and towards road trucks as indicated above. Changing freight transport from road to rail or water is however difficult and costly in most places, due to absence of rail connections or waterways, crowded railway systems, and need for additional transport from port or railway station. In terms of getting trucks off the road and improving congestion and air pollution it is fairly attractive, since small reductions in traffic volume

Table 6.3. CO_2 emissions per tonne kilometre for different freight transport modes

Freight transport mode	Average CO_2 emissions (grams per tonne kilometre)	Remarks
Inland shipping	31	
Ocean shipping	14	Varies from 8 for bulk tankers to 25 for container ships and 124 for refrigerated cargo ships
Rail	23	Mix of electric and diesel trains
Road	123	Varies from 92 for heavy trucks to 400 for light trucks

Source: EEA, TERM 2007 and 2003: indicators tracking transport and environment in the European Union.

can have a relatively large effect. Switzerland is currently undertaking a large programme of expansion in its rail infrastructure to shift environmentally damaging transalpine road freight transport towards rail[21].

More efficient fuel use

Passenger vehicles

New cars in Europe have become 30% heavier during the last 30 years[22], while fuel consumption has improved by 25% or so. So much of the progress in producing more efficient engines and transmissions has been cancelled out by heavier cars and more horsepower. In the USA fuel consumption of new cars has not improved for the last 25 years[23] and the share of SUVs (Sports Utility Vehicles like four wheel drives) with much higher fuel use has increased enormously. How can this trend be reversed?

Technically, drastic reductions in fuel use are possible. Just look at the cars on the market today. Fuel consumption of one of the best new passenger vehicle for sale (Toyota Prius hybrid) is about 1 litre for 20 km[24,25], while the average fuel consumption of all cars in the USA in 2005 was about 1 litre for 7 km[26]. In terms of CO_2 emissions per kilometre, a Toyota Prius emits about 105 g/km, compared with the average new car sold in Europe in 2005 emitting 161 g/km[27] (a Lamborghini Diablo about 520 g/km).

By improving aerodynamics, reducing size, reducing weight (with lighter materials such as aluminium and plastic), reducing power (smaller engines), further improvements in engines and overall design, and choosing (clean) diesel and hybrid systems (gasoline and electric combined; see Box 6.4), new cars by 2030 could be about 50% more fuel efficient than the best car for sale today[28]. Hybrid cars have a big role to play. For a significant reduction in transport CO_2 emissions their market share of newly sold cars would have to grow to something like 75% in 2030. Additional costs of hybrid and diesel hybrid vehicles by 2030 are estimated at US$3500–4500 per vehicle, but they are earned back within the lifetime of the vehicle (see also Table 6.5 below).

If fuel efficiency of new vehicles is halved by 2030, total transport energy use and emissions as a result of efficiency improvement only could then be reduced by 5–10% compared to business as usual[29]. Not a big contribution to the necessary reduction of emissions, but that is because cars on average last 10–20 years and changes in the fuel efficiency of the whole car fleet are slow. A 5–10% reduction in oil imports could be a useful contribution to improving energy security. The good news is that at oil prices above US$60 per barrel these measures together earn themselves back well within the lifetime of the vehicle.

Are more aggressive fuel efficiency improvements for the car fleet as a whole possible? A realistic way to speed up penetration of fuel efficient vehicles would be to take older, inefficient cars out of circulation. Such 'buy back' programmes have been applied at limited scale and have low cost per tonne of CO_2 avoided.

Box 6.4 Hybrid cars

Hybrid vehicles get their higher fuel efficiency from the following features (see diagram below):

- Using an electric motor to drive the vehicle at low and constant speed
- Using the gasoline or diesel engine only to recharge the battery and provide additional electric power when accelerating
- Capturing energy when braking and storing that in the batteries

The gasoline/diesel engine only runs to produce electric power, which makes it much more efficient. Particularly in city traffic with frequent stops and idling, the brake energy recovery and engine switch off make hybrids so efficient.

Hybrid electric engines are also applied in rail locomotives, buses, trucks, and submarines.

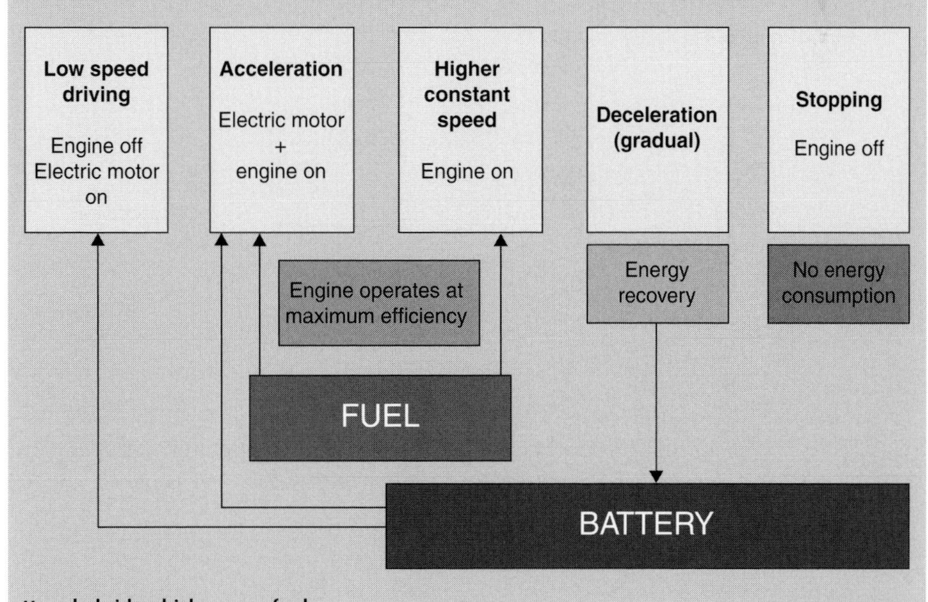

How hybrid vehicles save fuel.
Source: King Review of low-carbon cars, part I, UK Treasury, 2007.

Policy matters

The problem is this scenario of strong efficiency improvement cannot happen without fuel efficiency standards legislation. Experience in North America (frozen fuel efficiency standards between 1980 and 2006 and increased purchases of gas-guzzling SUVs), Japan (an increase in fuel use between 1985 and 2000), and Europe (a slowly declining fuel use and a failing voluntary agreement with automakers to reduce CO_2 emissions) shows that stronger policies are required. Fuel prices alone are also insufficient. Average European fuel consumption of newly sold cars is about 20% lower than in North America[30], as a result of taxes leading to three times higher fuel prices. With such differences in fuel prices one would have expected much better fuel efficiencies of European cars. However, when people decide on the purchase of a car, fuel efficiency is simply not an important factor, although differentiating (high European) car purchase taxes according to fuel efficiency can make some difference[31].

The European Union has introduced a legal CO_2 emissions standard of 130g CO_2 per kilometre to be achieved by 2012–2015 and is discussing lowering these standards for 2020. Figure 6.7 compares standards in various countries and states.

Freight transport

Freight transport fuel efficiency should not be measured in kilometres per litre as for passenger vehicles, because it is the tonnage that matters. The correct unit is tonne

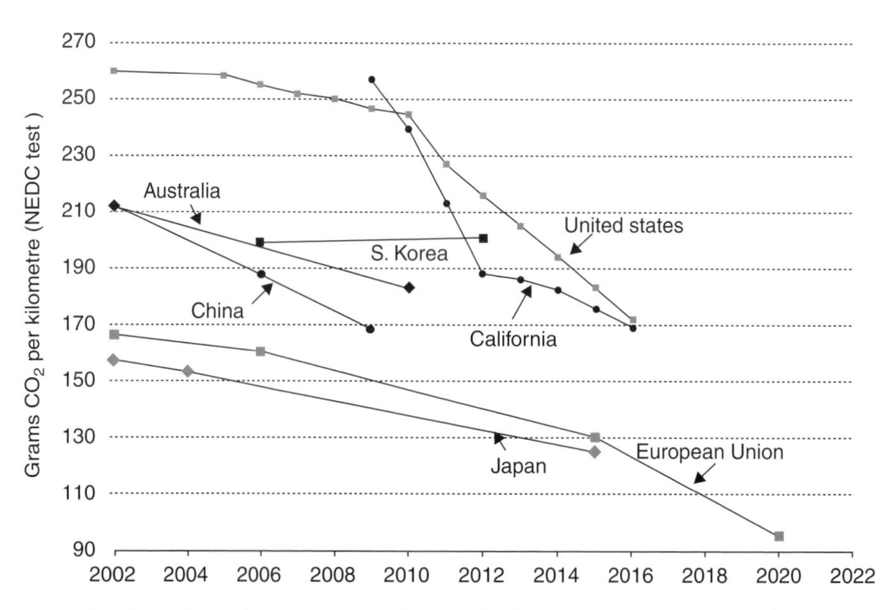

Figure 6.7 **Actual and projected greenhouse gas emission standards for new passenger vehicles by country.**
Source: ICCT, Passenger vehicle greenhouse gas and fuel economy standards: a global update, August 2008.

kilometres per litre or for emissions grams CO_2 per tonne kilometre. Table 6.3 shows the emissions of various freight transport modes. What efficiency gains can be made?

Technically a lot can be done to make trucks more fuel efficient by improving aerodynamics, tyres, engines and transmission, electrifying heating and lighting in stand-still mode, and introduction of hybrid electric systems. Hybrid electric trucks are particularly interesting for delivery vehicles, buses, etc. that make frequent stops where recovery of breaking energy makes sense. For long-haul trucks the advantages are small. In addition, limiting maximum speed through built-in speed controllers also can contribute significantly. There is an interesting behavioural factor here. In many countries drivers are paid by the distance they cover, so they have incentives to move as fast as possible. This is bad for fuel efficiency and road safety. Changing the payment system would make roads safer and reduce fuel use.

Altogether by 2030 freight truck fuel consumption and CO_2 emissions per tonne km could be reduced by 20–30%[32] compared to today. This does not include the gains that could be made by increasing the weight allowed (to be accompanied by lowering speed for safety reasons). Emissions per tonne kilometre in Australia where larger trucks are allowed are 20–50% lower than in Europe or North America[33]. Improved logistics (minimizing empty trucks) through ICT systems can also make a contribution, although only substantially higher transportation costs are likely to create meaningful incentives. Congestion is increasing fuel consumption, so there is a strong correlation between managing congestion and reducing CO_2 emissions.

In terms of policy for improving truck fuel efficiency governments have mostly relied on financial incentives, such as fuel taxes and toll charges. The reasoning is that freight companies are very motivated to reduce costs in light of the competition. So far only Japan has introduced fuel efficiency standards for new trucks, and those may be needed to speed up the penetration of advanced technical options.

The potential for improvements in fuel efficiency of freight transport by ships, rail, and aircraft is mixed. For shipping a combination of technical and operational measures could deliver CO_2 emission reductions of more than 30% compared to business as usual, which would still lead to an increase of total CO_2 emissions of shipping. The use of alternative fuels that could realize further reductions is discussed below. Opportunities for freight rail efficiency improvement are poorly studied and contributions to CO_2 reduction would probably be marginal. Air freight is growing strongly. Opportunities for fuel efficiency improvements are limited because of the slow turnover of airplanes. Fuel use and CO_2 reductions of 10–20% compared with business as usual are possible when CO_2 prices go up to 50–100 US$/tonne CO_2 avoided.

Change the fuel

Fuel efficiency improvements only have limited influence in the short term. Other measures are needed to significantly reduce oil consumption, air pollution, and CO_2 emissions. Changing the fuel is the only big option left. What are the alternatives?

Table 6.4. **Main producers of biofuels in 2005**

Country	Ethanol (Mtoe)	Biodiesel (Mtoe)	Total (Mtoe)
USA	7.5	0.22	7.72
Canada	0.12	0.00	0.12
EU	0.48	2.53	3.01
Brazil	8.17	0.05	8.22
China	0.51	0.00	0.51
India	0.15	0.00	0.15
WORLD	17.07	2.91	19.98

Source: IEA WEO 2006.

Biofuel

The most discussed alternative is biofuel, i.e. alcohol or diesel fuel produced from plants. There are a number of processes for producing biofuels (see Figure 6.8). Fuel alcohol is produced commercially today from sugar cane, maize, wheat, and sugar beet. The sugar is biologically converted to alcohol, which then has to be separated into concentrated form. Alcohol can be used in blends with gasoline (up to 25% alcohol without the need for engine adjustment) or in pure form (which requires engine adjustment). The fuel alcohol policy in Brazil triggered automakers to develop so-called flex-fuel vehicles (FFVs[34]) that automatically adjust engine settings, according to the composition of the fuel. This has given an enormous boost to fuel alcohol production, since cars no longer depend on specific fuelling stations (see also Chapter 4). FFVs are not yet available in all countries however.

Diesel is produced from animal fat, waste vegetable oils, and oilseeds, such as soybean, oil palm fruits, rapeseed, and cottonseed. In 2005 the global production of biofuels was about 20 Mtoe, or less than 0.1% of all transport energy (but about 2% of fuel use in the EU, see Table 6.4). Brazil (mainly sugar cane alcohol), the USA (mainly maize based alcohol), and the EU (mainly biodiesel) were the biggest producers[35]. Biodiesel is normally blended with regular diesel – up to 20% – so that no engine adjustment is required. It can also be used in pure form, if engines are adapted.

There are several other biofuel processes under development, producing so-called 'second generation biofuels'[36]. These are derived from cellulosic materials such as straw, other crop residues, grasses, or wood chips. Alcohol is produced through a biological process using specific bacteria that can break down cellulose into sugar. Alternatively chemical processes can be used to break down cellulose, leading to biodiesel. A third process gasifies the biomass and makes a synthetic diesel or other hydrocarbon fuel via a chemical process, the so-called Fisher Tropsch synthesis. None of these processes has so far reached the stage of commercial production, but prospects for large scale commercial deployment before 2020 are good.

Yet another process under development uses oil producing algae that are grown with sunlight and nutrients, after which oil is separated from the algae and processed as diesel. At small scale these algae systems have shown a high productivity. There are still

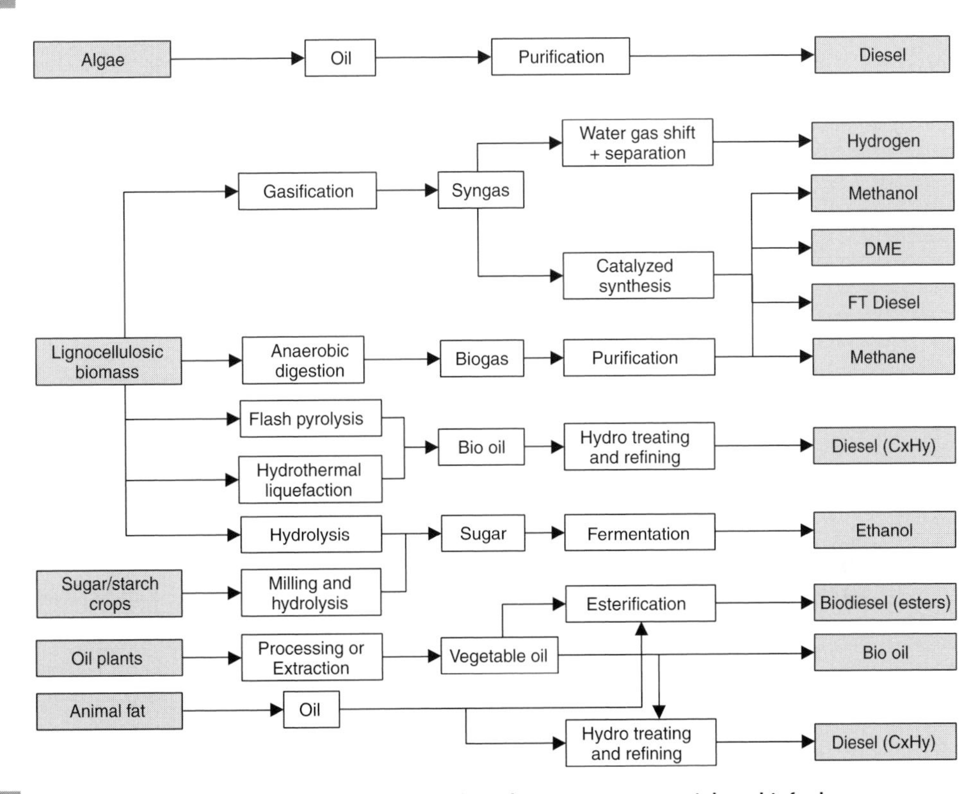

Figure 6.8 **Overview of conversion routes from biomass raw materials to biofuels.**
Source: adapted from IPCC Fourth Assessment Report, Working Group III, fig 5.8.

problems to be solved, particularly to get the sunlight to the algae when equipment is being scaled up. The algae form a thick 'green soup', where light is not penetrating easily. This is the same phenomenon that makes 'algal bloom' in polluted lakes lead to oxygen depletion and dying off of fish. The amount of energy algae systems can produce per hectare could be 10 to 30 times that of bioenergy crops. This would be a great advantage. Several big oil companies, such as Shell and BP, are investing in further development of this technology[37].

In Chapter 5 bioenergy was discussed in general. Questions were raised about the net carbon reductions and sustainability of bioenergy. For biofuels it is worse. The additional processing needed for biofuel requires additional energy compared to burning biomass. And the price paid for transport fuels is relatively high so that competition with food production becomes more of a problem.

Let us first look at the net carbon reduction as a result of using biofuel, compared to gasoline and diesel. Figure 6.9 shows results from two different studies. No land use change emissions due to conversion of forests, natural vegetation, or grassland were taken into account. Only energy use and emissions from producing and processing the crops were considered. Calculations like this are complex and large uncertainties still exist. As far as the results are concerned, sugar cane alcohol from Brazil is performing best with a net CO_2 emission reduction of about 80%. Alcohol from maize and wheat only achieves a

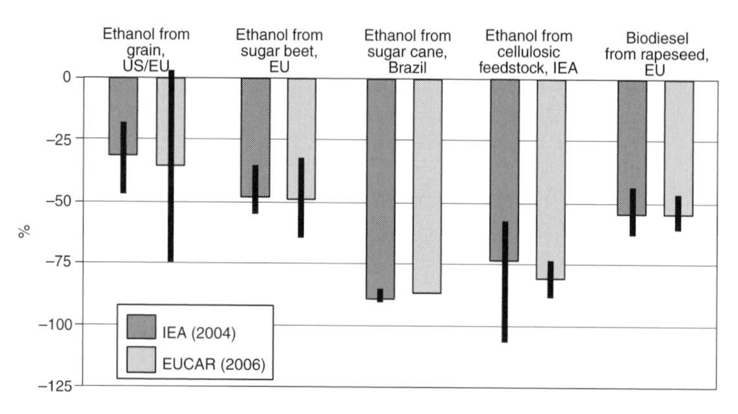

Figure 6.9 **Reduction of overall life cycle greenhouse gas emission reductions from biofuels compared to vehicles running on conventional fuels.**
Source: IPCC Fourth Assessment Report, Working Group III, fig 5.10.

30% reduction, but poor systems are getting close to zero. Biodiesel lies somewhere in between with about a 50% reduction. The prospects for cellulosic alcohol are good: around 75% CO_2 emission reduction can be expected.

However when land use change is assumed to take place, things can change dramatically. Whether or not land use change happens is a complex question. The plot where bioenergy crops are grown could have been used for other crops, in which case there is no displacement. In another location however new cropland might have been created from natural vegetation or forest to produce food that was displaced by bioenergy crops. In case of degraded land, where regular agriculture is not possible, crops do not displace food production. But refraining from bioenergy cropping could have meant recovery of the degraded land through natural vegetation with a fair amount of carbon stored in trees, shrubs, and soil. From currently available studies no reliable picture on the carbon emissions from land use change can be obtained.

The conclusion from this is that the first generation biofuels are contributing to reducing oil imports, but probably not much to CO_2 reduction. There is even a risk that some of them, particularly alcohol from grains and biodiesel from palm oil, produce more CO_2 than gasoline or diesel. A lot will depend on how fast the second generation biofuels that have a better performance become commercially available. The question now is whether this second generation biofuel can become available soon enough for the EU to maintain its 10% biofuel requirement for 2020.

To safeguard food production and biodiversity protection, a lot of effort is being put into the development of sustainability criteria and certification systems. The EU has adopted a target for biofuels of 5.75% by 2010 and 10% by 2020 and has made it clear that qualifying biofuels have to meet sustainability standards. Standards are under development within the EU and elsewhere[38]. Important criteria are the net carbon reduction, the risk of displacing food production, and the risk of destroying biodiversity. There are however other, more positive factors that should not be forgotten: creation of income for small farmers, opportunity to provide rural areas with modern energy services, and energy security.

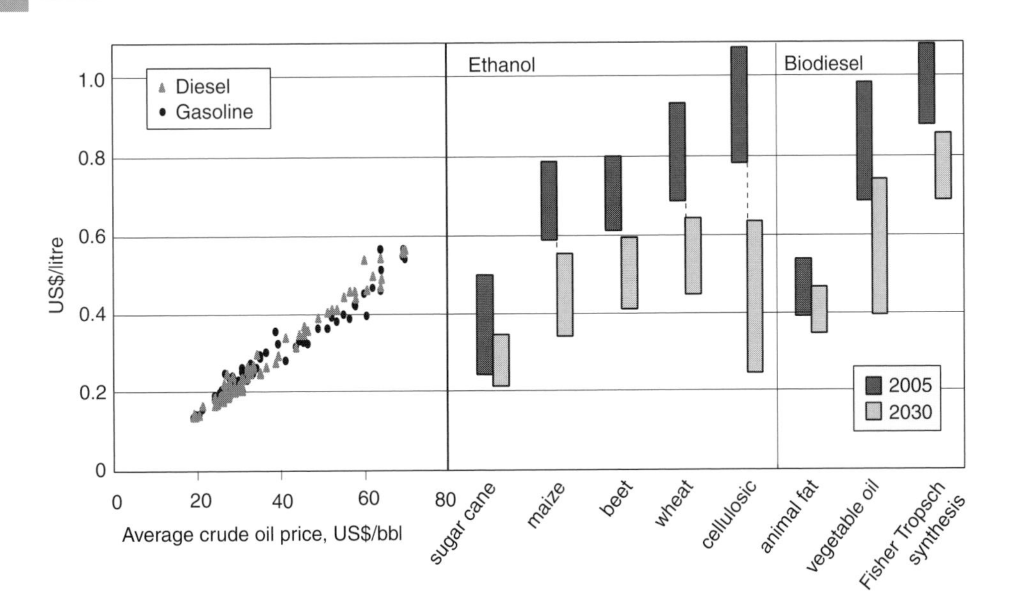

Figure 6.10 **Comparison between current and future biofuel production costs (in US$/litre) versus gasoline and diesel prices (at refinery gate, so excluding fuel taxes).**
Source: IPCC Fourth Assessment Report, Working Group III, fig 5.9.

For a long time biofuels have been more costly to produce than gasoline and diesel. This depends on the oil price of course. For oil prices above US$50 per barrel, alcohol from sugarcane is competitive with gasoline and biodiesel from animal fat with regular diesel. For oil prices of US$100 per barrel, alcohol from grains and vegetable biodiesel also are competitive. Expected cost reductions by 2030 would make most biofuels, including the second generation ones, competitive with gasoline and diesel for oil prices of more than US$80 per barrel[39] (see Figure 6.10).

To realize these expected cost reductions, investments in production need to be made to realize economies of scale. So over the next 20 years or so additional costs have to be borne to get there. What are these costs and who is paying them? What you currently see in terms of policy instruments being used is a combination of quota (like the EU targets), subsidies (mainly to farmers to produce bioenergy crops cheaper), and excise tax exemptions on biofuels sold.

Calculations of costs per tonne of CO_2 avoided are highly dependent on the oil price assumed and on the amount of CO_2 avoided by using biofuels to replace gasoline or diesel. One way to estimate these costs is to look at subsidies provided and CO_2 avoided. This results in costs per tonne of CO_2 avoided of 160 to more than 4000US$/tonne for a range of countries in 2006[40]. This definitely is not cost-effective. Interestingly, per barrel of oil avoided these costs are in the same range as oil prices of 60–100US$/barrel[41]. So current subsidy schemes may be beneficial from an energy security or agricultural point of view, but they are far from cost-effective for CO_2 reduction.

Another way to calculate cost per tonne of CO_2 avoided is to calculate the additional costs of biofuel production compared to gasoline and diesel, without considering subsidies. That leads to very different outcomes. For US$25/tonne CO_2 avoided (and assuming an oil price of US$60/barrel) biofuel could cost-effectively replace 5–10% of transport fuels[42]. For higher oil prices this would increase. The conclusion can only be that current subsidy systems pump way too much money into this sector.

How much can biofuels contribute to reduction of oil imports and CO_2 emissions? It will be clear from the discussion above that uncertainties are high. The most realistic scenarios available show a biofuel share of 3% of the total transport fuel by 2030 in the business as usual case. With ambitious policy this could grow to 5–10%[43]. This would translate into a global CO_2 reduction of 0.6–1.5 $GtCO_2$/year in 2030 at costs below US $25/tonne CO_2 avoided, although these costs will be higher if carbon loss from land use change (as discussed above) is factored in.

Electricity

The prospects for all-electric vehicles (driven by an electric motor that runs on a large battery that needs to be recharged) are a bit unclear[44]. There is a niche market for small electric vehicles like golf carts, distribution vehicles, warehouse trolleys, limited edition passenger vehicles, and electric sports cars and an emerging market for electric two-wheelers (30 million now operating in China alone, see Box 6.5).

The electric two-wheeler market has a lot of growth potential in developing countries, because it is one of the first things households purchase when they can afford it. In China, India, Africa, and many other developing countries the number of two-wheelers is expected to at least double between now and 2030[45]. Where electricity is available (which is not the case for many rural areas in India and Africa), this can then be an important option. In terms of oil consumption or CO_2 emissions two-wheelers contribute only a small percentage however. From an air pollution point of view electric vehicles could make a bigger contribution: two-wheelers are relatively big polluters.

Box 6.5 E-bikes in China

Electric motorbikes have become very popular in China. In 2005, 10 million e-bikes were sold in hundreds of different models, either bicycle or scooter style (see picture). They typically have a range of 40–50 km on a single charge and can be charged at standard electrical outlets. In most cities they may be used in bicycle lanes, without a driver's license. With city planning in many Chinese cities leading to the move of residential areas to the outer parts of the city, bicycle use has been declining strongly and motorbikes and e-bikes have replaced them.

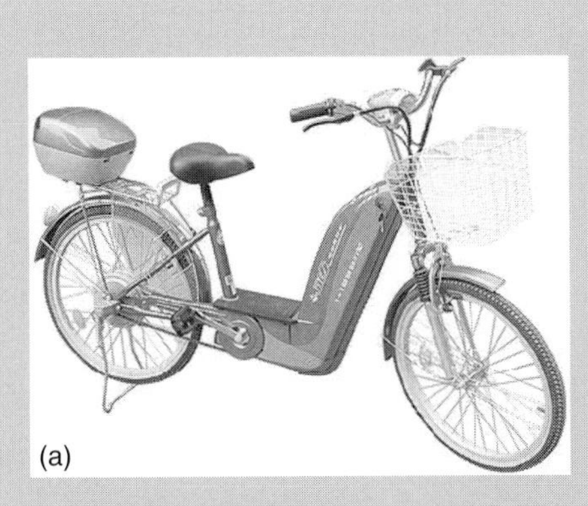

Bicycle style and scooter style electric bikes.
Source: http://www.forever-bicycle.com.

Success factors for the e-bikes have been:

- Low costs (about a third of a motorbike per kilometre due to lower fuel costs)
- Setting of national standards for e-bikes, ensuring good quality
- Promotion of e-bikes by city governments (not all cities have been in favour though) to help improve air quality

(Source: Weinert JX, Ma C, Cherry C (2006) The Transition To Electric Bikes In China: History And Key Reasons For Rapid Growth. Springer Transportation 34 (3), 301 – 318)

Currently there is a small number of serious electric passenger vehicles types for sale[46] in very limited numbers and with a limited range (the best up to about 350 km, because of battery limits[47]). Although most people do not drive their car further than that each day, not being able to drive further is a strong barrier. Progress on the front of better batteries with higher loads and low weight is slow. Much is expected of lithium ion batteries, now used extensively in laptop computers and applied in the TESLA Roadster electric vehicle, but their costs are still high.

How much could electric vehicles reduce CO_2 emissions? The origin of the electricity is obviously very important. For the average US electricity fuel mix, all-electric cars emit typically in the range of 60–130 gCO_2/km, compared to the US fleet average of 230 tsgCO_2/km and hybrid cars around 100 gCO_2/km. The climate change advantages of all-electric vehicles in high carbon electricity areas are therefore limited. In countries with low carbon electricity they perform much better of course. Costs are still high. There are no reliable estimates of the global contribution that all-electric vehicles could make to the reduction of CO_2 emissions. From an air pollution point of view they have strong advantages of course.

Plug-in hybrids

The road to all-electric vehicles (if reached at all) more likely goes via hybrid electric vehicles. A mixture between a classic hybrid and an electric car is the so-called ‘plug-in

Table 6.5. Cost comparison of hybrid and plug-in hybrid cars

Near-term incremental costs	Conventional hybrid	Plug-in hybrid (with a 40-mile all-electricity range)
Battery	US$2000	US$17500
Other	US$1500	US$1500
Annual fuel savings	US$480	US$705
Payback (years)	7.3	27.0
Long term incremental costs		
Battery	US$600	US$3500
Other	US$1000	US$1000
Annual fuel savings	US$480	US$705
Payback (years)	2.9	6.4

Source: Plug-In Hybrids: an Environmental and Economic Performance Outlook, ACEEE, 2006.

hybrid' vehicle. They operate as hybrid vehicles, but the batteries can be charged from the grid. In this way the vehicle can operate on battery power for a larger fraction of the time (initially probably 50–100km). In the US this would mean CO_2 emissions could be 15% lower than for conventional hybrid vehicles, while in California it could be 30% lower[48]. In countries with a low carbon electricity supply the advantages would of course be greater. However, battery costs are currently very high, making these vehicles too costly (see Table 6.5). Many automakers have announced their introduction, but so far they are not available commercially. Before 2030 costs are expected to come down sufficiently for plug-in hybrids to penetrate the market. Their contribution to emission reduction in the 2030 timeframe will be very limited.

Hydrogen

Hydrogen can be used to operate fuel cells, a type of battery that generates electricity through the reaction of hydrogen with oxygen from air. The process is extremely clean: only water comes out of the tailpipe. Electric motors drive the car. The CO_2 reduction that can be achieved completely depends on the source of hydrogen. As explained in Chapter 5, hydrogen is used extensively in the oil and chemical industry. It is produced from coal or gas. By using CO_2 capture and storage (CCS, see Chapter 5) near zero carbon hydrogen can be obtained. In principle biomass can be used as a source, leading also to near zero hydrogen, albeit at much higher costs. Hydrogen from electric decomposition (electrolysis) of water with renewable energy would also be near zero carbon, but costs are currently even higher. This could only change if hydrogen were produced with excess renewable power at off-peak times.

With low carbon hydrogen from gas or coal with CCS, the CO_2 emission is about 10–20% of that of regular gasoline cars[49], i.e. 30–50 gCO_2/km. Other air pollutants are not emitted.

All major automakers have prototypes of hydrogen fuel cell vehicles (HFCVs). There are three major problems still to be resolved before commercial production could be considered: the supply of hydrogen to fuel stations, the hydrogen storage in the vehicle, and the fuel cell reliability and cost. Hydrogen pipelines exist in areas with oil and petrochemical complexes. Extending that to whole countries would be costly, but feasible. The biggest problem is the fact that there are no hydrogen cars, and there will be no HFCVs unless there is a hydrogen infrastructure, which is down to the government. Hydrogen is clean, but is has a low energy density. This requires compressing hydrogen to very high pressures to keep the volume of the fuel tank to a reasonable size for a good driving range. Pressurized hydrogen tanks are now available that use three times the pressure of a compressed natural gas tank in cars, allowing a driving range of up to 450km. Another approach being developed is to absorb hydrogen in a special metal hydride powder, allowing larger amounts to be stored at lower pressures. Fuel cell development still has some way to go to get to cheaper and more reliable fuel cells.

Significant penetration of HFCVs in the vehicle fleet is not expected before 2030. In the period thereafter however contributions could be significant.

So what can be achieved in terms of reduction of energy use and CO_2 emissions?

In Table 6.6 the potential contributions by 2030 are shown from the various measures to reduce energy use and CO_2 emissions with costs up to US$100/tonne CO_2 avoided[50]. The potential is expressed as the so-called 'economic potential' (see Box 6.6).

Box 6.6 Economic versus market potential

Economic potential is the mitigation potential, calculated with payback times for investments as used in public sector projects (low discount rates) and assuming that market barriers are removed through policy intervention.

Market potential is the mitigation potential based on private payback times for investments used in business and household decisions (big discount rates) and occurring under real market conditions, including policies and measures currently in place, noting that barriers limit actual uptake.

The difference between the two is that market potential assumes all sorts of barriers, limiting the uptake of measures, i.e. not everything that is economically sensible is being done. Economic potential only looks at the question that makes economic sense at a certain carbon price, if barriers are removed by policy actions. Normally there is a pretty large difference between those potentials: economic potential is higher than market potential.

The total emission reduction potential is at least about 1.6–2.3 $GtCO_2$, equivalent to a reduction of 15% of the expected emissions without policy. Public transport and biking facilities in cities will add to this, but no reliable estimates of the reductions are available

Table 6.6. **Global economic potential for reduction of CO_2 emissions from transport by 2030 with costs up to US$100/tonne CO_2 avoided**

Measure	Total oil consumption reduction (% from BaU)	CO_2 emissions reduction compared to BaU ($GtCO_2$/year)	Other benefits
Reduce demand	Low	Low	Congestion can benefit more
Modal shift passenger transport	Moderate	Moderate	Congestion can benefit considerably
Modal shift freight transport	Negligible	Negligible	
Efficiency passenger road transport	10	0.75	
Efficiency freight road transport	2–5	0.1–0.4	
Biofuel	5–10	0.1–0.4	Sustainability constraints could reduce this amount
Electricity	Low	Negligible	
Hydrogen	Negligible	Negligible	
More efficient airline transport	n/a	0.28	
Freight shipping	n/a	0.3–0.4	
Rail	Negligible	Low	
TOTAL		1.6–2.3	

Source: based on IPCC Fourth Assessment Report, Working Group III, ch 5.

unfortunately. This also means oil consumption can be reduced by a significant amount. Congestion will be reduced and air pollution from transport as well.

The real big reduction in the transport sector can only come after 2030, with hydrogen fuel cell vehicles and possibly electric vehicles in combination with a largely decarbonized energy system, second generation biofuels from sustainable biomass, as well as structural changes in city planning and public transport systems.

How do we get it done?

As indicated above, policy action is needed to make reduction potentials a reality. Table 6.7 summarizes the most effective policy approaches for the various segments of the transport sector. If you look at the mix of policies, there is a strong emphasis on regulation and infrastructure. It reflects the experience that with financial incentives alone transportation problems cannot be managed effectively.

Table 6.7. Summary of policy approaches that have proven to be effective in managing transport problems

Segment	Effective policy approaches in industrialized countries	Effective policy approaches in developing countries
Reducing passenger transport demand	• Teleworking • City gentrification • Tax air travel	• City planning • Tax air travel
Reducing freight transport demand	• Increase cost of freight transport (taxes, road fees) • Truck road use restriction	• Industrial zoning • Increase cost of freight transport (taxes, road fees) • Truck road use restriction
Modal shift passenger transport	• Make driving and parking more expensive and time consuming (congestion charges, fuel tax, restricted areas, parking charges) • Pay-as-you-drive for road taxes (shift costs from one-time to operational) • Provide good public transport	• Maintain bicycle/walking provisions • Provide efficient, clean, and affordable public transport (e.g. Bus Rapid Transit Systems; intercity bus systems)
Modal shift freight transport	• Develop rail/water infrastructure	• Maintain/develop rail/ water infrastructure
Fuel efficiency improvement	• Set fuel efficiency standards • Make road/vehicle taxes dependent on CO_2 emissions • Subsidize hybrid vehicles • Scrap old vehicles	• Set fuel efficiency standards • Make road/vehicle taxes dependent on CO_2 emissions • Ban inefficient second hand car imports • Subsidize hybrid vehicles • Scrap old vehicles
Biofuel	• Set quota • Mandate sustainability certification • Support R&D second generation biofuels, incl for jet fuel	• Set quota • Mandate sustainability certification • Support R&D second generation biofuels
Electric/hydrogen fuel cell vehicles	• Provide hydrogen infrastructure • Support R&D (fuel cell vehicles)	• Promote e-bikes (allowing the maintenance of bicycle facilities, subsidies) • Support R&D (fuel cell bikes)

Source: based on IPCC Fourth Assessment Report, Working Group III, ch 5.

In addition to these specific policies, general economy wide policies have an important role to play. This particularly applies to carbon taxes and so-called cap and trade systems. These general polices are discussed in Chapter 11, but remarks need to be made here regarding their application to the transport sector. Cap and trade systems basically limit the amount of GHG emissions from the sector (or subsector, such as aviation), thereby creating a price for CO_2. There is a lot of debate as to whether this can sufficiently control emissions from transport, since car users would probably only notice an increase in fuel price, which is not so effective in changing the vehicle fleet. Fuel efficiency standards and more structural changes in public transport systems are more effective. In aviation the situation is better, because the cap and trade system could work through the aviation companies, who could be held accountable for their respective emission quotas. They can then either reduce emissions by using more efficient airplanes, use biofuel (in the future), or purchase allowances on the emission trading market. The EU has decided to include the aviation sector in the EU Emission Trading System.

Policies can only have an impact when they are carefully integrated in a coherent package. Different parts of the transport system require different policy instruments. And keeping a focus on the combined effect of policies to deal with congestion, air pollution, reducing oil imports, and CO_2 emissions is crucial.

Notes

1. World Business Council on Sustainable Development, Sustainable Mobility, 2004.
2. ICCT, Air pollution and greenhouse gas emissions from ocean-going ships, 2007.
3. Climate for transport: indicators tracking transport and environment in the European Union, EEA report 1/2008
4. http://www.fra.dot.gov/us/content/25.
5. IEA, WEO 2008.
6. IPCC Fourth Assessment Report, Working Group III, fig 5.1.
7. Worldbank, Cities on the move: a Worldbank Urban Transport Review, 2002.
8. IPCC Fourth Assessment Report, Working Group III, box 5.1.
9. IEA WEO 2007.
10. IPCC Fourth Assessment Report, Working Group III, ch 5.2.2.
11. IPCC Special Report on Aviation and the Global Atmosphere, 2000.
12. http://www.imo.org/Safety/mainframe.asp?topic_id=1709&doc_id=10268.
13. UN, World Urbanization Prospects: the 2007 Revision Population Database.
14. http://www.economist.com/markets/rankings/displaystory.cfm?story_id=8908454.
15. IPCC Fourth Assessment Report, Working Group III, ch 5.5.1.5.
16. See note 15.
17. IPCC Fourth Assessment Report, Working Group III, ch 5.5.1.2 and table 5.16; CEMT, Cutting Transport Emissions: what progress, OECD, 2007, ch 5.5.
18. IPCC Fourth Assessment Report, Working Group III, fig 5.13.
19. IPCC Fourth Assessment Report, Working Group III, ch 5.5.1.1.

20. IPCC Fourth Assessment Report, Working Group III, table 5.6.
21. IEA Energy Review Switzerland, 2007.
22. WBCSD, Sustainable mobility, p81.
23. http://pdf.wri.org/automobile-fuel-economy-co2-industrialized-countries.pdf.
24. http://www.eta.co.uk/car_buyers_guide gives 23 km per litre, based on test data; in real-use cars have about a 15% higher fuel consumption.
25. Converson factors: from km/l to miles per gallon (US): multiply by 2.35; from km/l to miles per gallon (UK) multiply by 2.82.
26. Davi, Diegel, Transport Energy Data Book, ONRL, 2007, table 4–2.
27. EEA, TERM, 2007, table 7.1.
28. IPCC Fourth Assessment Report, Working Group III, table 5.9a.
29. IPCC Fourth Assessment Report, Working Group III, ch 5.4.2.1.
30. See note 23.
31. ECMT, Cutting transport CO_2 emissions, OECD, 2007.
32. Grezler A. Heavy duty vehicle fleet technologies for reducing carbon dioxide: an industry perspective. In Sperling D, Cameron J (eds) Climate policy for transportation, Springer, Dordrecht, 2008.
33. Schipper L. http://pdf.wri.org/automobile-fuel-economy-co2-industrialized-countries.pdf.
34. IPCC Fourth Assessment Report, Working Group III, box 5.2.
35. IEA WEO 2006 (table is from there).
36. Childs B, Bradley R. Plants at the pump: biofuels, climate and sustainability, WRI.
37. http://www.shell.com/home/content/aboutshell/swol/jan_mar_2008/algae_13022008.html.
38. UN Energy, Sustainable bioenergy: framework for decision makers, 2007; Eickhout B et al. Local and global consequences of the EU renewable directive for biofuels, Netherlands Environmental Assessment Agency, 2008.
39. Biofuel production costs have gone up after the publication of this graph in the IEA World Energy Outlook 2006, while fossil fuel prices soared also. So for 2008 oil prices the biofuel cost bars are higher and the 'break-even point' as reflected in the text more or less stays the same.
40. OECD. Biofuels: is the cure worse than the disease? OECD Paris, SG/SD/RT (2007) 3.
41. Metcalff GE. Using tax expenditures to achieve energy policy goals, National Bureau of Economic Research, 2008; http://www.nber.org/papers/w13753.
42. IPCC Fourth Assessment Report, Working Group III, ch 5.4.2.3, based on IEA WEO, 2006 and IEA Energy Technology Perspectives 2006.
43. IPCC Fourth Assessment Report, Working Group III, ch 5, Exec summary.
44. See http://www.ieahev.org/.
45. WBCSD Sustainable Mobility, figure 2.9.
46. See http://www.autobloggreen.com/2007/02/07/the-top-ten-electric-vehicles-you-can-buy-today-for-the-most-pa/.
47. A range of 300 km would require batteries with a weight of more than 400 kg (IPCC, 2007, ch 5.3.1.3).

48. Kliesch J, Langer T. Plug-in hybrids: an environmental and economic performance outlook, ACEEE, Report T061, September 2006.
49. IPCC Fourth Assessment Report, Working Group III, fig 5.11
50. \$50/t CO_2 avoided is equivalent to an increase in gasoline prices of about 13 \$c/litre or 50\$c per gallon.

7 Buildings

What is covered in this chapter?

Buildings are a big user of energy, through the building materials and through heating, cooling, lighting, and use of equipment in the buildings. They contribute almost 20% to global greenhouse gas emissions, when emissions from the electricity used in buildings are included. At the same time the opportunities for energy savings and CO_2 reduction are enormous. And most of these savings pay for themselves. Modern techniques now allow net zero energy buildings to be built. This chapter will investigate these possibilities and try to find out why these opportunities have not been taken advantage of and what could be done about that.

Developments in the buildings sector

Buildings are the basic infrastructure of human societies. Housing is a fundamental human need. Unfortunately many people on this planet do not yet have an adequate house. One out of three people living in cities in developing countries lives in a slum[1]. The average number of people in Pakistan per room is three, while this is 0.5 for many countries in Europe and the USA[2]. World population will grow by several billion people over the next 50 years. They all need proper housing. Factories, offices, schools, shops, and theatres also require buildings. And these buildings require energy: energy to build them, to heat and cool them, to cook food, to heat water, and to run the appliances and equipment used in buildings (see Box 7.1).

Box 7.1 **China's building boom**

China is currently adding about 2 billion (= two thousand million) square meters of building floor space every year, about half for residential and half for commercial buildings, a growth rate of about 7%. Most of these new buildings are in cities (urban population is now about 40%, will rise to 60% by 2030). An important driver for the increase in residential buildings is the official government target of increasing the living space per urban person to 35 m^2 by 2020 (now 26 m^2) and the trend towards smaller households (from 4.5 people in 1985 to 3.5

now and to 3 by 2030). Building codes were introduced in 1986, revised in 1995, differentiated according to climate, and revised again in 2006. Compliance with building codes is poor: from 60% in the North to 8% in the South.

The buildings sector used about 35% of total primary energy (including electricity and heat from central supply) in 2005, two-thirds from traditional biomass. This is expected to be only about one-third in 2030. Natural gas and electricity use in the buildings sector are expected to grow by about 6% per year till 2030. Natural gas supply to cities for household use is a government priority. Heating, cooling, and appliances are the biggest energy consumers. About 80% of urban households own an air conditioner now and almost all urban households have a refrigerator, a washing machine, and one or more televisions. Appliances are generally less efficient than comparable European models. Policies on efficiency standards for appliances and phasing out of residential electricity subsidies are in place.

(Source: IEA,WEO 2007)

Energy

The building sector uses almost 40% of the final energy[3] (final energy = energy as used, not including the losses due to electricity production). This share varies from region to region: about 20% in Australia and New Zealand to more than 50% in Sub-Saharan Africa. Residential buildings are responsible for about three-quarters of this energy use, varying from slightly more than 50% in North America to about 90% in most developing countries[4]. Commercial buildings are responsible for one-quarter. Buildings currently use more than 50% of all electricity generated[5].

Residential energy use per person differs enormously across countries. An average Ethiopian used less than one-hundredth of the energy of a Western European or North American in 2003[6]. And an average Chinese person used about one-third the energy of a Western European and a quarter of a North American. Traditional biomass (firewood, crop residues, cow dung, etc.) is still a very big energy source for household heating and cooking in developing countries. In China it provided 65% of all final energy used in the building sector in 1999 (and 80% in rural areas)[7].

What is the energy used for?

In industrialized countries heating and cooling typically use something like 40% of all residential energy, appliances 30%, and lighting and water heating about 10% each. In developing countries these shares are very different. In colder climates heating is by far the biggest use and given limited use for lighting and appliances, water heating is also pretty important.

For commercial buildings the picture is different: appliances, computers, and other equipment take a much higher share (could be 40–50%), lighting could be in the order of 20%, and a relatively lower share is accounted for by heating and cooling (more like 20%). Again, there is a marked difference in developing countries, where space heating

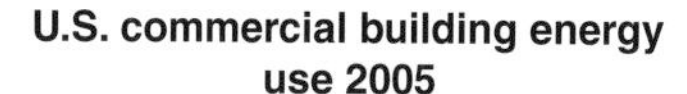

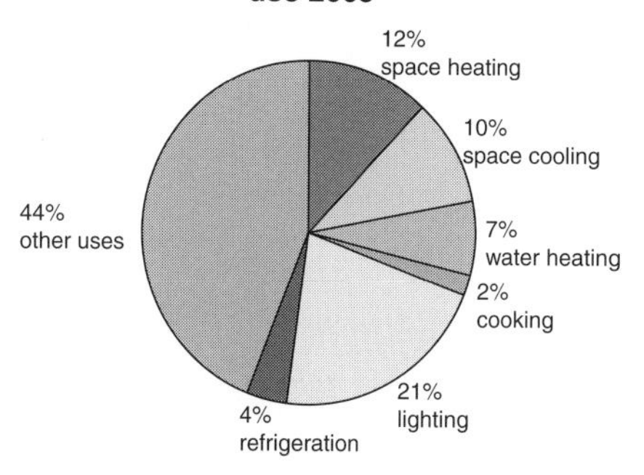

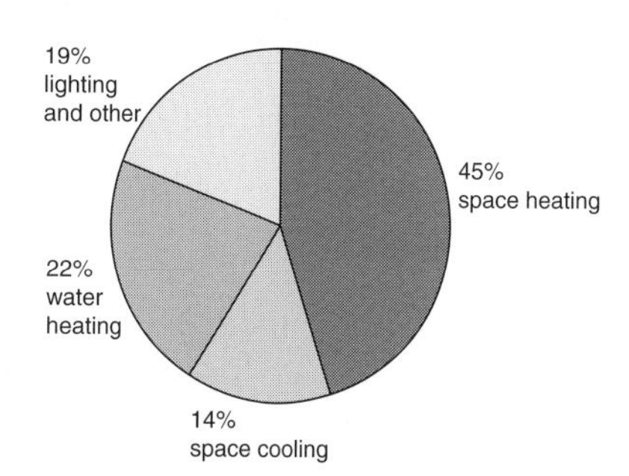

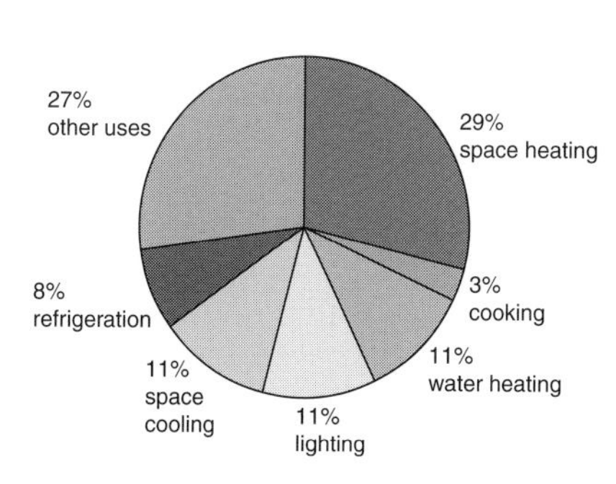

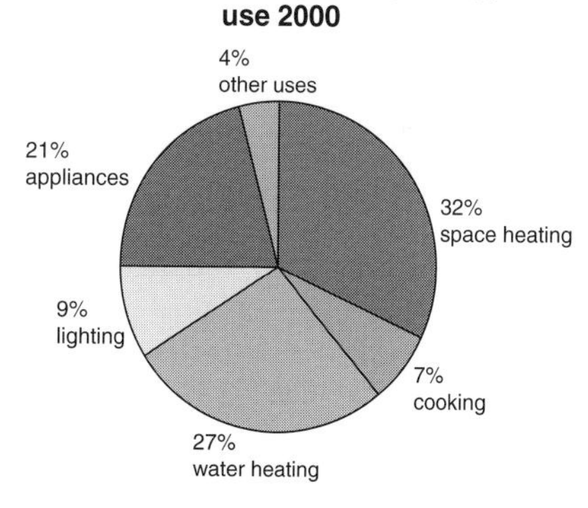

Figure 7.1 Share of energy used for different purposes in residential and commercial buildings in the USA and China.
Source: IPCC Fourth Assessment Report, Working Group III, fig 6.3.

and cooling is the dominant usage, followed by water heating. Figure 7.1 gives some breakdowns for the USA and China.

Climatic conditions of course have a major impact. Figure 7.2 shows the different energy use patterns for different climatic zones in the US. The shift from heating to cooling needs is clearly visible.

How do buildings compare?

There are large differences in energy use between buildings in the same climatic zone. The average heating energy use per unit of floor space in Germany for instance is about 220 kWh per square metre per year. In Central and Eastern Europe the average is 250–400 kWh/m^2/year. Buildings designed according to the 'passive solar house' concept use about 15 kWh/m^2/year. A selection of existing office buildings in Malaysia, Singapore,

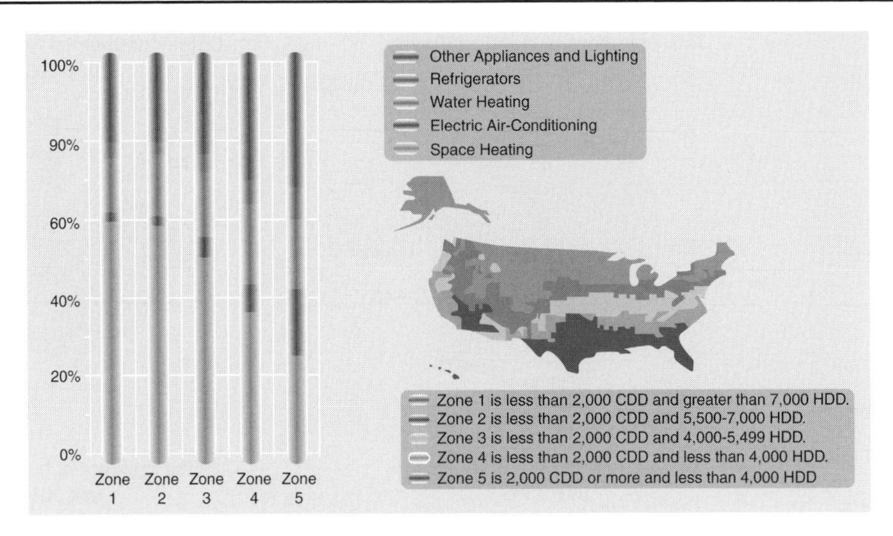

Figure 7.2 **Energy consumption shares in US residential buildings.**
Source: UNEP, Buildings and climate, 2007, fig 2.15. See Plate 13 for colour version.

Thailand, and the Philippines showed a range of 80–250 kWh/m^2/year[8]. This means there is a huge potential for reducing energy demand.

Future energy use

Energy use in the building sector is projected to increase by about 40% in the period until 2030. Most of this growth is expected to happen in developing countries, where large expansions of the housing and building stock will be needed. Building energy use in industrialized countries will stabilize or even decline. The share of electricity in the energy use in buildings is expected to double, making it by far the most important energy source for buildings. Centrally supplied heat (from urban heating networks) will remain relatively small (less than 5% of the energy used).

What are the greenhouse gas emissions?

Direct greenhouse gas emissions from the building sector are responsible for 8% of global greenhouse gas emissions or 3.9 GtCO_2-eq per year (for 2004 numbers, see Figure 7.3). CO_2 from energy covers 80% of the emissions, CH_4 10%, and the rest is N_2O and fluorinated gases[9]. Here only the emissions of HFCs are included, because emissions of the fluorinated gases CFCs and HCFCs (accounting for 1.3 GtCO_2-eq/year) do not fall under the Kyoto Protocol. They are being phased out under the Montreal Protocol however. Table 7.1 gives a summary.

When all emissions from electricity and heat used in buildings, but produced elsewhere, are also included then the share of the building sector goes up to 22%[10].

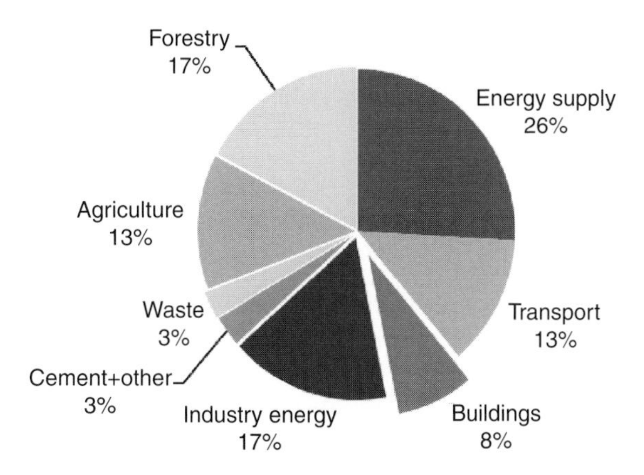

Figure 7.3 **Buildings sector contribution to direct global greenhouse gas emissions in 2004. Indirect emissions from the sector are about 14%, bringing the total of direct and indirect emissions to 22%.**
Source: IPCC Fourth Assessment Report, Working Group III, ch 1.

Table 7.1. **Contribution of different greenhouse gases to emissions from buildings: data for 2004**

Gas	Source	Emission ($GtCO_2$-eq/year)
CO_2	Heating, cooking	3.2
CO_2	Externally supplied electricity and heat	7.1
CH_4	Gas and biomass burning	0.4
N_2O	Gas and biomass burning	0.1
HFCs	Refrigeration, air conditioning, and insulation	0.2
CFCs and HCFCs (non-Kyoto gases)	Refrigeration, air conditioning, and insulation	1.3

Source: IPCC Fourth Assessment Report, Working group III, ch 1 and IEA, WEO 2007.

As for the share of energy use, residential buildings are responsible for about three-quarters of emissions on average. Regional differences are large. Developed countries are responsible for 70% of emissions, developing countries 30%. Sub-Saharan Africa covers only 6% of the total.

Without new policies total building sector emissions are projected to increase by 50–100% by 2030.

How can we reduce energy use and greenhouse gas emissions?

Energy use by and greenhouse gas emissions from buildings can be reduced in the following ways:

- by reducing energy needs
- by using energy more efficiently

- by changing the energy source
- by changing materials that emit fluorinated gases
- by changing behaviour

Reduce energy needs

The energy needs of a building are to a large extent determined by its design. Orientation to the sun, daylight entry, shading, insulation, and use of natural ventilation are some of the critical variables that determine the heating and cooling requirements. They are set at the design stage. As indicated above there are enormous differences in energy use between buildings in different places today. Several houses have been built according to the so-called 'passive house standard'[11], where energy use for heating and cooling is 75–90% lower than in a standard new-built house[12]. Typical heating energy requirements are 15 kWh/m^2/year. This is achieved with maximizing use of incoming solar radiation through glass windows in winter and minimizing it in summer, storing incoming solar heat in thick walls, very good insulation, airtight design with mechanical ventilation with heat recovery, natural ventilation, and proper orientation to the sun (see Figure 7.4).

Insulating properties of several building elements have improved enormously over time. Replacing windows with double or triple insulating windows reduces the heat loss by 45–55%. Coated double glazed windows only have 25–35% of the heat loss of regular double glazed windows[13]. Reflecting glazing can reduce incoming solar radiation by 75%. The newest windows have the capacity to become more reflecting when temperatures go up[14].

Ventilation systems have become much more advanced. Uncontrolled ventilation in buildings in cold climates can be responsible for half the total heat loss of a building. Advanced controlled ventilation systems can reduce this heat loss by a factor of 5–10. In warm climates cooling requirements can be reduced enormously by making use of natural ventilation, assisted by some small fans and exhaust ventilators. In California such houses

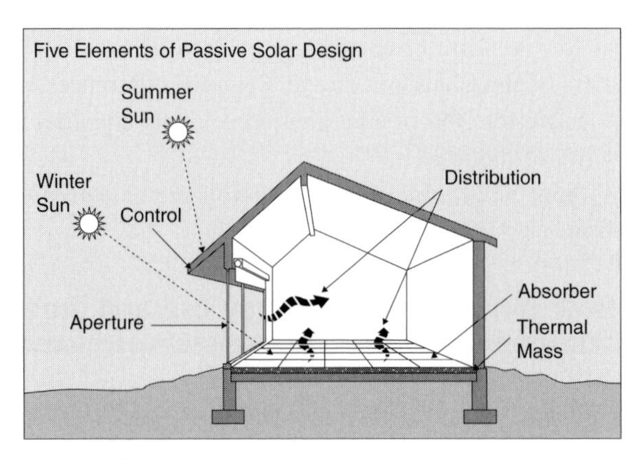

Figure 7.4 **Schematic drawing of passive solar design.**

Source: US Department of Energy, http://apps1.eere.energy.gov/consumer/your_home/designing_remodeling/index.cfm/mytopic=10270.

Figure 7.5 **Use of insulated daylight panels to reduce the need for electric lighting.**
Source: http://www.inhabitat.com/2006/09/06/green-building-101-design-innovation/.

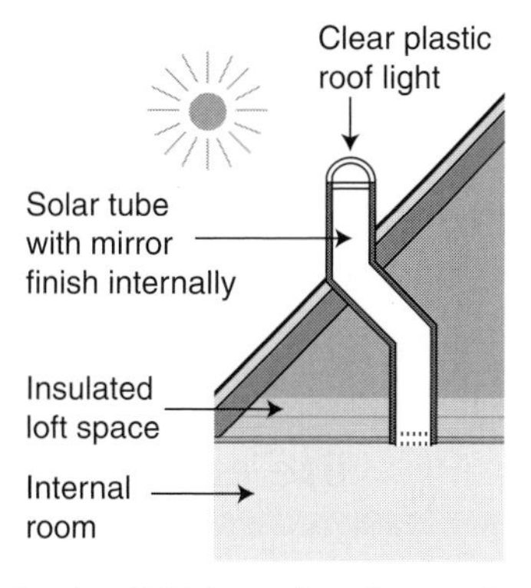

Figure 7.6 **Solar tube to pipe light to enclosed rooms in a building.**
Source: www.lowenergyhouse.com.

are able to keep temperatures below 26°C with night-time mechanical ventilation only for 40% of the time[15].

Better use of daylight can save a lot of lighting energy. Many office building are designed in such a way that in most workplaces electric light is a necessity, even on a sunny day. Through proper design of office buildings 40–80% of lighting energy can typically be saved by making better use of daylight (see Figure 7.5).

Advanced technologies are now available to 'pipe' light from outside into enclosed rooms in a building[16] (see Figure 7.6).

Costs

Halving heating and cooling needs compared to current building standards is possible without net additional costs. The saved energy pays for the extra measures taken. Still, in many instances the additional upfront investment required or aesthetical considerations by architects are a reason not to take these economically rational decisions.

Many of the design features described here are of course only achievable in a new building, and from now until 2030 many new houses will be built. According to the UN '*An estimated 21 million new housing units are required each year, in developing countries, to accommodate growth in the number of households between 2000 and 2010. 14 million additional units would be required each year for the next 20 years if the current housing deficit is to be replaced by 2020*[17].'

And that is only part of new construction. In developed countries urban renewal projects will lead to knock-down of old buildings and construction of new ones, together providing a big opportunity for energy saving and CO_2 emission reduction. It would even be attractive to demolish older, energy inefficient buildings well before their economic lifetime, because the energy embedded in the construction of a building is normally only 15–20% of the total energy used over the lifetime[18].

For existing houses and commercial buildings the possibilities for energy conservation are somewhat more limited and more expensive, but a lot can be done at low cost. Insulation of walls of existing buildings by filling of wall cavities with spray foam or rock wool, of floors with insulating foils, and of roofs and lofts with foam or rock wool can be done in many buildings. Care should be taken to use available climate friendly blowing agents when applying foam, because HFC or HCFC blowing agents would add to the GHG emissions[19].

Use energy more efficiently

Heating, cooling, lighting, and running refrigerators, washing machines, TVs, computers, etc. require energy in the form of electricity, gas, oil, coal, or (traditional) biomass. How efficient is that energy used? And how much can CO_2 emissions be brought down?

Space heating

Space heating in industrialized countries and urban areas of developing countries is done with gas, oil, or electricity. Except when very low carbon electricity is available, electric heating is inefficient and leads to high CO_2 emissions. First turning fossil fuel into electricity, losing about 60% of the energy, and then converting electricity into heat, again with a substantial loss of energy, is not a good idea. And most electricity is produced with fossil fuels, guaranteeing high CO_2 emissions from electric heating.

Modern gas fired building heating installations have reached an efficiency of more than 97% due to advanced burner design and recovery of waste heat. On average, installations being used today have an efficiency of 60–70%. With an average lifetime of a central heating boiler of about 15 years, big reductions in energy use and CO_2 emissions can be achieved by replacing those with advanced high efficiency installations. The newest, most efficient boilers earn the additional cost back well within their lifetime. It even makes sense to replace boilers before the end of their economic lifetime.

Heat pumps

In places where there is no way of replacing electric heating with modern gas fired systems, heat pumps can be used to improve energy efficiency. The heat pump is a sort of 'reverse refrigerator' that transfers heat into the house from the surrounding air or the soil. Since the soil is relatively warm in winter time compared to air, it is attractive to draw the heat from the soil. Heat pumps can also work the other way around in summer, cooling the building. Since the soil remains substantially cooler than the air, 'pumping' the heat into the soil in summer is more energy efficient. Given the energy losses when producing electricity, the overall efficiency of heat pumps is lower than that of modern gas fired heaters, except in cases where low carbon electricity is available. By doubling as air conditioners, heat pumps can also eliminate emissions of fluorinated gases from traditional air conditioners[20] (see Figure 7.7).

District heating

An efficient way to heat a building is to use waste heat from a power plant via a district heating system. The power plant then becomes a combined heat and power plant (CHP, see also Chapter 5). Of course the heat will have to be transported via a pipe network. This limits the scope for district heating to a radius of about 50 km around a power plant. There are many cities where that condition applies and district heating with CHP is applied in many cities already.

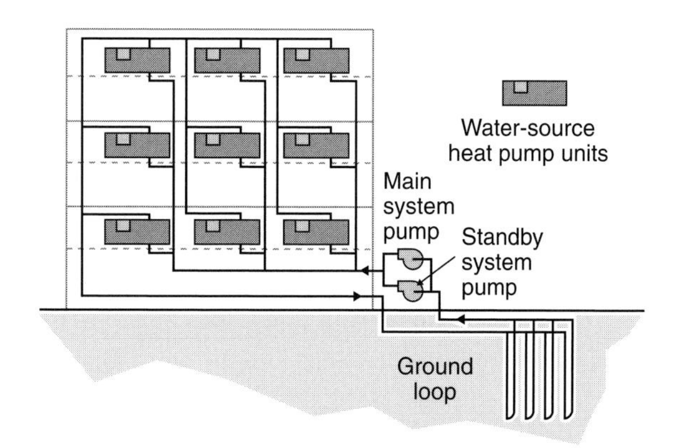

Figure 7.7 **Ground source heat pump system for building heating and cooling.**
Source: redrawn from http://www.geo4va.vt.edu/.

Micro CHP

Yet another alternative that is being installed in some places is the so-called micro combined heat and power installation (micro CHP). It is an installation that produces both electricity and heat for a small building. They are normally gas fired. Because of their very high efficiency, the overall efficiency for heat and electricity is often better than electricity from the grid and a separate heater. In terms of CO_2 emissions this also holds, except in cases where grid based or decentralized electricity is from renewable sources.

Rural areas

In rural areas of developing countries the situation is quite different. About 3 billion people in rural areas depend on wood, charcoal, crop residues, cow dung, and coal (particularly in China) for heating and cooking, although many of these people live in tropical areas where no heating is required. This practice causes severe indoor air pollution and causes disease and premature death. In terms of contribution to greenhouse gas emissions the picture is mixed. Much of the heating fuel is renewable, although wood consumption in many areas is not sustainable[21].

There are not many low carbon alternatives for rural energy in the short term. A lot of work has been done on the development of efficient cook stoves. Results are mixed. Efficiency has been shown to be 10–50% better, but penetration is limited, new cook stoves were not always working properly, and costs were not always low enough. The impact on women and children (reduced time for fuel gathering and less indoor pollution) is bigger than on CO_2 emissions. Biogas installations (see Chapter 5) do have a good potential to provide renewable cooking fuels. Capacities of these installations are generally not sufficient however to cover heating needs. Solar and small electric cookers have some potential. For the time being more efficient heating stoves are the only short term solution.

Air conditioning

Full mechanical air conditioning is becoming the norm for cooling of buildings. In urban areas of developing countries it is one of the first things households want to have if they can afford it. In cooler areas of industrialized countries it is also becoming more common to have air conditioning, where this used not to be necessary. The world production of small air conditioners for instance increased by 25% between 1998 and 2001.

Air conditioners come in a range of sizes and types, from small room size wall mounted units, to so-called split system units for homes and small buildings to large cooling devices for use in larger residential and commercial buildings. Their energy efficiency generally improves with size. Big centrifugal chillers are about 2–3 times as efficient as small room air conditioners. Further improvement of energy efficiency is possible[22].

Figure 7.8 **Window-mounted air conditioners in apartment building.**
Source: Shutterstock.com, © Phaif, image #15585142.

Air conditioners generally use a halo-carbon refrigerant. More than 90% of the installations use HCFC-22. As this substance will be phased out in the near future under the Montreal Protocol, a shift to HFCs, a powerful greenhouse gas controlled by the Kyoto Protocol, is noticeable[23] (see also Chapter 2). Leaks in some installations and repair or service work lead to an emission of HFCs of about 0.2 $GtCO_2$-eq per year. Alternatives in the form of refrigerants that have a zero or lower global warming contribution are available. For supermarket cooling systems the combined effect of choosing more energy efficient cooling equipment and a change of the refrigerant can lead to a 60% reduction in overall CO_2 equivalent emissions[24].

There are other, much lower energy alternatives by moving from air conditioners to low energy cooling techniques. One approach is to mechanically assist air flows through buildings using cooler night-time air or using (cool) underground inlet ducts. Another is to cool the inlet air by evaporating water directly in it or cool the incoming air with an evaporative cooling driven heat exchanger. Energy savings in the order of 90% compared to traditional air conditioners are possible. In areas with hot and humid air the drying of the air by over cooling consumes a lot of energy. Desiccants can reduce this energy use by 30–50%. Figure 7.9 shows a schematic diagram of a building where several energy reducing measures have been taken, including a small sized centrifugal chiller as the main cooling machine on the roof.

Light[25]

Lighting consumes roughly 20% of global electricity, 10% of the total energy use in residential buildings, and 20% in commercial buildings. The total energy used for lighting is about one-third used in residential and two-thirds in industrial and commercial buildings. Lighting is responsible for about 1.9 $GtCO_2$ per year, which is 70% of the emissions from all passenger vehicles. Traditional so-called incandescent lamps represent 80% of lamps sold, 30% of all lighting energy, but only 7% of delivered

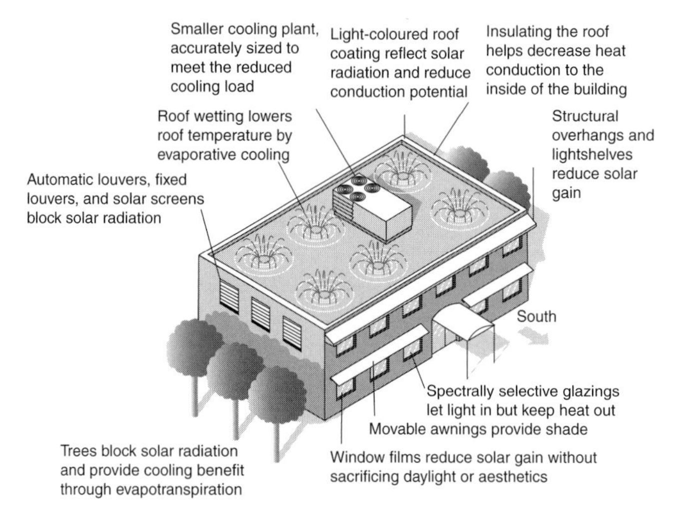

Figure 7.9 **Combination of measures to reduce the cooling requirements and energy use for cooling, including a small size centrifugal chiller on the roof.**
Source: Madison Gas and Electric, http://www.mge.com/business/saving/madison/pa_14.html.

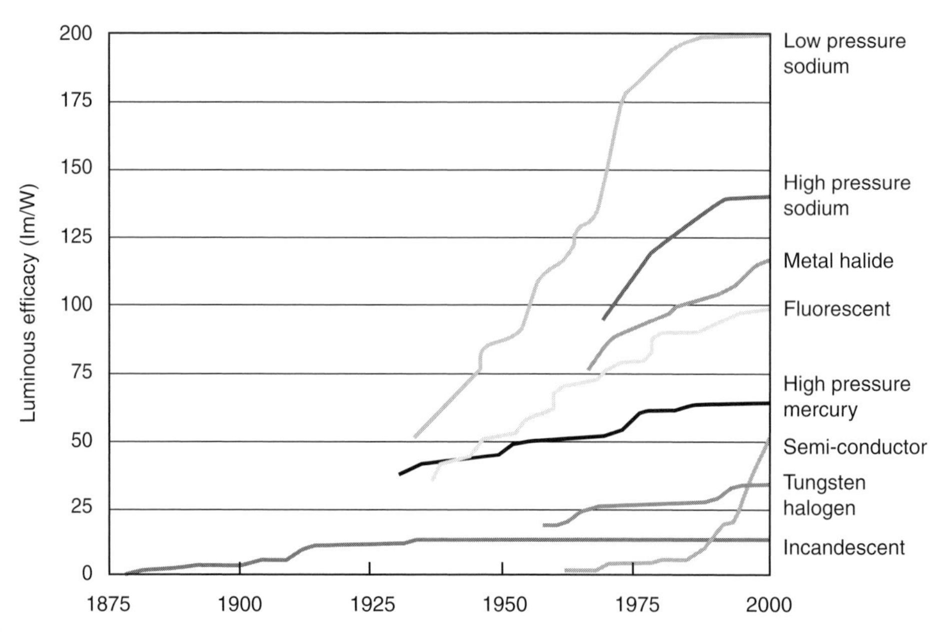

Figure 7.10 **Efficiency of different lamp types over the years. Efficiency is expressed in light delivered (lumen) per Watt.**
Source: Light's Labour's Lost, International Energy Agency, 2006.

light. This is a complex way of saying they are very inefficient but are still widely used. Penetration of more efficient fluorescent tubes, compact fluorescent, and halogen lamps is limited: in the best European country about one in three light bulbs in households was efficient. For a comparison of the efficiency of lamps see Figure 7.10.

The potential for reduction of energy use and emissions is considerable. Reductions of 75–80% are possible in residential buildings, primarily by shifting from incandescent light bulbs towards compact fluorescent lamps (CFLs) and (in the future) LED (light emitting diode) lamps. The use of sensors to switch light on when people are present (and daylight is not enough) and off when they have left is an important way to assist people to save on energy.

In commercial buildings, where lighting is already more efficient, a further 50% improvement is possible through use of more efficient lamps, sensors, and use of local, so-called task lights. And this on top of a further 20–40% reduction by minimizing the need for lighting by designing buildings to make better use of daylight.

About one third of the world population depends on kerosene, paraffin, or other hydrocarbon fuel for lighting. Only 1% of all lighting is provided in this way, but it represents 20% of the lighting related CO_2 emissions and 3% of the world's oil consumption. Together with efforts to provide electricity to the 1.6 billion people that do not have it now, efficient fluorescent lamps allow people to use a minimum of electricity, which is an expensive commodity for many poor people. Bringing the costs of these CFLs down is therefore of prime importance.

Appliances

In 11 large OECD countries the electricity used by refrigerators, freezers, ovens, washing machines, dryers, computers, etc. (in short: household appliances) is more than 40% of the total residential primary energy use. In developing countries this share is much lower, although in several countries, for example amongst China's urban population, the penetration of electrical appliances is increasing strongly. In commercial building the share of equipment in the total energy use is normally higher than in residential buildings. Figure 7.11 gives an overview for the average electricity used by appliances in US households.

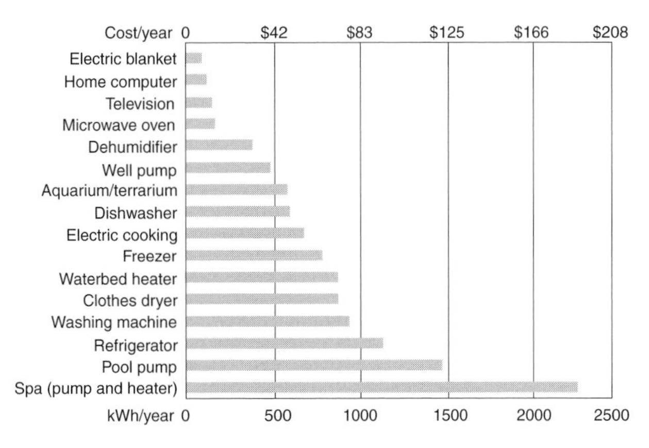

Figure 7.11 **Energy use of a typical appliance per year and its corresponding cost based on national averages for US households. For example, a refrigerator uses almost five times the electricity the average television uses.**

Source: US Department of Energy, http://www1.eere.energy.gov/consumer/tips/appliances.html.

How much efficiency improvement is possible?

Efficiency of appliances has improved considerably over the years. Refrigerators sold in the US today use less than 400 kWh per year, while those sold in the late 1970s used about 1800 kWh. Due to the lifetime of appliances there is a significant difference between the average appliance in use and the new models on the market. In the UK in 2005 average energy use of washing machines was 1.24 kWh per washing cycle. The best machines for sale used about 0.85 kWh[26]. And then there is the difference in appliances for sale today: the most efficient use 50–80% less energy than the worst ones, as is for instance shown by the energy labels used in the EU (see Figure 7.20).

Unfortunately efficiency is not the only thing. The volume of refrigerators tends to increase with income and is influenced by cultural aspects: in the US they are much bigger than in Europe. In the US the best standard size refrigerators use less than 400 kWh per year; in Europe the figure is about half this, because of smaller size.

So-called 'standby power', the electricity consumed when appliances are switched off but still in sleeping mode, is becoming a big contributor to electricity consumption. In the US it is now more than all refrigerators combined, due to the sheer volume of appliances that are kept plugged in (see Box 7.2).

Box 7.2 Global efforts to combat unneeded standby and low power mode consumption in appliances

Standby and low-power-mode (LoPoMo) electricity consumption of appliances is growing dramatically worldwide, while technologies exist that can eliminate or reduce a significant share of related emissions. The IEA estimated that standby power and LoPoMo waste may account for as much as 1% of global CO_2 emissions and 2.2% of OECD electricity consumption. The total standby power consumption in an average household could be reduced by 72%, which would result in emission reductions of 49 million tCO_2 in the OECD. Various instruments – including minimum energy efficiency performance standards (MEPS), labelling, voluntary agreements, quality marks, incentives, tax rebates, and energy efficient procurement policies – are applied globally to reduce the standby consumption in buildings, but most of them capture only a small share of this potential. The international expert community has been urging a one Watt target. In 2002, the Australian government introduced a 'one-watt' plan aimed at reducing the standby power consumption of individual products to less than one watt. To reach this, the National Appliance and Equipment Energy Efficiency Committee has introduced a range of voluntary and mandatory measures to reduce standby – including voluntary labelling, product surveys, MEPS, industry agreements, and mandatory labelling. As of mid-2006, the only mandatory standard regarding standby losses in the world has been introduced in California, although in the USA the Energy Policy Act of 2005 directed the USDOE to evaluate and adopt low standby power standards for battery chargers.

(Source: taken from IPCC Fourth Assessment Report, Working group III, box 6.4)

Further efficiency improvements are possible, through innovation and by removing inefficient appliances from the market.

Change the energy source

Changing to low carbon energy sources is an obvious way to reduce CO_2 emissions. As far as electricity is concerned the options are discussed in Chapter 5, including PV cells mounted on building roofs and small scale wind turbines. PV cells integrated in building materials will be considered below as will solar water and space heating. Combining renewable energy generation by buildings with energy needs reduction and energy efficiency improvements can lead to so-called 'net zero energy' buildings that produce all the energy that is needed.

PV integrated building materials

Photovoltaic panels, mounted on the roof of a building, are now a common thing. There are many building materials on the market however where PV cells are integrated in the building material itself. PV Roof tiles for flat roofs and PV slates and shingles for slanted roofs are commercially available (see Figure 7.12). South facing facades of buildings are ideal for PV integrated wall tiles, but also for PV sunshades (see Figure 7.13).

Solar water heating

Solar water heaters absorb heat from the sun, either in an insulated dark flat panel (flat panel type) or in pipes that are insulated with a double vacuum wall like a thermos can

Figure 7.12 PV integrated roof slates.

Source: www.newagesolar.com.

Figure 7.13 **PV integrated sunshades as part of the building design.**
Source: Power Glaze, www.romag.co.uk.

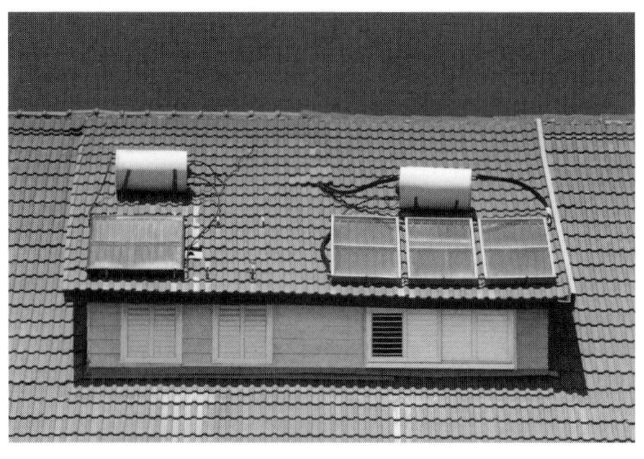

Figure 7.14 **Flat panel solar water heaters providing 80% of the hot water needs of the house.**
Source: © mtsvn/shutterstock.com, image # 14253103.

(vacuum tube type, see Figures 7.14 and 7.15). For swimming pools unglazed plastic collectors are often used, particularly in the USA.

China is now by far the biggest market for solar water heaters, with about 65% of all installed capacity in the world. The EU has 13%, followed by Turkey (6%), Japan (4%), Israel (3%), and Brazil (2%)[27]. Local building codes are a very strong driver for installation of solar water heaters. In Israel for instance there are strict national regulations and in a number of other countries stimulation programmes and municipal building codes have contributed to a significant penetration (see Figure 7.16.). The total installed capacity of solar heaters is of the order of 220 million m^2. Annual growth rates are of the order of 20%. The heat produced however is still less than 5% of all heat used in the buildings sector[28]. More than 50 million households worldwide have a solar heater system.

Costs of solar water heaters in China are typically 200–300US$ each, while systems in Europe vary from US$700 to US$2300. Prospects for solar water heating as a

Figure 7.15 **Vacuum-tube solar water heater.**

Source: www.himfr.com.

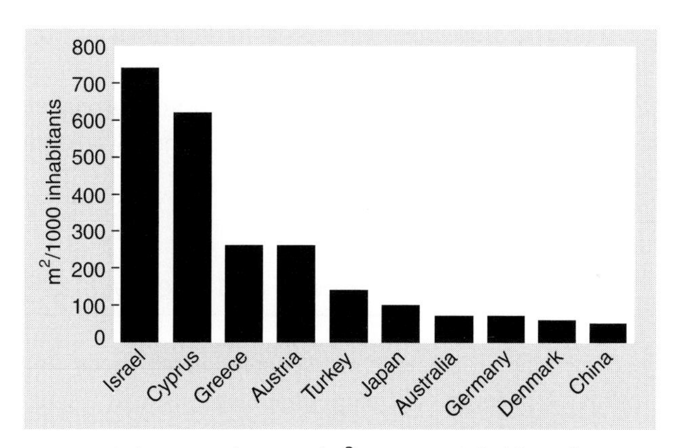

Figure 7.16 **Solar water heaters (m^2 per 1000 inhabitants).**

Source: REN21, Renewables 2005.

contribution to CO_2 emission reduction are modest at a global scale, although growth of these systems could be strong when adequate policies are put in place in many countries. Especially in tropical and subtropical developing countries the need for drastic expansion of the housing stock provides excellent opportunities at low costs.

Solar space heating and cooling

Passive solar heating has already been discussed above. The same principle as for solar water heaters can be used to provide additional solar space heating, albeit with much larger solar collectors, which is the reason why this technology is not applied widely yet.

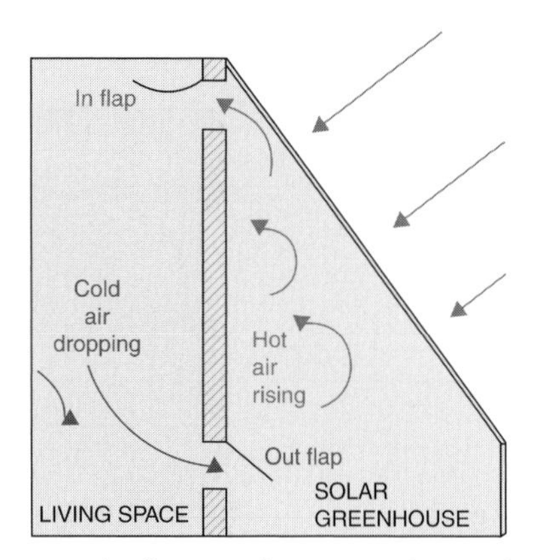

Figure 7.17 Schematic diagram of solar greenhouse attachment for space heating.
Source: http://jc-solarhomes.com/passive_solar.htm.

A more attractive form of solar space heating is the attachment of glass extensions (a greenhouse actually) to buildings that act as a greenhouse and capture heat. By controlling air flow from the glass extension the adjacent house can be (partially) heated (see Figure 7.17).

Solar heating and cooling can benefit from seasonal storage. Excess heat captured in summer can be stored for instance in the groundwater under the building. In winter this warm water can be used for heating again

Zero-energy and Energy-plus buildings

By combining all elements of reducing energy needs through the passive house concept (see above), energy efficiency improvements, and use of solar energy, buildings can be constructed that use no external energy or are even net energy producers. The US Department of Energy database contains seven examples of commercial zero or net positive energy buildings in the US[29]. Box 7.3 describes a new energy-plus office building in Paris. There are additional costs involved, but part of these will be earned back due to lower energy bills.

Box 7.3 **Energy Plus office building in Paris**

The 'Energy Plus' office building, to be located outside of Paris, is designed to produce all its own energy for heating, lighting, and air conditioning. This zero-energy building, according to the designers, will be the greenest office building ever created. It will accomplish this by having more solar panels on its roof than any other building – producing enough energy to power the entire building and still feed extra back into the grid. Its unique cooling system will take cold water from the river Seine and pump it around the building – eliminating the need for a traditional air conditioner. The 70 000 m^2 building will also utilize cutting edge insulation, reducing amount of electricity consumption per square meter of office space per year to 16 kilowatts, the lowest in the world for a building of its size. The building is expected to house up to 5000 people. It's expected to cost approximately 25–30% more than a traditional office building. It was designed by Skidmore Owings Merrill, the architectural firm behind New York's upcoming Freedom Tower.

(Source: http://www.metaefficient.com/architecture-and-building/the-energy-plus-building-produces-all-its-own-power.html)

Change behaviour

Behaviour is an important driver of energy use and greenhouse gas emissions in buildings. Setting the temperature, switching off lights, purchasing lighting and appliances, decisions to invest in insulation or PV panels, buying green electricity, etc. are all human decisions that determine energy use and emissions. We know from research that, for similar houses (i.e. similar design, insulation, and other features) and composition of families, energy use can vary by a factor of 2.

Most of the actions that people can take to reduce energy use in buildings have a net benefit. In other words, they save money. However, only a small percentage of people react 'economically' to these existing financial incentives for installing insulation or energy efficient heating and cooling equipment or appliances. There are many reasons for this seemingly irrational behaviour, which is not so irrational actually. Lack of motivation, lack of time, lack of information, and competing issues that people have to attend to are important. There are also limitations to what individuals can do. People that rent a home or an apartment have only a limited influence on the insulation of the building and the efficiency of the heating and cooling facilities. Scarcity in the housing market often reduces the choices and location is often more important. Figure 7.18 gives an example, based on research in the UK, of willingness and ability to act. When ability is low, attempts to change people's behaviour will of course fail. And when willingness is absent, prospects are not good either.

Willingness and ability are not enough to change behaviour. It is well known that people who say they are very concerned about climate change are not doing all the things they could. So changing behaviour is about creating the additional incentives to turn willingness into action. Information campaigns have traditionally been the preferred instrument to change behaviour. They often were focused on motivating people, in other words increasing their willingness. That explains the limited success of such campaigns. If action is not made easier, behaviour will not really change. There is another complicating factor: consumers are not all the same. There are distinct groups with different values and preferences: environmentally conscious people, trendsetters, rationalists, ill informed followers, conservationists, hedonists, etc. They react differently to campaigns. Effectiveness of behavioural change campaign is also affected by culture. In Japan for instance information campaigns seem to work much better (see Box 7.4).

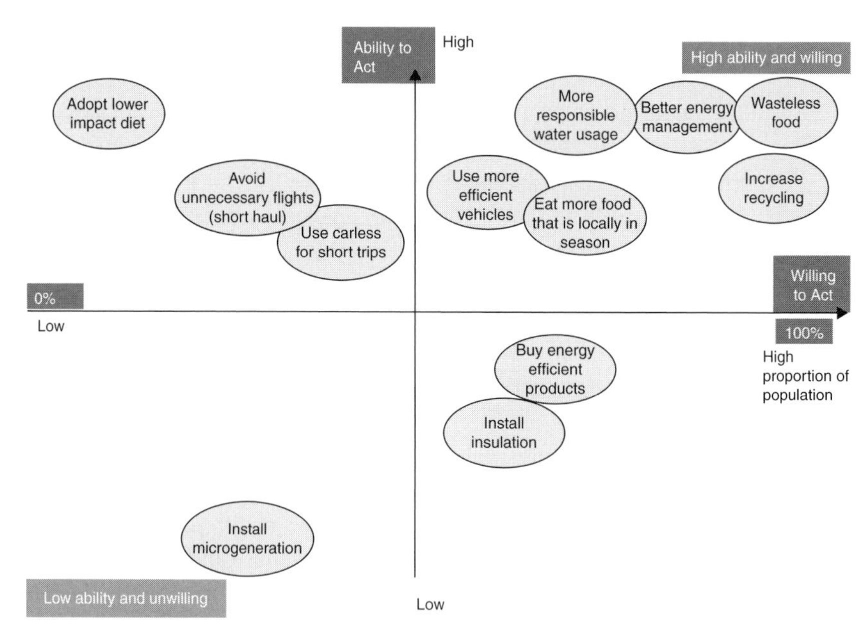

Figure 7.18 **Willingness and ability of people to change environmental behaviour.**
Source: UK DEFRA, A framework for pro-environmental behaviours, January 2008.

An important lesson from research and practice on behavioural change is that 'hard' measures, such as appliance standards, building codes, automatic power off features, bans on certain energy wasting equipment, and things like motion detecting light switches, free compact fluorescent lamps and subsidies on efficient appliances, supported by 'soft' information instruments, work best. In that way information about the need for change is combined with a practical and easy way to actually change behaviour[30].

Box 7.4 Japan Cool Biz campaign

In 2005, the Ministry of the Environment (MOE) in Japan widely encouraged businesses and the public to set air conditioning thermostats in offices to around 28 °C during summer. As a part of the campaign, MOE has been promoting summer business styles ('Cool Biz') to encourage business people to wear cool and comfortable clothes, allowing them to work efficiently in these warmer offices. In 2005, an MOE survey of 562 respondents showed that 96% of the respondents were aware of 'Cool Biz' and 33% answered that their offices set the thermostat higher than in previous years. Based on this result, CO_2 emissions were reduced by approximately 460 000 tonnes in 2005, which is equivalent to the amount of CO_2 emitted from about 1 million Japanese households for 1 month. MOE will continue to encourage offices to set air conditioning in offices at 28°C and will continue to promote 'Cool Biz'.

(Source: IPCC Fourth Assessment Report, Working Group III, box 6.5)

How does this all fit together?

Many studies have been performed in specific regions on how much reduction of energy and CO_2 emissions can be realized and at what cost levels. Because studies assume different combinations of measures, different electricity and fuel prices, different economic criteria when calculating cost, and not all regions are adequately covered, only a rough estimate of the global potential can be given. Overall, total emissions can be reduced by about 30% in 2030, compared to what they would have been otherwise, at zero costs or at a profit ('negative costs'). An additional 10% can be reduced for costs up to US$100/tonne CO_2 avoided. Both numbers are an underestimate, because most studies have looked at only a part of the attractive options available and ignored many of the higher cost options since so much can be done at low costs. That corresponds to a minimum of about 4.5 and 5.6 Gtonnes of CO_2 per year by 2030, at zero and US$100/tonne, respectively. These are the reductions achievable for the total building stock. Given the 50–100 year lifetime of buildings a lot of the reductions have to be achieved through retrofitting of existing buildings. For new buildings, about a 75% reduction can be realized, compared with current practice, at little or no extra costs[31]. Accepting 20–30% higher initial costs would bring zero energy buildings within reach as discussed above. Pushing the new construction to very low or zero energy use is needed to bring the overall building emissions down.

The potential differs from region to region. Most of the reductions can be found in developing countries, in light of the expected population growth and the building activity in these countries. Of the total reduction potential developing countries cover about 45%, OECD countries about 35%, and former Soviet Union countries about 20% (a high share compared to the size of the population, caused by a long neglect of energy conservation in these former centrally planned economies).

How to realize this large potential?[32]

With the large potential for reductions at negative cost, the building sector seems to be ideal for realizing energy and CO_2 reductions without specific policy. There is such a strong economic argument, things should happen automatically, shouldn't they? The reality is very different. The savings that can be made are not happening and even with specific policy actions it is extremely difficult to get measures implemented. Why is that?

Barriers

The most important reasons are summarized in Table 7.2. Financial barriers to a large extent have to do with the problem of making higher initial investments acceptable.

Table 7.2 Barriers that hinder the penetration of energy efficient technologies and practices in buildings

Barrier categories	Definition	Examples
Financial costs/benefits	Ratio of investment cost to value of energy savings	Higher upfront costs for more efficient equipment Lack of access to financing Energy subsidies Lack of internalization of environmental, health, and other external costs
Hidden costs/benefits	Cost or risks (real or perceived) that are not captured directly in financial flows	Costs and risks due to potential incompatibilities, performance risks, transaction costs, etc. Poor power quality, particularly in some developing countries
Market failures	Market structures and constraints that prevent the consistent trade-off between specific energy efficient investment and the energy saving benefits	Limitations of the typical building design process Fragmented market structure Landlord/tenant split and misplaced incentives Administrative and regulatory barriers (e.g. in the incorporation of distributed generation technologies) Imperfect information
Behavioural and organizational non-optimalities	Behavioural characteristics of individuals and organizational characteristics of companies that hinder energy efficiency technologies and practices	Tendency to ignore small opportunities for energy conservation Organizational failures (e.g. internal split incentives) Non-payment and electricity theft Tradition, behaviour, lack of awareness, and lifestyle Corruption

Source: IPCC Fourth Assessment Report, Working Group III, table 6.5; *Source Carbon Trust, 2005.*

Individual decisions are often driven by initial capital investment rather than overall costs that include energy costs during the use of the building, lack of financial incentives for delivering excess PV electricity back to the grid, and lack of attractive financing for energy efficiency investments. The relative costs of more energy efficient or renewable energy options are often a disadvantage as a result of low or subsidized (fossil fuel) energy prices.

Hidden costs primarily emerge from uncertainty about the performance and reliability of alternative options, the cost of collecting the necessary information or of getting approval for alternative solutions.

A typical example of a market failure is the so-called 'split incentive' situation, where owners/landlords have little incentive to put in additional investment to save energy, while tenants (that have a good incentive) are not in a position to make the investments. Other examples are regulations that prohibit the installation of some energy saving or renewable energy options, or policy priorities to keep rents affordable (meaning limiting the capital investments). It also covers lack of information about energy use, options for reduction and costs, or lack of time to investigate how measures can be taken; this applies to architects, builders, and owners.

Behavioural and institutional barriers include the issues of personal choice and behaviour mentioned above, as well as real world issues such as non-payment and corruption, preventing rational decisions to be made (see above).

Policies

In light of the multitude of barriers it is no surprise that an effective policy to realize energy and CO_2 reduction needs to be based on multiple policy instruments, each addressing specific barriers. For a sector with a large number of decision makers (down to individual home owners or tenants) effectiveness of policy instruments is a function of reaching these decision makers. A package of many different kinds of information, financial incentive, and other measures still would only reach a fraction of these decision makers. In such a situation regulatory approaches are usually the most effective. The most effective are building codes and legislation requiring utilities to invest in energy savings and to pay adequately for electricity delivered back to the grid by decentralized solar PV. They can address a whole range of barriers at the same time[33]. For reducing fluorinated gas emissions from air conditioners and refrigeration regulation is also an effective approach.

Building codes

Building codes come in two different styles: the *prescriptive style* where specific provisions for insulation, windows, and heating/ cooling systems are prescribed; and the *performance style*, where standards for the energy performance of whole buildings are specified, leaving flexibility for architects and builders. The first type is easier to enforce, making it attractive for countries with limited enforcement expertise, but provides no incentive for further improvements. The second allows for optimizing

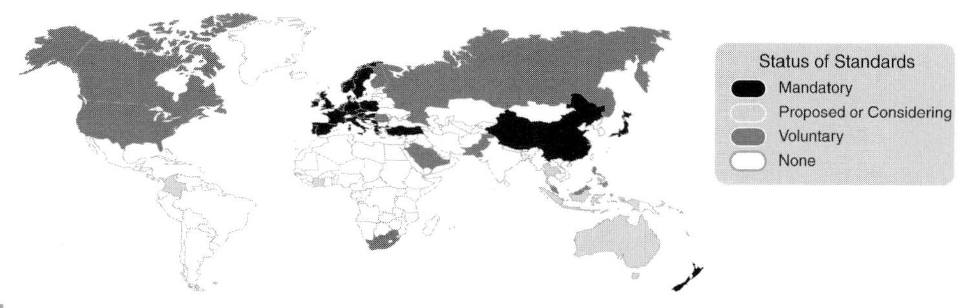

Figure 7.19

Status of building codes around the world.
Source: UNEP. Buildings and Climate Change: Status, Challenges and Opportunities, 2007.

design in light of the specific situation and gives more room for introducing new technologies, but requires more sophisticated expertise to ensure compliance.

Building codes are often limited to new buildings, although requirements in relation to remodelling of existing buildings can extend their influence. They also need to be renewed regularly, to adopt the latest developments in building technology, use of renewable energy, and energy savings. And they need to be enforced, which is not always done (see Figure 7.19 and Box 7.5). This is a well-known weak spot. Harmonizing building codes across countries, which is for instance done in the EU, is a very effective way to push possible energy and emission savings.

Box 7.5

EU Directive on Energy Performance of Buildings

One of the most advanced and comprehensive pieces of regulation targeted at the improvement of energy efficiency in buildings is the European Union Directive on the Energy Performance of Buildings (European Commission, 2002). The Directive introduces four major actions. The *first action* is the establishment of 'common methodology for calculating the integrated energy performance of buildings', which may be differentiated at the regional level. The *second action* is to require member states to 'apply the new methods to minimum energy performance standards' for new buildings. The Directive also requires that a non-residential building, when it is renovated, be brought to the level of efficiency of new buildings. This latter requirement is a very important action due to the slow turnover and renovation cycle of buildings, and considering that major renovations to inefficient older buildings may occur several times before they are finally removed from the stock. This represents a pioneer effort in energy efficiency policy; it is one of the few policies worldwide to target existing buildings. The *third action* is to set up 'certification schemes for new and existing buildings' (both residential and non-residential), and in the case of public buildings to require the public display of energy performance certificates. These certificates are intended to address the landlord/tenant barrier, by facilitating the transfer of information on the relative energy performance of buildings and apartments. Information from the certification process must be made available for new and existing commercial buildings and for dwellings when they are constructed, sold, or rented. *The*

last action mandates Member States to establish 'regular inspection and assessment of boilers and heating/cooling installations'. It is estimated that CO_2 emission reductions to be tapped by implementation of this directive by 2010 are 35–45 million tCO_2-eq at costs below 20EUR/tCO_2-eq, which is 16–20% of the total cost-effective potential associated with buildings at these costs in 2010.
(Source: taken from IPCC Fourth Assessment Report, Working group III, box 6.3)

Demand side management

In the USA, Demand Side Management (DSM) programmes, run by electric utilities, have been very successful. They operate on the basis of regulatory requirements imposed on utilities to first invest in energy savings, before expanding power plant capacity. At first sight that looks to be against the interest of these utilities. Why would they put money into selling less electricity? The crucial element is the rule that energy saving investments can be recovered via the electricity tariffs. So customers pay for it, but less than what they would have paid if investments had been put in new power plants. These programmes are implemented through utility based incentive programmes or direct investments in energy savings in buildings. This policy approach is spreading to other countries now. The UK has introduced the Energy Efficiency Commitment legislation for instance[34].

Appliance standards and labelling

Legally based appliance standards are in place in many countries. The US programme applied in 2004 to 39 residential and commercial products. Experience with this programme is very positive: costs are low (in the order of US$2 per household), and standards are effective (estimated reduction of 10% in 2020 compared to business as usual and more than US$1000 savings per household). Standards can speed up the improvement of energy efficiency, provided they are regularly strengthened. In that respect the Japanese 'Top-Runner' programme is very interesting. Performance of the best-in-class equipment is automatically becoming the standard 3 years later. This is a built-in mechanism to stimulate innovation by companies[35].

Labelling of appliances, heating/cooling and lighting equipment, and whole buildings is becoming quite popular. Figure 7.20 shows how efficiency of refrigerators in the EU improved over time, and how consumer preference shifted. Labels make it easier for people who are motivated to buy an efficient appliance. It does not change behaviour of those who are not sensitive to energy conservation.

Financial incentives

Supplementary policies are needed to take care of barriers that cannot be removed through building codes. The obstacle of higher initial investments, for example, one of

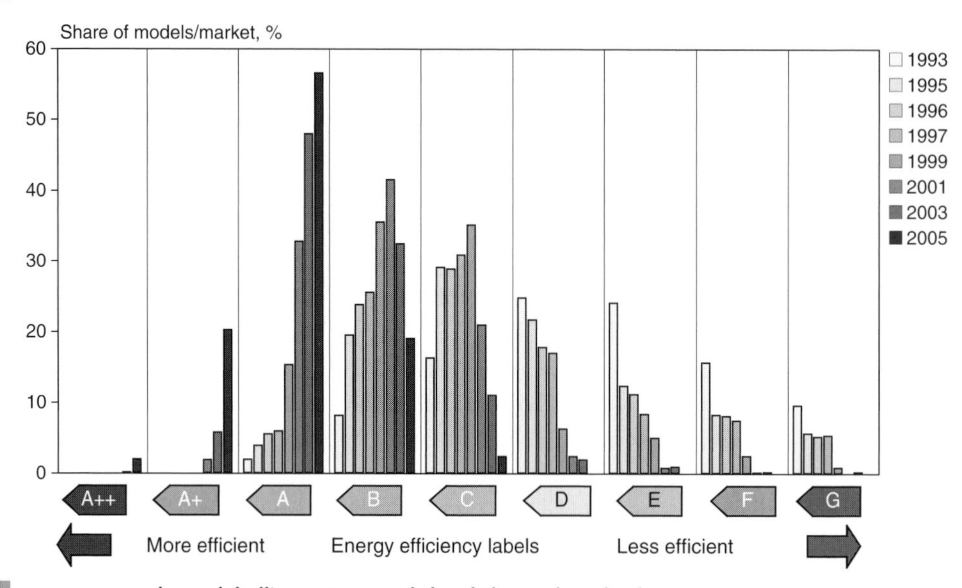

Figure 7.20 **The EU labelling system and the shift in sales of refrigerators over time.**
Source: IPCC, Fourth Assessment Report, Working Group III, figure 6.5.

the more important factors that make people resist strengthening building codes and refraining from cost-effective energy saving measures in existing building. This can effectively be addressed through financial incentives. They can take the form of upfront subsidies (often called rebates), taxes on energy based on the carbon content to make energy and CO_2 savings financially more attractive, or tax deductions (tax credits). All of these are widely used.

Surprisingly, many countries still subsidize fossil fuel based energy. So the first priority should be to remove these subsidies. Given the political sensitivity of removing subsidies, alternative forms of support for poor households, such as installing energy saving features free of charge to lower electricity bills, would be needed.

Feed-in tariffs for solar PV panels, making it attractive to deliver electricity back to the grid, are used in many countries. So-called 'net metering' is becoming popular. If you deliver electricity from a solar PV equipped building back to the grid, the meter turns backwards; meaning you receive as much for a kWh delivered back as for a kWh consumed from the grid (see also Chapter 11).

In several countries low interest mortgages are available for energy saving investments in buildings. Effectiveness of these financial incentive policies varies.

The specific design and the presence of other policies have a large impact on effectiveness. In terms of cost effectiveness caution is warranted. Government expenditures can be in the region of US$30–100 per tonne of CO_2 avoided[36], although the savings by owners and tenants could still make these policies cost effective for the national economy as a whole. Avoiding complex and overlapping incentives helps to make these policies more effective.

Energy Service Companies

A somewhat different approach that is showing good results in the commercial buildings sector is the promotion of so-called Energy Service Companies (ESCOs). These companies contract with businesses to reduce energy consumption and get paid on the basis of achieved results. This is an ideal way to take the burden of energy conservation out of the hands of busy managers of small and medium sized companies and institutions. In the USA the turnover of ESCOs in 2006 was of the order of US$2 billion[37].

The building sector challenge

With an abundance of technical options to reduce greenhouse gas emissions from buildings at low to negative costs, the real challenge is to find effective ways to realize this potential. A tailored approach with a mixture of instruments is needed. But above all the focus of policy should be shifted towards regulatory instruments. These also hold the best opportunities to induce behavioural change, if 'soft' information instruments are closely aligned with and supporting the introduction of 'hard' instruments.

Notes

1. UN Habitat, State of the World's Cities 2008–2009, 2008.
2. http://www.nationmaster.com/graph/peo_per_per_roo-people-persons-per-room.
3. Price et al. Sectoral trends in global energy use and greenhouse gas emissions, LBL, report LBNL-56144, July 2006.
4. See UNEP, Buildings and Climate Change: status, challenges and opportunities, Nairobi, 2007, fig 2.10.
5. IEA, WEO 2006, ch 9.
6. See Earth Trends, Residential energy data 2003; http://earthtrends.wri.org.
7. Tonooka Y et al. Journal of Asian Architecture and Building Engineering, vol 1(1), February 2002, 1–8.
8. See note 4.
9. Here only the emissions of HFCs are included, because other fluorinated gas emissions of CFCs and HCFCs do not fall under the Kyoto Protocol. CFCs and HCFCs are good for 1.3GtCO_2eq/year. They are being phased out under the Montreal Protocol however.
10. IPCC Fourth Assessment Report, Working Group III, ch 1 and IEA WEO, 2007.
11. This is different from 'zero net energy' houses, that generate some of their own energy; see section on 'Changing the energy source'.
12. For mid-latitude (40–60 degrees North) regions the 'Passive House standard' is 15kWh/m^2/year maximum energy use for space heating and cooling and 120kWh/m^2/year for all appliances, water heating, and space heating/cooling together, see http://www.passivhaus.org.uk/index.jsp?id=668.

13. see Smith PF. Architecture in a climate of change: a guide to sustainable design, Elsevier, 2nd edition, 2005, page 65.
14. IPCC Fourth Assessment Report, Working Group III, ch 6.4.2.2.
15. IPCC Fourth Assessment Report, Working Group III, ch 6.4.2.3 and 6.4.4.
16. IPCC Fourth Assessment Report, Working Group III, ch 6.4.10.
17. http://www.habitatforhumanity.org.uk/lea_need.htm.
18. See note 4, page 8.
19. See for a detailed description of climate friendly insulation the IPCC Special Report on Safeguarding the Ozone Layer and the Global Climate System, 2005, chapter 7.
20. IPCC Fourth Assessment Report, Working Group III, ch 6.4.3.
21. In estimates of global CO_2 emissions it is assumed that 90% of traditional biomass is from sustainable sources; this means that 10% of it is not and should be counted when emissions are calculated.
22. IPCC Fourth Assessment Report, Working Group III, ch 6.4.4.3.
23. HCFC-22 is also a greenhouse gas. It has a Global Warming Potential that is slightly higher than that of HFC-134a (the most used replacement). A shift to HFCs therefore does not improve the warming effect of emissions much.
24. See IPCC Special Report on Safeguarding the Ozone Layer and the Global Climate System, 2005, chapter 5.
25. IEA, Light's Labour's Lost, 2006 and IPCC Fourth Assessment Report, Working Group III, ch 6.4.9.
26. See UK National Energy Foundation, http://www.nef.org.uk/energysaving/labels.htm.
27. see Renewable Energy Network 21, Renewables 2007 Global Status Report, 2008, www.ren21.net.
28. IEA Solar Heating and Cooling programme, Solar Heat Worldwide, 2008.
29. See http://zeb.buildinggreen.com/.
30. See DEFRA, A framework for pro-environmental behavior, 2008; CE, Energy conservation behavior, 2006; Policy Studies Institute, A green living initiative, http://www.green-alliance.org.uk.
31. IPCC Fourth Assessment Report, Working Group III, ch 6.5.
32. IPCC Fourth Assessment Report, Working Group III, ch 6.7.
33. IPCC Fourth Assessment Report, Working Group III, ch 6.8.
34. IPCC Fourth Assessment Report, Working Group III, ch 6.8.3.1 and 6.8.3.6.
35. Joakim Nordqvist, Evaluation of Japan's Top Runner programme, AID-EE, 2006.
36. IPCC Fourth Assessment Report, Working Group III, ch 6.8.3.3.
37. IPCC Fourth Assessment Report, Working Group III, ch 6.8.3.5.

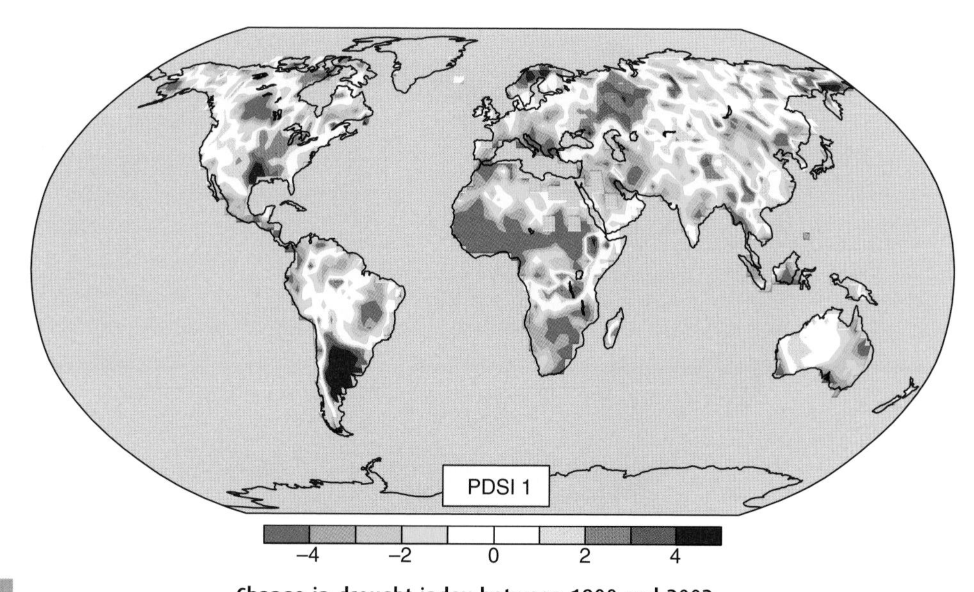

Plate 1

Change in drought index between 1900 and 2002.
Source: IPCC Fourth Assessment report, Working Group I, figure1 from box FAQ3.2.

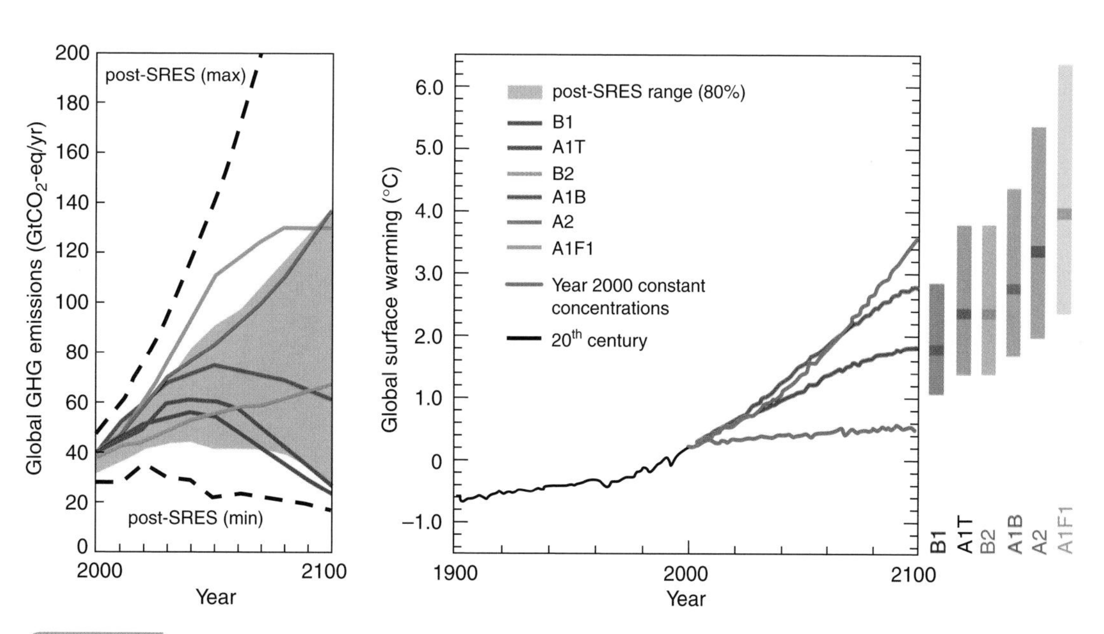

Plate 2

(Left panel) Scenarios for global greenhouse gas emissions, according to IPCC; (right panel) projected global mean temperatures belonging to the scenarios in the left panel.
Source: IPCC Fourth Assessment report, Synthesis Report, figure SPM.5.

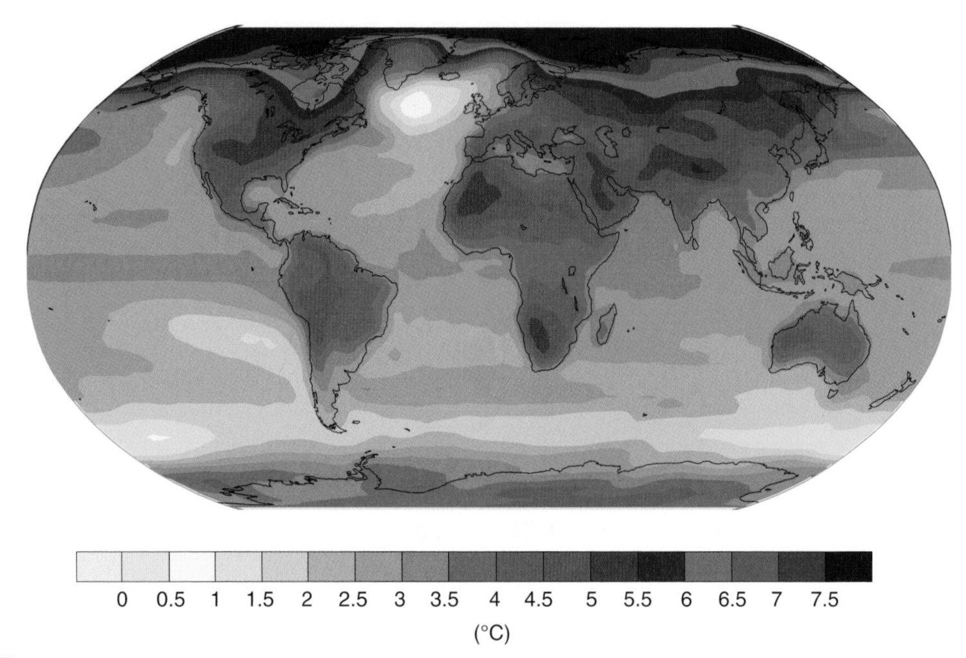

Plate 3 Projected surface temperature changes for the period 2090–2099, compared to 1980–1999. The average of different models is shown for the IPCC SRES A1B scenario (a middle of the range one).
Source: IPCC Fourth Assessment report, Synthesis Report, figure SPM.6.

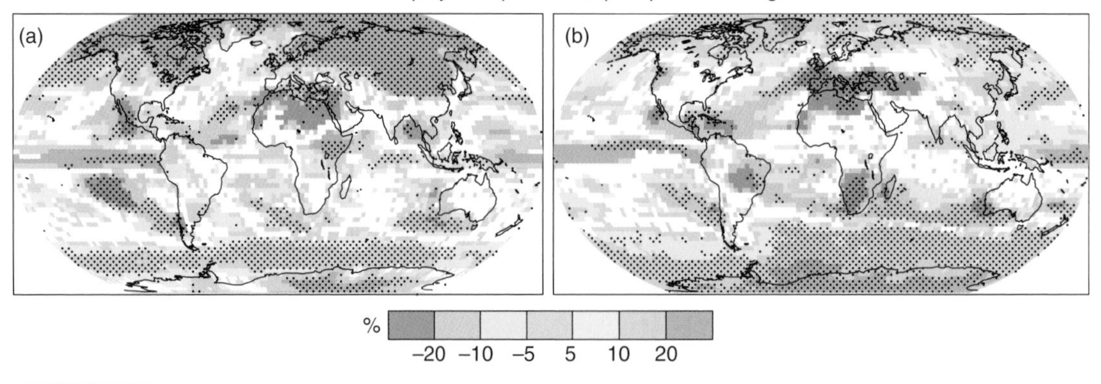

Plate 4 Relative change in precipitation for the period 2090–2099, compared to 1990–1999. (a) December to February. (b) June to August. Model averages are shown for the IPCC SRES A1B scenario (a middle of the road one). White areas are where less than 66% of the models agree about increase or decrease. In stippled areas more than 90% of the models agree.
Source: IPCC Fourth Assessment report, Synthesis Report, figure 3.3.

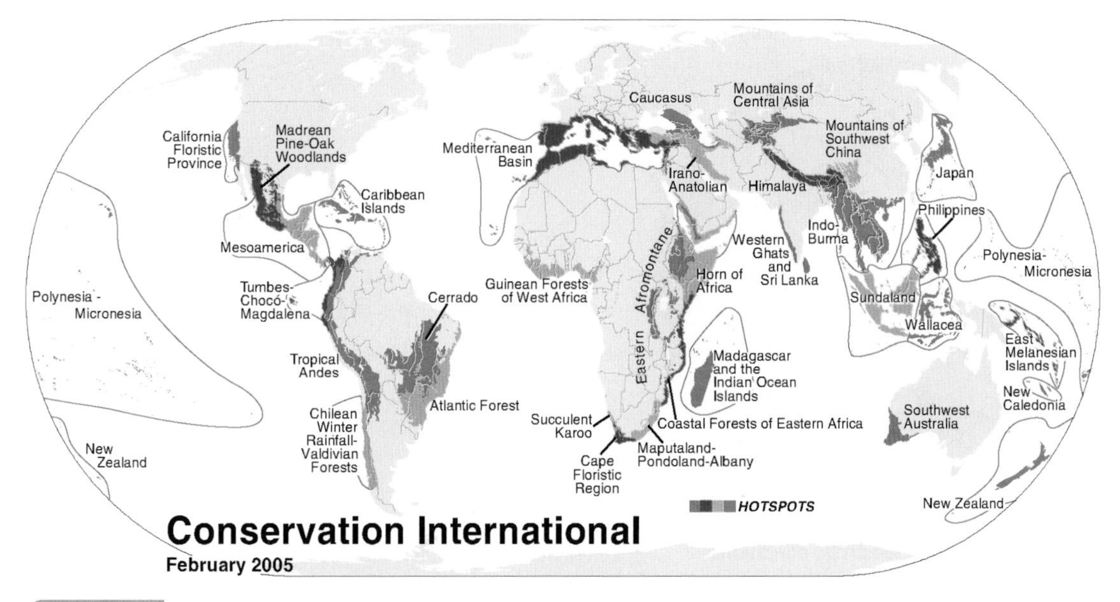

Plate 5 **Biodiversity hotspots.**

Source: Conservation International, www.biodiversityhotspots.org.

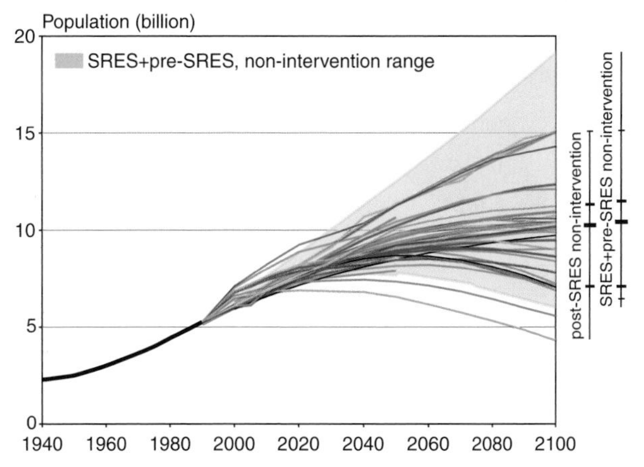

Plate 6 **Global population projections as reported in the IPCC Special Report on Emission Scenarios (SRES+ pre-SRES; light shaded area) and in the IPCC Fourth Assessment Report, Working Group III, chapter 3 (post-SRES non-intervention, dark shaded area).**

Source: IPCC Fourth Assessment Report, Working Group III, figure 3.1.

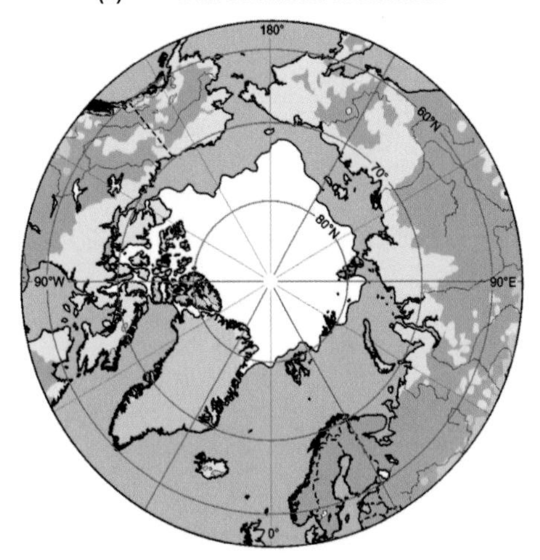

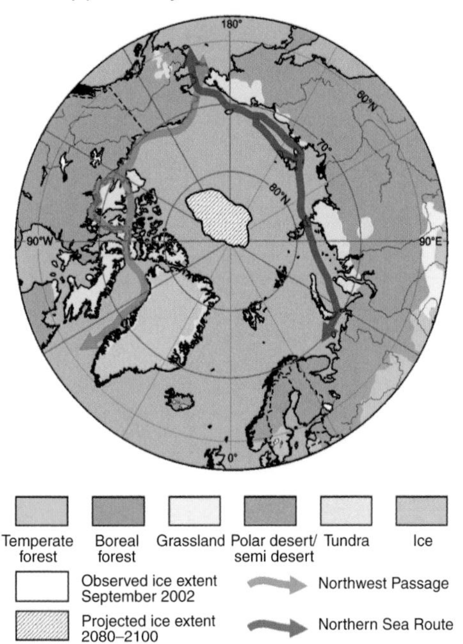

Plate 7 Arctic sea ice and vegetation of Arctic and neighbouring regions. (a) 2005 conditions. (b) Projected for 2090–2100 under an IPCC IS92a scenario. Note the sharp decline in sea ice and tundra area.

Source: IPCC Fourth Assessment Report, Working group II, figure TS.16.

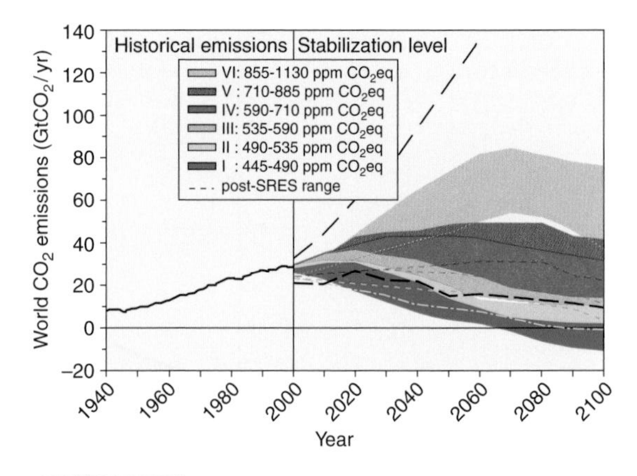

Plate 8 CO_2 emission reductions required to achieve stabilization of greenhouse gas concentrations in the atmosphere at different levels, compared to 2000 emission levels. The wide bands are caused by different assumption about the emission trends without action (so-called 'baselines') and different assumptions about timing of reductions. Emission trajectories are calculated with various models (see also Box 3.2).

Source: IPCC Fourth Assessment Report, Synthesis Report, figure SPM 11.

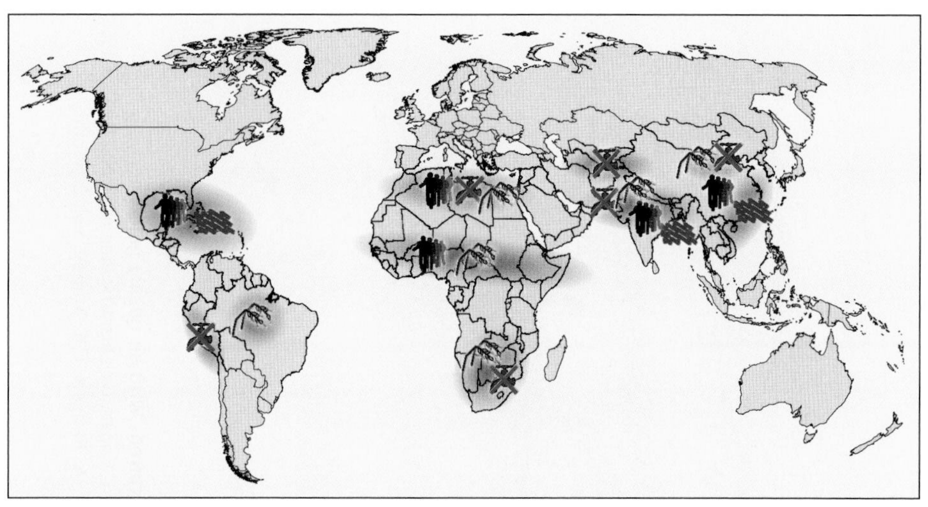

Conflict constellations in selected hotspots

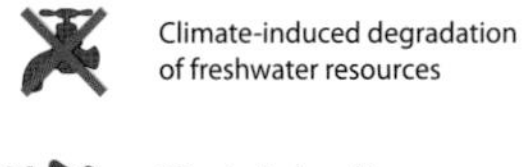
Climate-induced degradation of freshwater resources

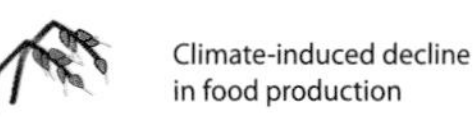
Climate-induced decline in food production

Hotspot

Climate-induced increase in storm and flood disasters

Environmentally-induced migration

Plate 9 **Potential areas where violent conflicts could emerge as a result of climate change.**
Source: German Advisory Council on Global Change, World in Transition: Climate Change as a Security Risk. Summary for Policy Makers. Berlin, 2007.

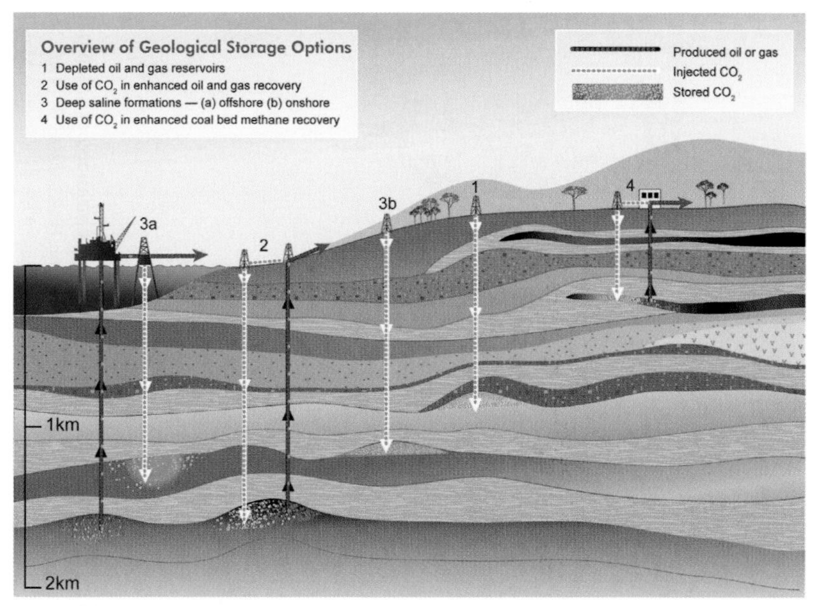

Plate 10 **Methods for storing CO_2 in deep underground geological formations. Two methods may be combined with the recovery of hydrocarbons: EOR (2) and ECBM (4).**
Source: IPCC Special Report on CO_2 Capture and Storage, figure TS.7.

Plate 11 **Cities with annual mean concentrations of small particles (PM10) at least 50% above the current WHO air quality guideline.**
Source: Cohen, A. J., Anderson H. R., Ostro B. *et al*., Mortality impacts of urban air pollution. In: Comparative Quantification of Health Risks: Global and Regional Burden of Disease Due to Selected Major Risk Factors, eds. M. Ezzati, AD Lopez, A. Rodgers, CJL Murray, vol. 2. World Health Organization, Geneva, 2004. p. 1374.

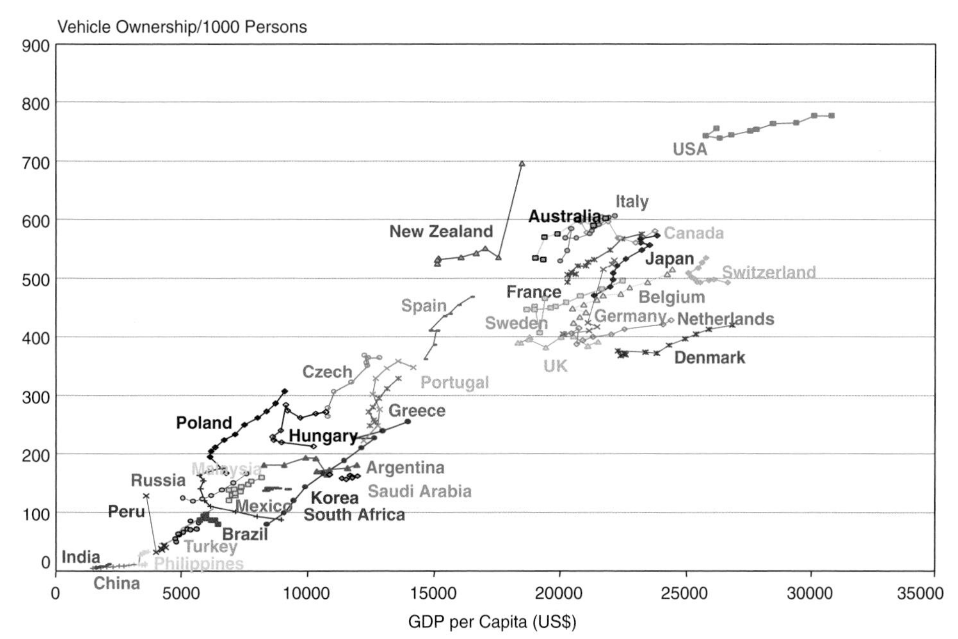

Plate 12 **Comparison of vehicle ownership between countries as a function of per capita income. Data for the period 1990–2000 (with some differences for specific countries).**

Source: IPCC Fourth Assessment Report, Working Group III, figure 5.2.

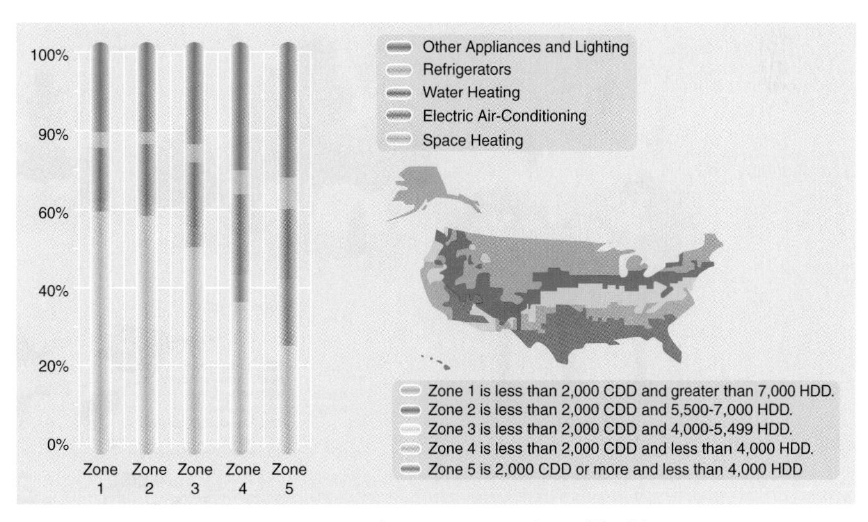

Plate 13 **Energy consumption shares in US residential buildings.**

Source: UNEP, Buildings and climate, 2007, fig 2.15.

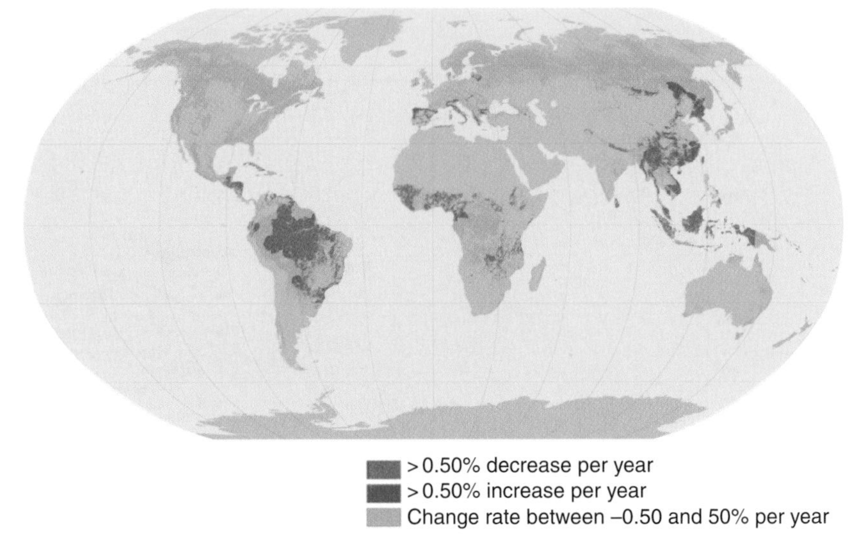

Plate 14 **Net change in forest area between 2000 and 2005.**
Source: FAO, Global Forest Resource Assessment 2005.

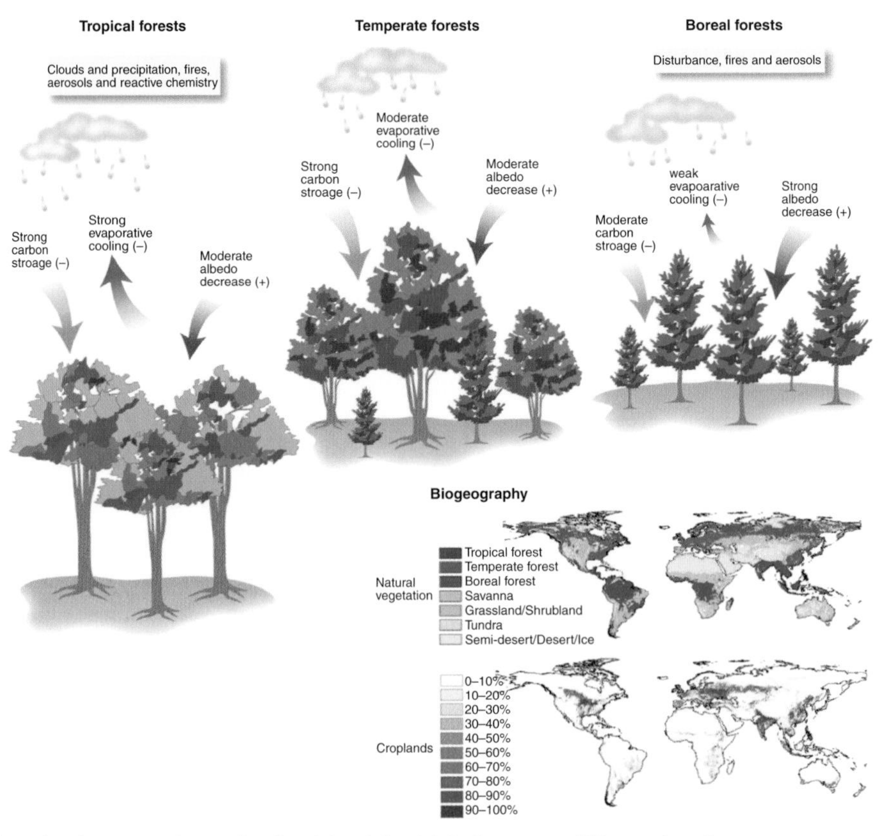

Plate 15 **The role of evaporation and reflectivity (albedo) in forests at different locations.**
Source: Bonan, *et al*. Forests and climate change: Forcings, feedbacks and the climate benefits of forests, Science, vol. 320, June 2008, p.1444.

8 Industry and waste management

What is covered in this chapter?

Products manufactured by industry form an essential part of modern economies. Industrialization is a step in the development of countries that brings jobs and better living standards. Industrial production will therefore keep rising and will increasingly be located in developing countries. Interestingly, the most modern installations are often found in developing countries. Since industry contributes about 20% to global greenhouse gas emissions, any serious attempt to reduce global GHG emissions will have to involve industry. Using less energy intensive industrial goods like steel by making lighter cars for instance and using wood for constructing buildings instead of steel, concrete, and bricks is one way to go. Most of the emissions reduction will have to come from more efficient production (less greenhouse gas emissions per unit of product), shifting to low carbon energy sources, and using CO_2 capture and storage to remove CO_2 at the smokestack. For the most important processes the reduction opportunities are discussed. Government policies are needed to make these reduction opportunities a reality. Experience with various policy instruments shows that for big reductions in emissions more stringent instruments, such as cap and trade and regulations, will be needed. Voluntary agreements do initially help to raise awareness amongst participants and to encourage corporate responsibility, but delivering major emission reductions through voluntary agreements is not possible.

Waste is an important emission source in industry and for household and commercial waste. There are strong interactions via recycling of paper, glass, and metals. That is why greenhouse gas emissions and waste are discussed in this chapter together. Waste contributes a few per cent to global emissions. Greenhouse gas emission reduction often goes hand-in-hand with proper waste management for sanitary reasons.

Trends in industrial production

Industry covers a large number of products that are essential for modern economies: food products, building materials like cement, concrete and construction wood, iron and steel,

Table 8.1. **Production of steel (2006) and cement (2005)**

Country	Steel production (Mt/year)	Share of global (%)	Cement production (Mt/year)	Share of global (%)
China	419	34	1064	47
EU	210	17	230	10
Japan	116	9	74	3
USA	98	8	99	4
Russia	71	6	45	2
South Korea	48	4	50	2
India	44	4	130	6
Ukraine	41	3	n/a	n/a
Brazil	31	2	39	2
Turkey	23	2	38	2
World	1242		2284	

Source: IEA Sectoral approaches to greenhouse gas mitigation: exploring issues for heavy industry, 2007.

aluminium and other metals, glass, ceramics, fertilizers, chemicals, paper and cardboard, oil products, cars and other transport means, computers and computer chips, electrical equipment, machinery, and many others. As a result of increased population and economic growth the production capacities of these various industries have increased tremendously. Since 1970, global production of cement increased about threefold, while aluminium, paper, ammonia (for fertilizers), and steel production approximately doubled These are the most energy intensive industries, contributing most to global greenhouse gas emissions.

Much of the production of these energy intensive goods is now located in developing countries. China is the world's largest producer of steel, cement, and aluminium. Developing countries together produced 42% of steel, 57% of nitrogen fertilizer, 78% of cement, and 50% of aluminium in 2003[1]. Production is concentrated in a limited number of countries. China, the EU, Japan, USA, Russia, South Korea, and India account for 82% of the steel and 74% of the cement production in the world (Table 8.1). Many industrial goods are traded globally. Of all aluminium produced, about 75% is traded. For steel it is about 30% (not counting products made with steel); for paper products about the same. For many other industrial products like metals, chemicals, or paper, plants are located where raw materials are readily available, leading to large trade volumes of the manufactured products. For heavy and bulky materials like cement where raw materials are readily available in many places, trading is limited (about 5%). Many manufactured products with limited energy contents (and relatively small emissions) are produced in places with low labour costs. International competition therefore plays a role for a limited set of energy intensive products and that has implications for emission reduction policies.

Since many of the plants in developing countries are relatively new, they are often the most efficient. The reason is that cost minimization is a dominant issue in these internationally competing industries and efficiency (of energy or raw material use) is directly affecting costs. For steel, cement, aluminium, and fertilizer, energy costs are typically 10–20% or more of total costs. For chemicals, paper, ceramics, and glass it is in the order of 5%, still a significant amount and worth reducing. For products like transportation equipment, textiles, food, electrical equipment, and machinery it is less than 2%[2], and incentives for efficiency improvement are less. Energy use per tonne of product has therefore gone down substantially over time in those industries where energy costs are high (see Figure in Box 4.1).

Globally, large companies dominate the energy intensive industry sector. Cement production in China is an exception: there are more than 5000 plants with an average production of not more than 200 000 tonnes/year. In developing countries small and medium sized companies (SMEs) can have a significant share in production, such as in the metals, chemical, food, and paper industries. These SMEs often use older, less efficient technologies and do not have the capacity to invest in modern equipment and emission controls.

Demand for industrial products is expected to increase strongly: for cement a doubling by 2020 and a fourfold increase by 2050.

Trends in waste management

Waste can be separated into industrial waste, which is a by-product of manufacturing, and household/commercial waste, which is the remains of consumption (often called post consumer waste). They are very different in nature: industrial waste is very process specific and can consist of hazardous materials, while post-consumer waste is mostly organic material, wastewater, paper, plastic, metals, and textiles. Construction waste is usually counted under industrial waste. Treatment of waste is also different. In industry recycling of waste streams is an economic necessity. Sending waste off-site for treatment can cost a lot of money. For post-consumer waste collective treatment of waste water and solid waste is a matter of improving health conditions. Keeping as much valuable material out of the solid waste stream as possible is attractive for use as raw materials in industrial glass, paper, and steel production and is widely practised. The small quantities of hazardous waste from households and offices are kept separate as much as possible to avoid spreading these substances in the environment.

Post-consumer waste is increasing with increasing income. In low income countries it is less than 100kg per person per year. In high income countries it is more than 800kg. Total solid waste volumes have therefore increased significantly. Currently they are about 900–1300 million tonnes per year globally[3]. The way solid waste is treated varies enormously across countries. In total more than 130 million tonnes (10–15%) is incinerated, often with energy recovery[4]. Roughly 50% is put in landfills (controlled or uncontrolled) and the rest is recycled. Waste water is increasing with income as well, not least because 40% of the world population still has no sewerage connection, septic tank, or latrine in their homes.

To improve health conditions, this situation needs to be addressed urgently.

Greenhouse gas emissions

The industry sector accounts for emissions of about 9.5 $GtCO_2$-eq per year (about 20% of the total). Waste management adds another 1.3 $GtCO_2$-eq or 3% of the total (see Figure 8.1). This excludes the emissions of the electricity used inside industry plants but generated outside (called indirect emissions). These emissions are counted towards the energy supply sector and are about 2.5 $GtCO_2$-eq/year. The share of industry in a country's total emissions varies considerably, even amongst industrialized nations (see Figure 8.2). If indirect emissions are included industry is responsible for about two-thirds of China's total CO_2 emissions.

About 5% of industry and 95% of waste management emissions are from non-CO_2 greenhouse gases: in industry mostly fluorinated gases and some N_2O; in waste management largely CH_4 and a little N_2O. Solid waste landfills generate most of the CH_4 from waste. Waste water treatment generates N_2O and CH_4.

The contribution of specific industry sub-sectors is shown in Figure 8.3.

Future emissions

Greenhouse gas emissions from industry are projected to increase by 20–65% until 2030. For waste management the increase is about 30%, ranging from about zero for N_2O from waste water treatment to about 50% for CH_4 emissions from landfills[5].

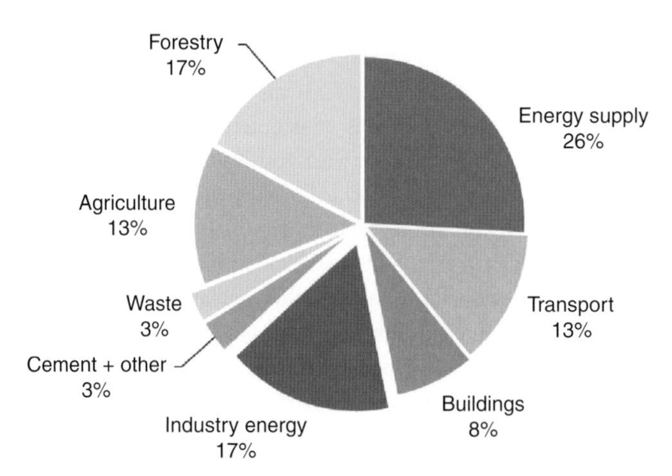

Figure 8.1 Industry and household/office waste management sector emissions. These are direct emissions only (i.e. excluding the emissions from electricity used in the plant but generated outside). Emissions are separated into energy related emissions from industry, cement and other non-energy related emissions, and waste management emissions.

Source: IPCC Fourth Assessment Report, Working Group III, ch 1.

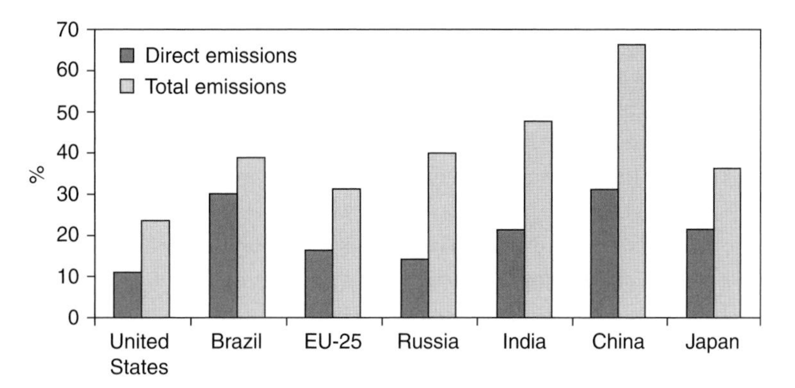

Figure 8.2 **Share of industry in total CO_2 emissions. Both direct and indirect emissions are shown.**
Source: Houser et al. Levelling the carbon playing field, Peterson Institute for International Economics and WRI, 2008.

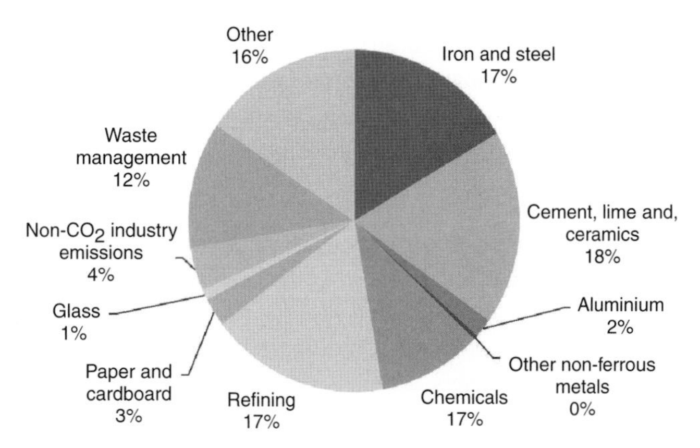

Figure 8.3 **Contributions of subsectors to industry and waste management greenhouse gas emissions in 2004/2005. Includes direct CO_2 and non-CO_2 emissions only.**
Source: IPCC Fourth Assessment Report, Working Group III, ch 7 and IEA Energy Technology Perspectives, 2008.

Opportunities to reduce emissions

Emissions reduction in the industry sector can in principle be achieved in three ways:

1. Replacing energy intensive products with low emission alternatives (e.g. replacing steel and concrete for buildings with wood)
2. Reducing the amount of industrial products consumed (e.g. by producing lighter cars requiring less steel)
3. Reducing the emissions per unit of product by modifying the production process

For waste management this 'hierarchy' of options is slightly different: (1) reducing waste volumes; (2) recycling waste; or (3) managing it with lower emissions of greenhouse gases per unit of waste.

Quantitative data on the first two industry options are scarce. However, there is a clear trend towards reducing weight per unit of product in automobile manufacturing, computers, TV sets, packaging, and many other products. It is a matter of becoming more efficient with raw materials (and saving costs) as well as shifting to lighter and cheaper materials. Examples include using thinner material in aluminium cans and steel tins and replacing steel in automobiles with lighter metals and plastics. Due to the strong increase in demand the total amount of material used (and therefore the emissions from production) keeps going up.

Data on the emissions per unit of product are available for many countries and production processes, allowing international comparisons. Very often these comparisons are made in energy use per unit, which can give a very different picture from emissions per unit (see Box 8.1). For the industrial processes that produce most greenhouse gas emissions (iron and steel, cement, and chemicals, together good for about three-quarters of the total emissions from the industry sector) the opportunities for emission reduction through process modifications will be discussed in detail. For some other processes the options will be summarized. In addition, there are many reduction options that apply across the whole sector. These will be discussed separately.

Box 8.1 Energy efficiency and carbon efficiency

Efficiency of industrial installations is often evaluated in terms of energy use per unit of product (energy efficiency). This is because energy costs are an important factor in operating these processes. When comparing installations from a climate change point of view the CO_2 emission per unit of product (or carbon efficiency) is more relevant. This requires one look at the carbon content of the sources of energy used, including the way the electricity is generated that comes from outside the plant.

Iron and steel

There are three different steel making processes (see schematic diagram in Figure 8.4):

1. Reduction of iron ore in blast furnaces, usually with coal (in the form of coke[6]) and conversion of the so-called 'pig iron' into steel in a Basic Oxygen Furnace. About 60% of the steel in the world is produced this way
2. Melting of recycled iron (so-called 'scrap') in Electric Arc Furnaces (35%)
3. Direct reduction of iron ore with natural gas and further processing it in an electric furnace (5%)

In terms of energy use and CO_2 emissions the traditional blast furnace/basic oxygen furnace process is the worst. Scrap melting (Electric Arc Furnace) only uses about 30–40% of the energy of the traditional process, with CO_2 emissions depending on the source of the electricity. The Direct Reduction/Electric Arc Furnace process (using natural gas) only produces 50% of the CO_2 emissions per tonne of steel compared to the traditional process.

Emissions per tonne of steel in different countries vary considerably, from about 1 to 3.5 tCO_2 per tonne of steel. This is caused by different production processes, sources of electricity, efficiency of equipment, and types of products. Figure 8.5 shows the average CO_2 emissions per tonne of steel for various countries. Both the direct emissions

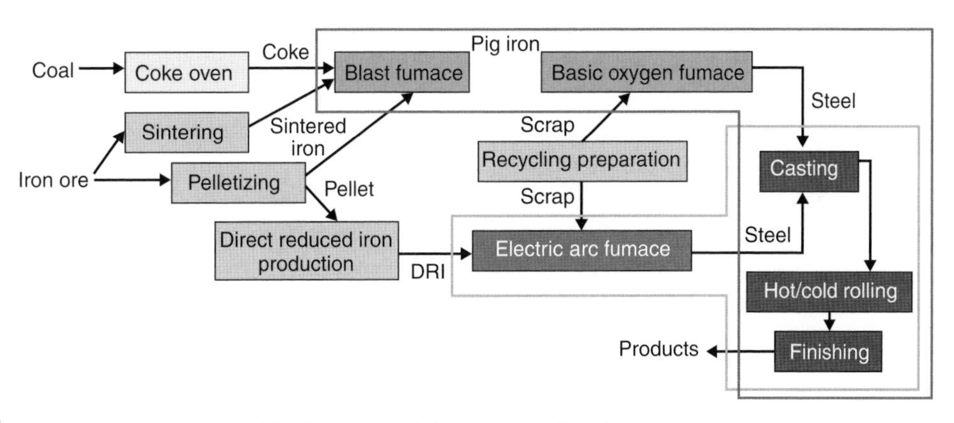

Figure 8.4 **Simplified diagram of the main steel making processes.**
Source: IEA, Assessing measures of energy efficiency performance and their application in industry, 2008.

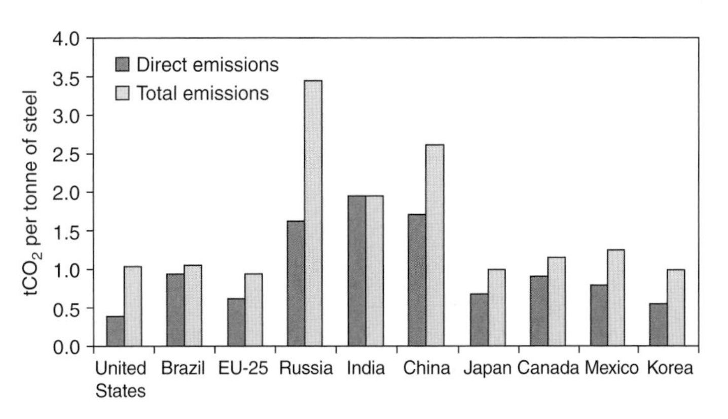

Figure 8.5 **Carbon intensity of steel production, expressed as tCO_2 per tonne of steel. Direct emissions only cover emissions produced at the steel plant. Total emissions also include emissions from coke production and electricity generation off-site.**
Source: Houser et al. Levelling the carbon playing field, Peterson Institute for International Economics and WRI, 2008.

(from the process itself), as well as the indirect emissions (from the production of coke and electricity generated off-site) are given here. The high emissions in India and China are caused by the fact that steel production is overwhelmingly of the Blast Furnace/Basic Oxygen Furnace type, because of insufficient recycled (scrap) iron.

Apart from shifting to production processes with lower emissions (i.e. those that use more scrap iron) there are many opportunities for improving the energy efficiency of the blast furnace and process steps. Adding these up gives efficiency improvement potentials like 15% for Japan and 40% for China when compared with best practices currently found in major steel producing countries[7]. More advanced energy efficiency options are being studied.

Another important way to reduce emissions is to shift from coal (in the form of coke) to a lower carbon reducing agent. Oil, natural gas, waste plastics, and biomass are being used. In Brazil charcoal is used in blast furnaces, but this is unlikely to be from a sustainable source, so net CO_2 emissions are in fact much higher. The use of hydrogen is being investigated for future use, which could bring down emissions considerably. For Electric Arc Furnace processes CO_2 can be reduced by moving to a low carbon electricity source. Recovery of combustible gas that is produced during coke and steel manufacturing can also contribute to emission reductions in places where that is not yet done.

Finally it is technically feasible in principle to apply CO_2 Capture and Storage (CCS, see also Chapter 5). Where applied it could reduce something like 85–90% of the CO_2 emissions. Costs of this reduction option in blast furnaces are relatively high (US$40–50/t$CO_2$ avoided) and CCS has therefore not yet been applied commercially in steel making. Small scale demonstrations are being done and plans exist for large scale demonstrations by 2015. In direct reduction (DRI) plants costs would be lower (US$25/t$CO_2$ avoided), but DRI capacity is still relatively small. By 2030 the CCS reduction potential is estimated at 0.1–0.2 GtCO_2/year, but this could grow to 0.5–1.5 GtCO_2/year in 2050[8].

The worldwide mitigation potential of all options by 2030 at costs of US$20–50/t$CO_2$ avoided is estimated at 15–40%, or 0.4–1.5 GtCO_2/year.

In the longer term, new, so-called 'melt reduction' processes are expected to deliver further reductions. These processes integrate the iron ore preparation, coke making, and blast furnace iron making steps. That increases the energy efficiency and also produces gases with higher CO_2 concentrations that make CCS more attractive. By replacing air with pure oxygen the CO_2 content of the gases can be further increased to make CCS even more attractive. By 2050 these new processes in combination with CCS could deliver an additional 0.2–0.5 GtCO_2/year reduction. The other long term option is to move to different methods of steel processing. Currently steel is first cast into slabs, which are later reheated to be rolled into steel plates and other steel products. By integrating these steps (so-called 'direct casting') significant energy savings can be made.

Cement

The principal component of cement, called clinker, is produced by heating limestone with some additives to high temperatures of about 1500°C. In the process CO_2 is released from the limestone, good for about half of the CO_2 emissions from cement manufacture. The energy used for heating the oven (called a kiln) is also a major source of CO_2 emissions. In the USA, China, and India the energy comes mostly from coal. In Canada, Brazil, and Europe large amounts of biomass are used. The type of kiln also has a big influence. So-called 'wet kilns', with a high moisture content, use 25–125% more energy than dry kilns. Wet kiln processes are predominantly found in Russia, India, China, and Canada. Europe, Japan, Thailand, and Korea mostly use dry kilns.

The additional process steps are also energy intensive. The limestone has to be dug out of a quarry and ground. After that it is pretreated (dried and ground). At the end of the process the clinker is cooled, ground, and other materials are added to get the final cement product (see diagram Figure 8.6).

Emissions per tonne of cement vary from country to country (see Table 8.2).

Since clinker production is the major source of emissions, the clinker content of cement to a large extent determines the emissions per tonne. Standard, so-called Portland, cement contains 95% clinker. In blended cement some of the clinker is replaced by alternative materials, such as fly ash from coal fired power plants, waste material (slag) from blast furnaces, and natural volcanic minerals (pozzolanes). This results in lower CO_2 emissions per tonne of cement. Blended cements are used widely in Europe, but hardly at all in the USA. Replacement of clinker contributes about 30% to current reduction potential, based on best available technologies[9].

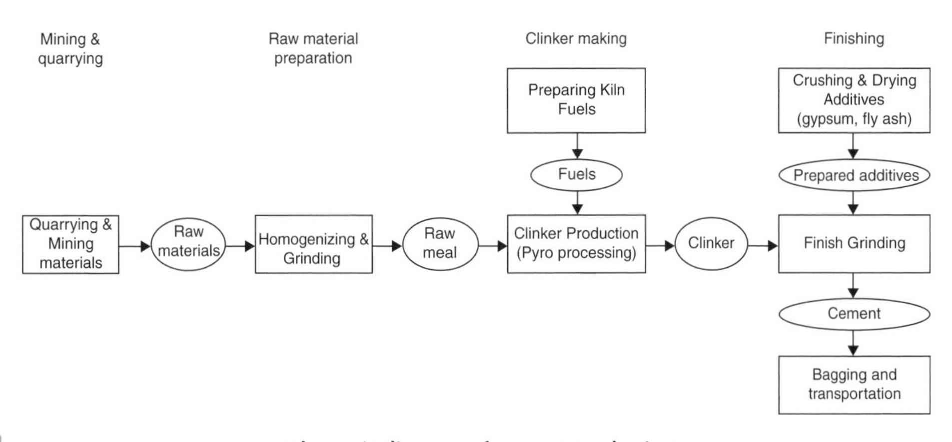

Figure 8.6 **Schematic diagram of cement production.**

Source: Ecofys, Sectoral Approach and Development, Input paper for the workshop 'Where development meets climate', 2008, http://www.pbl.nl/en/dossiers/Climatechange/Publications/International-Workshop-Where-development-meets-climate.html.

Table 8.2. Selected emission intensities of cement (2000 data)

Country	Average emissions (tCO_2/t cement)
Europe	0.70
Japan	0.73
South Korea	0.73
China	0.90
India	0.93
USA	0.93

Source: IPCC Fourth Assessment Report, Working Group III, ch 7.4.5.1.

The efficiency of the kiln and the fuel used to heat the kiln are also important in reducing emissions. As indicated above, the energy use per tonne of cement in dry kilns is lower than in wet kilns. Compared to best available technologies, emissions per tonne of cement can be reduced by about 40%. Shifting from coal to waste materials, including tyres, plastics and biomass, can contribute up to 20% to emission reduction. More efficient use of electricity and lowering the carbon content of electricity used in the process (often generated off-site) can contribute the rest (in the order of 10%).

Cement kilns produce gas streams with high CO_2 concentrations, originating from fuel and limestone. This makes cement plants a good candidate for CO_2 capture and storage (CCS, see also Chapter 5). Since costs would be high (initially more than US\$100/tCO_2 avoided, over time to be reduced to US\$50–75), CCS in cement plants has not yet been applied, nor are there any large demonstration units. With increasing CO_2 prices it is estimated that about 0.25 $GtCO_2$ could be reduced economically by 2030 with CCS in the cement industry at costs of US\$50–100/tCO_2 avoided[10].

For the whole cement sector the estimate of the worldwide mitigation potential at costs up to US\$50/tCO_2 avoided is about 10–40% of the emissions in 2030, or 0.5–2.1 $GtCO_2$/year.

Chemicals and petroleum refining

The chemical industry is very diverse. It covers tens of thousands of products, with annual production varying from a few tonnes to more than 100 million tonnes. The industry covers thousands of companies. Plants are often integrated with petroleum refineries, because oil products are an important raw material. There are more than 700 refineries in 128 countries.

A small number of processes are responsible for about 70% of the energy use in the chemical industry:

1. Ethylene (used mainly for producing plastics), produced by so-called steam cracking (high temperature heating with steam) of oil or gas. Important by-products are propylene (also used for plastics), and aromatic hydrocarbons like benzene. Emissions are about 0.2 $GtCO_2$/year.
2. Methanol, used as an industrial solvent, antifreeze and basis for gasoline additives, produced mainly from natural gas

3. Ammonia, used mainly as a raw material for nitrogen fertilizers. It is produced by reacting nitrogen with hydrogen (produced from gas or coal)

Most of the emissions are in the form of CO_2. Of all the fossil fuel used about half is burned for heating purposes. The other half is used as so-called feedstock for the processes and converted into products. Since many of these products are burned or decomposed with a certain delay, eventually all of the feedstock ends up as emissions of CO_2.

There are some chemical processes that produce significant quantities of non-CO_2 GHGs as by-products from the production process. An important one is N_2O emissions from plants that produce raw materials for the manufacture of nylon and nitric acid. Another major contribution is fluorinated gas (HFC-23) as a by-product of the manufacture of a liquid used in air conditioners (HCFC-22).

Emission reduction opportunities in ethylene manufacture are twofold: (1) energy efficiency improvements in the various stages of the process (cracking, separation); and (2) feedstock choice, affecting the energy required for the cracking process. Energy use per tonne of ethylene has been reduced by about 50% since 1970. This can be further improved by at least 20% for cracking and 15% for the separation processes by applying higher temperature furnaces, combined heat and power gas turbines, and advanced refrigeration systems.

In ammonia production, reduction opportunities are found in efficiency of energy use, the choice of feedstock for making hydrogen, and the application of CCS. Energy efficiency of ammonia plants has already been improved so much that the most recent plants are performing close to the theoretical minimum energy consumption levels (see Figure 8.1 above). Replacing and upgrading existing plants remains to be done.

Hydrogen, one of the main inputs for ammonia manufacture, is produced from natural gas (77% of ammonia production), gasified coal (14%, mainly in China), or oil products (9%). The amount of CO_2 produced by the hydrogen manufacture process makes a big difference in the total CO_2 emission per tonne of ammonia (which varies from 1.5 to 3.1 tCO_2/t ammonia). Moving to a low carbon hydrogen source is therefore an important reduction measure. Adding CO_2 capture and storage (CCS) is the cheapest way to do that, because in the hydrogen plant the CO_2 has already been separated from hydrogen and the expensive capture step can thus be skipped (see Chapter 5). Costs are estimated to be about US$25/ tCO_2 avoided, which is much lower than producing low-carbon hydrogen from biomass or from electrical decomposition of water. The effect of this reduction option is somewhat limited by the partial use in many ammonia plants of the CO_2 stream for producing urea, a popular fertilizer.

Refineries

Petroleum refineries cannot easily be compared across countries, because there are too many differences in crude oil type, set of products, and equipment. There are however significant opportunities for energy efficiency improvement. Refineries use 15–20% of the energy in crude oil for their operation, leading to current emissions of about

1.9 $GtCO_2$-eq/year. The reduction potential from improving energy efficiency is estimated at 10–20%, representing about 0.3 $GtCO_2$/year. CCS provides additional opportunities of around 0.1 $GtCO_2$/year by 2030.

Non-CO_2 greenhouse gases

Non-CO_2 emission reduction potential is considerable. For many sources in the chemical industry emission reductions of 50–90% are achievable by 2030 at costs lower than US\$20/tCO_2-eq avoided. N_2O from nitric and adipic acid and caprolactam manufacture for instance can be reduced by more than 80% at practically zero cost. More than 80% of HFC-23 emissions can be destroyed by incinerators at costs of less than US\$1/tCO_2-eq. Because of the high Global Warming Potential (GWP) of HFC-23 (see Chapter 2) the small amount destroyed represents a significant amount in terms of CO_2-eq (see Chapter 12 for a discussion of this very cheap option and the implications for the Kyoto Protocol implementation). In total about 0.2–0.3 $GtCO_2$-eq/year can be reduced at relatively low cost.

Altogether, the chemical and petroleum refinery industry can reduce at least 1 $GtCO_2$-eq/year, about 75% at costs below US\$20/tCO_2-eq avoided.

In the longer term significant CO_2 emission reductions can be expected in the chemical industry by shifting to biomass as feedstock, instead of petroleum products, and by using biological or enzymatic processes that can operate at lower temperatures. Reduction percentages of up to 60% would be possible by 2050.

Other industries

Manufacturing of aluminium, magnesium and other metals, paper and cardboard, glass, bricks and ceramics production, and food processing can contribute significantly to the industry reduction potential.

Aluminium

Aluminium production is a highly energy intensive process. Bauxite aluminium ore is refined to aluminium oxide in a high temperature oven. Then the aluminium oxide is reduced to aluminium metal with carbon electrodes in a hot 'reverse battery', filled with molten fluoride containing minerals. This process produces large amounts of CO_2, just as in iron ore reduction, but also perfluorinated carbon compounds (PFCs) with a very high GWP. Reduction opportunities lie in more efficient use of energy. The average amount of electricity consumed per unit of product has gone down about 10%

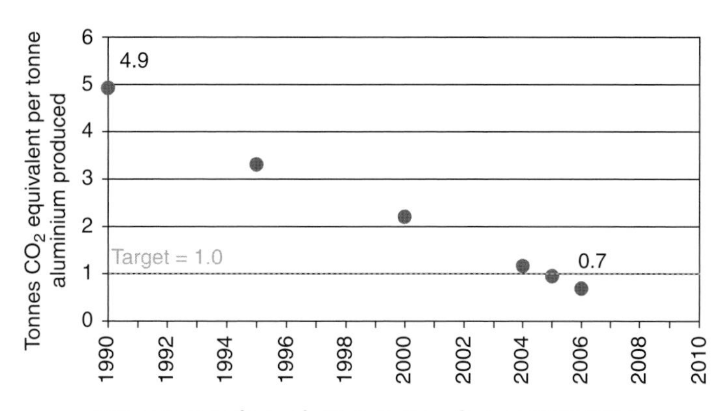

Figure 8.7 **PFC emissions from aluminium manufacture 1990–2006.**
Source: International Aluminium Institute, 2007 Sustainability report.

over the last 25 years but more is possible. PFC production per tonne of aluminium has gone down by about 85%, stimulated by a voluntary programme implemented by the aluminium industry (see Figure 8.7). Costs of these measures have been low to zero. So-called secondary aluminium smelters use recycled aluminium and have much lower emissions. Increasing the recycling rates (currently about 50%) is therefore an attractive reduction measure. The reduction potential from aluminium manufacturing by 2030 is about 0.1 $GtCO_2$-eq/year. In the longer term processes with non-carbon electrodes could further reduce emissions by 0.1–0.2 $GtCO_2$/year.

Other industries

Other energy intensive industries, like paper and cardboard (usually called pulp and paper), food processing, and glass manufacturing have good mitigation opportunities as well. Energy efficiency improvement is of course a primary one. Management of process waste is another. Anaerobic waste water treatment with methane recovery for energy, use of biomass waste as fuel, and gasification of wood pulping waste for fuel are prominent options to reduce greenhouse gas emissions. In these industries that use a lot of heat for their processes, combined heat and power (CHP) units can make a major contribution (see Chapter 5). Surplus electricity from these CHP units can be exported off-site. Table 8.3 gives a summary of the major mitigation measures. The total mitigation potential in these industries is at least 0.3–0.4 $GtCO_2$/year.

Non-CO_2 gas reduction potential from these other industries is about 0.1 $GtCO_2$-eq/year. SF_6 from magnesium production can be reduced by almost 100% at negative costs. Various fluorinated gases from semiconductors and LCD TV and computer screens manufacture can be reduced by at least 10% at zero costs through recycling and alternative compounds. HFCs, which were introduced as alternatives for ozone depleting fluorinated gases in foam production, refrigeration and air conditioning, or solvent applications, can be replaced with alternatives that have a low GWP or have no greenhouse gas effect (see Box 8.2).

Table 8.3. **Main mitigation opportunities in some energy intensive industries**

Industry	Main mitigation opportunity
Pulp and paper	Use of waste biomass fuel
	Combined heat and power
	Gasification of wood pulping waste (black liquor) for fuel use
	Increased recycling
Food processing	Energy efficiency improvement
	Combined heat and power
	Methane recovery from waste water
Glass	Energy efficiency improvement
	Switching from oil to gas heating
	CCS in combination with oxygen
	Increased recycling

Source: IPCC Fourth Assessment Report, Working Group III, Ch 7.4.

Box 8.2 **Replacing HFCs in industry**

Refrigeration equipment for frozen food processing and storage, industrial production of oxygen and nitrogen, and other cooling processes in industry predominately use ammonia or HCFC-22 as cooling agents. Since HCFCs are due to be phased out under the Montreal Protocol, a shift to HFCs (with high GWPs) is expected. Excellent alternatives exist however in the form of CO_2 (see note 1) or CO_2/ammonia mixtures as coolants. HFCs with very low GWPs, in combination with a leak tight system, could in some cases also be effective. Costs of such alternatives are about US$30–40/t$CO_2$-eq avoided.

Foam production for mattresses, furniture, and packaging is currently mostly done with HFCs as so-called blowing agents. Alternatives do exist however in the form of hydrocarbons or CO_2 that can completely replace HFCs at low costs.

In the electronics and other industries CFCs were originally used. After they were banned under the Montreal protocol, water and soap proved to be an excellent replacement for many applications. HFCs and PFCs have replaced CFCs for special purposes, but alternatives are also becoming available.

Note 1: CO_2 has a very low GWP compared to HFCs and HCFCs and given the limited quantities its contribution to overall warming from these applications is completely negligible.

(Source: IPCC Special Report on Safeguarding the Ozone Layer and the Global Climate System: Issues related to hydrofluorocarbons and perfluorocarbons, 2005)

Generic reduction options

Apart from the industrial processes we discussed above, there are other types of industrial processes and many small and medium enterprises that contribute a significant amount to industry emissions, mainly from the use of fossil fuel or electricity. They can also contribute to emission reductions, through energy efficiency improvement, fuel switching (either direct fuel use or fuel used in electricity generation), or recycling.

Electric motors are a prime example of what this could mean. In the EU and USA approximately 65% of all industrial electricity use is for electric motors (this includes the sectors that were discussed above). Typical numbers for the energy savings that can be realized by replacing motors with more efficient ones are 30–40%. Investments in such replacements are normally earned back very quickly, after which they produce net benefits. Compressed air systems, widely used in industry, are another example. In general, 20% of such systems are leaking, wasting a lot of energy. With simple measures considerable savings can be achieved. Steam boilers are used in many types of industry. Efficiencies of modern steam boilers are now in the order of 85%, while in practice most boilers are doing much worse. There are many other cheap ways of saving energy through insulation, heat recovery, recycling, proper maintenance of equipment, etc. In particular in developing countries energy savings of 10–20% can be achieved with simple measures. More advanced measures, requiring larger investments, can realize a 40–50% reduction in energy use. Most of these investments have a very short pay-back time.

Recycling of industrial waste materials has a significant potential to reduce emissions. The discussion on the steel industry above showed for instance that recycled steel as input in electric arc furnaces leads to much lower emissions per tonne of steel. Aluminium production from recycled aluminium waste requires only 5% of the energy needed for primary aluminium production. Increasing the recycling rate will therefore reduce emissions significantly. Waste paper as raw material for paper and cardboard manufacture saves energy. Increasing recycling rates to levels of 65% and above (as in Japan and parts of Europe) can realize significant CO_2 emission reduction. Many waste materials can be used as fuel in industrial boilers. If all waste materials were used, this could in theory lead to a 12% reduction of global CO_2 emissions; however, availability at the right place, transport costs, and user requirements will limit this potential considerably.

Renewable energy sources obviously are an important reduction option. In terms of primary energy sources this means use of biomass. The use of sugar cane waste (bagasse) is common in sugar mills. In the paper industry biomass waste is also widely used as an energy source. Increased use of biomass in industrial boilers as a reduction option depends on the local availability of biomass and the way the biomass is produced (see Chapter 9 for a more in-depth discussion). Renewable electricity is of course another good reduction option, as discussed in Chapter 5.

Management of post-consumer waste

Post-consumer solid waste management is schematically given in Figure 8.8. Most of the greenhouse gas emissions come from CH_4 from landfills due to biological conversion of organic waste materials. CO_2 is emitted from incineration and composting, but the fraction from organic (food and plant) residues does not count, because it is supposed to be neutralized by the uptake of CO_2 during growth[11]. CO_2 contributions are therefore small.

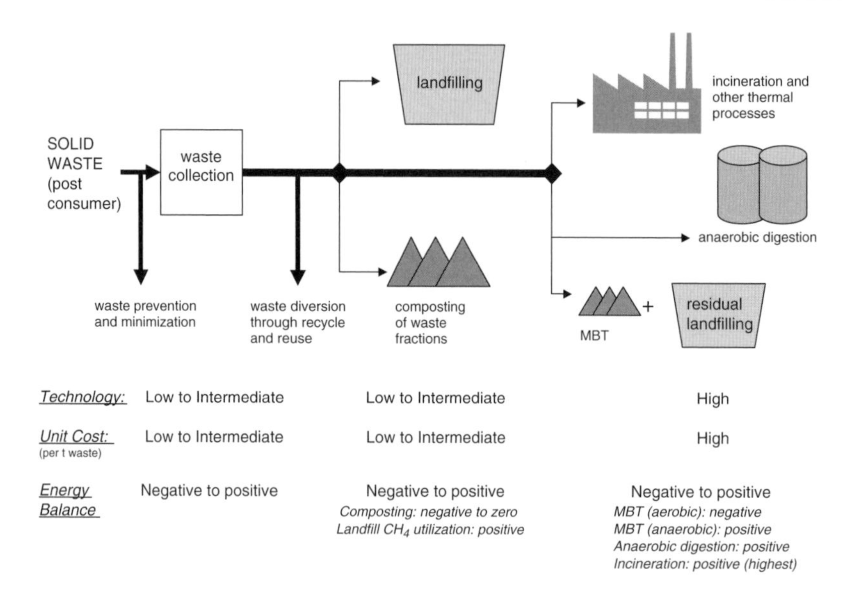

Figure 8.8 **Schematic diagram of post-consumer solid waste management options. MBT = mechanical biological treatment.**

Source: IPCC Fourth Assessment Report, Working group III, figure 10.7.

Reduction of CH_4 emissions from solid waste management can be realized in three ways:

- Waste minimization and recycling, so that less (organic) waste ends up in landfills
- Diversion of (particularly) organic waste to composting, mechanical-biological treatment, anaerobic digestion, or incineration
- Capture of CH_4 from landfills and use as a fuel

The total reduction potential in 2030 is about 0.4–1.0 $GtCO_2$-eq. Capture of CH_4 accounts for about half of that, with the other two approaches splitting the rest. About half of the potential can be obtained at negative costs, i.e. the benefits of the captured gas outweigh the costs of the measures. About 80% of the total costs less than US\$20/tCO_2-eq avoided[12].

For waste water management, measures to reduce emissions are first the provision of proper sewerage, septic tanks, and latrines. Water reuse and recycling and shifting to anaerobic waste water treatment can further reduce emissions. Reliable estimates of the potential are not available[13].

Overall reduction potential

The overall reduction potential (direct and indirect) for the industry and waste management sector is about 4.7 $GtCO_2$-eq/year in 2030, with a fairly large uncertainty margin of plus or

Table 8.4. Overall economic mitigation potential for the industry and waste sector

Mitigation option	Economic potential 2030 ($GtCO_2$-eq/year)	Cost range (US\$/tCO_2-eq avoided)
Iron and steel	0.4–1.5	20–50
Cement	0.5–2.1	<50
Chemicals and refining	1.0	75% of potential <20
Other industries	0.5–0.6	<100
Generic options	0.1–0.3	<100
Household/office waste management	0.4–1.0	<100
Total	2.9–6.5	<100

Source: IPCC Fourth Assessment Report, Working Group III, Ch 7 and 10.

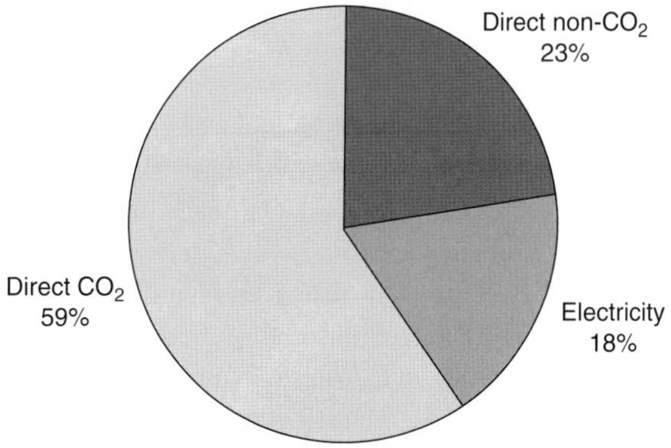

Figure 8.9 **Share of direct CO_2 and non-CO_2 and indirect CO_2 reduction in the economic reduction potential of the industry and waste management sector.**
Source: IPCC Fourth Assessment Report, Working Group III, Ch 7 and 10.

minus 1.8 $GtCO_2$-eq. This is at cost levels up to US\$100/tCO_2-eq. Of this potential about 30% can be obtained at costs lower than US\$20/t, about 90% at costs below US\$50/t. It means a reduction of 15–40% of emissions in 2030 without mitigation. The composition of the reduction potential is shown in Table 8.4.

The distribution between direct CO_2 and non-CO_2 reduction and indirect reduction from electricity use is shown in Figure 8.9.

Costs referred to above are strictly for the reduction measures taken and do not account for additional benefits (so-called co-benefits) achieved. Experience shows that energy efficiency programmes very often lead to improved maintenance and therefore reduced down-time of equipment, leading to better product quality, less waste, and better use of existing equipment. In a study of co-benefits in about 50 projects in several countries, costs of GHG reduction measures were cut in half when co-benefits were counted. Overall numbers for cost reduction from co-benefits are not available.

As indicated above, the reduction potential in industry in the longer term is higher than for 2030. By 2050 greenhouse gas emissions could be reduced by more than 60% compared to the baseline, when going up to cost levels of US$200/t$CO_2$-eq avoided. For comparison, by 2030 and costs <US$100/t$CO_2$-eq it is less than 40%.

How to make it happen?

Investment decisions in larger companies in industry are made on rational economic grounds. Given strong competition and global markets, companies cannot afford to do otherwise. This means that investments in greenhouse gas emission reduction are only made when there are economic benefits. Benefits can be a lowering of energy costs when investing in energy efficiency improvement. It can also be in the form of increased shareholder value, when a company takes the lead in climate change mitigation. BP experienced that when it undertook to lower its CO_2 emissions by 10% and Dupont's goal of cutting its GHG emissions by 65% was made part of its efforts to become a leader in sustainable growth.

Where CO_2 has a price, such as in the EU under the European Emission Trading System, economic benefits are obtained by lowering emissions in order to avoid purchasing additional emission allowances. Or, when there are regulations to use best available technologies, such as under the EU Integrated Pollution Prevention and Control Directive, the economic benefits of investing are avoiding penalties. Profitability of investments in industry is normally judged in terms of their pay-back time (the time needed to recoup the investment). Generally in industry only investments with pay-back times of not more than a few years are approved. The economic logic means that emission reduction investments that do not meet those private sector pay-back criteria are simply not made. Subsidies provided by governments will of course make investments more attractive.

The lifetime of facilities in industry is often tens of years, which slows the penetration of low-emission equipment and process plants (this is the so-called 'slow capital stock turnover'). Replacing an installation before the end of its economic life is economically difficult to justify, unless the alternative is very attractive.

Industry also rates the reliability of installations highly and is therefore reluctant to invest in new equipment that does not have a long track record, even when the pay-back time of the investment looks good. Banks are often reluctant to provide loans for new technologies, even if the company is convinced of its viability. Particularly in SMEs there are problems of lack of expertise or time to evaluate alternatives that slow down the acceptance of new technologies. In larger companies strategic consideration about mergers or acquisitions could take the attention away from economically justifiable investments.

Applying commercially available technologies across a whole industrial sector is a time consuming process therefore. In many developing countries there are additional problems with respect to technical capacity, availability of capital, and unattractive investment conditions. Modern technologies often have to be acquired abroad, which further complicates investments in modern low emission technologies (the so-called technology transfer problem[14]). This means that even profitable investments are not always made. In other words, many greenhouse gas emission reduction technologies are not taken up as much as the economic benefits would justify.

Policy instruments

Policy instruments are needed to make investments attractive. There are basically four approaches:

1. Make it more attractive to invest in profitable reduction measures
2. Make unprofitable investments profitable by creating incentives
3. Increase the price of greenhouse gas emissions to make more advanced technologies profitable (or require those through regulation)
4. Stimulate R&D to develop future mitigation technologies or make current ones cheaper

These policy instruments are discussed in Chapter 11, but the specific experiences in applying them in the industry and waste sector are outlined below.

Voluntary agreements

An instrument belonging to the first category is the so-called voluntary agreement between (a sub-sector of) industry and a government. They have been used in a number of industrialized countries since the early 1990s. They are essentially negotiated contracts, containing targets to be met by industry (often in terms of energy efficiency) and facilities and support to be provided by governments (for analyzing performance, information sharing, recognition, awards, etc). They vary in terms of the stringency of the targets, but more importantly in the verification and penalty provisions. Experience shows that agreements with a credible threat of regulations or taxes in case the agreement does not work, and with adequate government support, are the most effective (mainly in Japan and the Netherlands, see also Chapter 11). In such cases the effect is that pay-back criteria for low emission investments are somewhat relaxed. Generally speaking, voluntary agreements raise awareness in industry about the possibilities for GHG emission reduction and reduce barriers against low emission investments, such as the lack of information. However, for most voluntary agreements no difference from a business as usual improvement could be detected.

Industry initiated voluntary actions

Many companies have taken on actions related to reduction of greenhouse gas emissions on a strictly voluntary basis, without involvement of governments. Some, like Dupont, BP, and United Technologies Corporation, have achieved measurable reductions in energy use or emissions (see Box 8.3). Others have joined international initiatives, such as the Global Reporting Initiative (measuring and reporting of GHG emissions), the World Business Council for Sustainable Development's Cement Sustainability Initiative (CO_2 inventories and best practice sharing), the International Aluminium Institute's Aluminium for Future Generations programme (technical services, performance indicators, reduction objectives), and the International Iron and Steel Institute's voluntary action plan (measuring and reporting CO_2 emissions, general objectives for reduction of energy use and emissions).

Environmental NGO's like the World Wildlife Fund and the Pew Center on Global Climate Change increasingly work with companies to help them formulate voluntary actions.

Box 8.3 Some corporate achievements

Dupont is a chemicals company with 135 facilities in 70 countries, 60000 employees and about US$60 billion in sales. It formulated the following company-wide targets:

- Reduce greenhouse gas emissions by 65% below 1990 levels by 2010
- Hold total energy consumption at the 1990 level
- Supply 10% of energy from renewable sources

Energy use was indeed kept at the 1990 level through an aggressive energy efficiency improvement programme, involving company-wide training and energy audits. It also resulted in a net cost reduction by the way. By 2002 the emission reduction target had already been met. In 2004 reductions were 72%. It was achieved by eliminating N_2O emissions from adipic acid manufacturing (a raw material for nylon) and HFC-23 emissions from the production of HCFC-22 (a cooling liquid). With 80% of Dupont's emissions from N_2O and HFC-23, and energy related CO_2 emissions remaining constant, this helped them to meet the target.

In 1998 BP set a target of reducing company-wide direct greenhouse gas emissions (i.e. from operations only, not from burning the fuels BP produces) by 10% below 1990 values. In 2002 this target was met. In 2007 emissions had declined further. Most reductions were achieved through reduced flaring and venting and improvements in energy efficiency. They also resulted in net cost reductions.

(Source: IPCC Fourth Assessment Report, Working Group III, ch 7; Dupont testimony US Congress, http://oversight.house.gov/documents/20070523104438.pdf; BP Sustainability Report, 2001, 2006, 2007, http://www.bp.com/sectiongenericarticle.do?categoryId=9022214&contentId=7041069)

Most voluntary actions result in the implementation of economically profitable emission reductions. Since these are often not implemented under normal circumstances, this is a contribution to mitigation in the industry sector. However, as substantial emission reductions are needed in the future, voluntary actions and voluntary agreements will be unable to deliver these in the absence of strong government action.

Financial instruments

Financial instruments come in different forms: taxes, tax deductions, and subsidies.

Taxes on CO_2 emissions from industry have only been introduced in a limited number of countries so far. Norway has a CO_2 tax of about US$50/t$CO_2$. It mainly applies to the off-shore oil and gas industry, since many other industry sectors were exempted in exchange for committing to voluntary agreements on emission reduction. The Stattoil CO_2 capture and storage installation at its Sleipner gas production platform became attractive just because of this tax. Sweden has a carbon tax of about US$40/t$CO_2$, but industry pays only half. The same holds for Denmark, but with a much lower rate of US$14/t$CO_2$[15]. The UK introduced a general climate change levy (less than 1USc/kWh), but then created exemptions for industries that participate in voluntary agreements or emission trading. France has a modest tax on N_2O emissions from chemical industries. Industry in the Netherlands only pays energy/CO_2 tax up to a relatively small volume of electricity and gas usage. It is exempted from the rest. Germany introduced a similar eco-tax. The reason for all these exemptions (and the lobbying by industry for it) is the effect the taxes have on international competitiveness. Due to their limited impact, taxes have not achieved much in terms of moving to emission reductions with net positive cost. There is one clear exemption: the Norwegian tax on offshore oil and gas operations that led to the introduction of CCS. Taxes do contribute however to taking cheap reduction options more seriously.

Tax deductions are quite popular in many countries to encourage industry investments. They often however fail to discriminate between investments in reducing greenhouse gas emissions and other more traditional investments. So the impact of tax deductions is hard to assess. Effectiveness requires a quite precise description of investments the tax deduction applies to. Recent energy efficiency oriented tax deduction schemes in The Netherlands, France, and the UK are trying to do that[16].

Subsidies for investments by industry in energy efficient equipment in the form of grants or cheap loans are quite popular. In Europe, Japan, and Korea extensive programmes exist. Evaluations of such subsidy programmes generally show a positive influence on energy savings and corresponding emission reductions. They also show a positive influence on the development of markets for innovative technologies. The major drawback of these subsidy schemes is that they also benefit companies that would have made the investments anyway. That particularly applies to investments that deliver net profits. To make subsidy schemes more effective, targeting positive cost technologies is very important. In developing countries subsidy schemes for industry are generally lacking, making the role of commercial and development banks all the more important. Unfortunately these institutions are often not equipped or willing to give priority to loans for energy efficiency investments.

Cap and trade programmes

Cap and trade means setting maximum emission levels for companies and then allowing them to trade emissions allowances between themselves to fulfil their obligations.

A company that can reduce its emissions below the maximum can sell surplus allowances to others. A company that finds it too expensive to reduce its own emissions can buy allowances to fulfil the obligation. In doing so, a price for emitting a tonne of CO_2 will emerge. The more stringent the emission limits, the higher the price. If the market works well, i.e. companies do indeed trade surplus allowances and other companies do buy allowances, the overall cost of realizing the emission reduction will be minimized. The cheapest options are then done first.

The first big application of a cap and trade system happened in the USA in the 1980s under the Clean Air Act, when companies were allowed to trade SO_2 allowances. Greenhouse gas cap and trade systems are in operation in the EU (see Chapter 11 and Box 11.5), Norway (Norway will join the EU trading system soon), New South Wales (Australia), and several US States. Australia, New Zealand, Japan, and the USA are considering introducing national systems. Energy intensive industries are always included.

Cap and trade systems so far apply mostly to large installations. The EU ETS for instance covers over 11 500 energy-intensive installations across the EU, which represent about 40% of Europe's emissions of CO_2. These installations include combustion plants, oil refineries, coke ovens, iron and steel plants, and factories making cement, glass, lime, brick, ceramics, pulp, and paper. The chemical industry was only partially included through their large combustion unit or their integration with petroleum refineries. For the period after 2012 ammonia and aluminium plants and N_2O and PFC from some industrial sources are also included[17]. For smaller installations the administrative burden becomes bigger and so these are usually left out of the cap and trade system. In theory there are ways to bring these smaller entities into the system by allocating the allowances to suppliers of natural gas or other raw materials.

A big issue under any cap and trade system is the initial allocation of emission allowances. Most systems in operation today started on the basis of free allocation only, based on historic emission levels (so-called 'grandfathering'). This was very much pushed for by industry in light of international competitiveness. To some extent actual performance of the plant, in terms of previous reduction measures, can be taken into account. In the EU ETS this is done at the Member State level in the so-called national allocation plans. Normally there are provisions made for newcomers: if a new company wants to build a production plant there are guarantees that they can obtain the necessary emission allowances.

Gradually, a shift towards auctioning the allowances is visible. In the EU ETS during the period 2008–2012, Member States have the option of auctioning up to 10% of the allowances[18]. For the period after 2012 more auctioning will take place. For non-exposed industries 70% will be auctioned by 2020, increasing to 100% by 2027[19]. Most systems under consideration elsewhere also incorporate partial or full auctioning. Auctioning means companies have to buy the allowances at a market price, which adds to the cost of their operation. That is why globally competing companies that are subject to the EU ETS are lobbying hard to be given the allowances for free. There is a tendency however for them to exaggerate the impacts. As indicated above energy only forms a significant part of total operating costs in a limited number of industrial sectors.

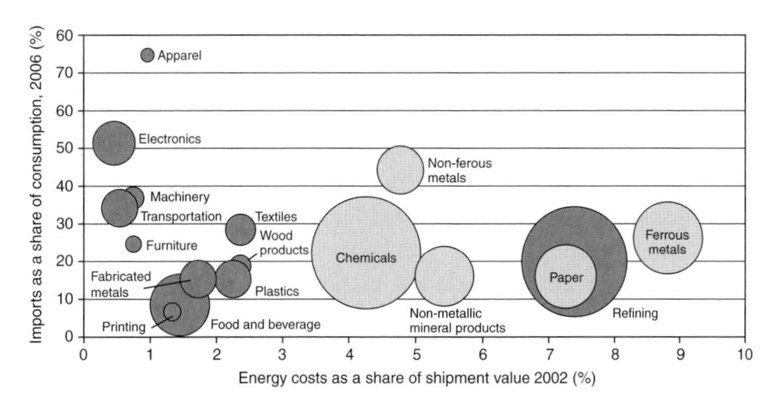

Figure 8.10 **US industry exposure to climate costs. Sectors that have high energy costs and are faced with large imports are the most vulnerable to competition. The size of the bubbles is proportional to the CO_2 emissions in 2002.**

Source: Houser et al. Levelling the carbon playing field, Peterson Institute for International Economics and WRI, 2008.

And the exposure to international trade also varies: it is high for aluminium, moderate for chemicals, oil products, paper and steel, but low for cement, glass, and ceramics. In response to industry concerns it has been decided to exempt certain industry sectors from auctioning of permits if they face high cost increases and are very exposed to trade[20]. Figure 8.10 shows the situation for US industry.

The most important lesson that was drawn from the EU ETS operation so far is the fact that more centralized allocation of allowances is needed. The system with 27 national allocation plans led to big differences in treating similar installations in different Member States. It has been decided therefore to replace this with one central allocation system, run by the Commission.

Regulation

Regulation on industrial greenhouse gas emissions is applied on a limited scale. In the EU the system of permitting large industrial installations, based on the Integrated Pollution Prevention and Control (IPPC) directive, covers N_2O and fluorinated gas emissions from some installations. It requires Best Available Technology standards to be applied (based on BAT reference documents issued by the European Commission[21]). In the EU and some other countries there are also regulations banning the use of HFCs, PFCs, and SF_6 in certain applications. China is using regulation to force the closure of a substantial number of old, inefficient cement plants[22]. It remains to be seen how much of this plan will be implemented. In general the notion of forced closing of outdated inefficient plants is a useful one to consider as part of a portfolio of policies to realize substantial emission reductions. It could for instance be part of a system of mandating Best Available Technologies in industry at the national level.

Technology policy

There are extensive policy efforts aimed at diffusion and transfer of modern efficient technologies and at developing new and better technologies. They normally are not seen as climate change related, but they in fact are. So it is good to discuss what these technology policies could mean for realizing deep reductions in greenhouse gas emission from industry.

An efficient plant, in terms of energy and raw materials use, can produce at lower costs than an inefficient plant. That means a better competitive position, which is good for the economic development of a country. That is why government policies to promote the application and development of modern efficient technologies exist.

Since technology policy also affects the energy, transportation, and buildings sector technology policies will be discussed in more detail in Chapter 11.

Air and water regulations

Controlling air pollution from sulphur oxides, nitrogen oxides, and fine particles can also reduce CO_2 emissions if focussing on energy efficiency improvements and fuel shifts (from coal to gas or renewable energy sources). So air pollution control policies can have an impact on reducing CO_2 emissions in industry. There is a tendency to integrate air pollution and climate change policy in order to maximize the win-win opportunities. Particularly in Europe, under the UN Convention on Long Range Transport of Air Pollution and under EU policy, this is being pursued. In many other places however air pollution is often controlled with add-on desulphurization units and particle filters that tend to increase energy use and CO_2 emissions.

Waste management

Waste management is to a large extent a government dominated and local industry. Competitiveness hardly plays a role. It is also dominated by health and environment considerations, which means that waste management rather than climate policies are the appropriate instrument. If we look at the biggest reduction opportunity, i.e. avoiding CH_4 from landfills, then it obvious that regulations on landfill construction and CH_4 capture are fundamental. But since the other half of the potential lies in avoiding waste and diverting it to other waste processing methods, policies also need to be focussed on that. The most effective approaches to reduce waste and encourage recycling are financial incentives (e.g. buy special waste bags for a price that reflects the disposal costs; no other bags allowed) and regulations (ban on recyclables in general waste, combined with collection of recyclables). In many places municipal governments are still relying on purely voluntary approaches through information and central recycling centres. These voluntary approaches are often reasonably successful, but fail to capture all of the potential[23].

More recycled material as input for steel and aluminium making, paper production, and plastic processing has immediate effects on industrial energy requirements and CO_2 emissions. Recycling also provides fuels for industrial boilers and cement plants, replacing fossil fuels. Effective waste management and recycling policies can therefore have a positive impact on realizing reduction of industrial emissions.

Future challenges

Big greenhouse gas emission reductions are possible in industry at reasonable costs. The greatest challenge is to realize these reductions, given that industry operates in a competitive environment. Many energy intensive products are traded internationally, if not globally. Forcing industry in some countries to drastically reduce emissions while competitors elsewhere are not facing restrictions will therefore not solve the problem. In practice this slows down the implementation of reduction options, because pushing companies to relocate and losing jobs is not an attractive proposition for politicians. The solution is either to develop international agreements covering most competitors, or to use trade mechanisms to protect domestic industries that face strict emission reduction obligations.

Waste management can make a large contribution when both industrial waste and waste water streams as well as so-called post-consumer waste are considered. The numbers usually point to waste treatment options such as capturing CH_4 from landfills as the most important ones. However, minimizing waste by reducing the material content of products and maximizing recycling are undervalued. At present we do not know enough about the reduction in energy use and greenhouse gas emissions from such dematerialization and 'cradle to cradle' approaches. More studies, in particular life cycle analyses that look at complete lifecycles of products, should provide better answers in the future.

Notes

1. IPCC Fourth Assessment Report, Working Group III, ch 7.1.2.
2. Houser et al. Leveling the carbon playing field, Peterson Institute for International Economics and WRI, 2008.
3. IPCC Fourth Assessment Report, Working Group III, ch 10.2.1.
4. IPCC Fourth Assessment Report, Working Group III, ch 10.1.
5. IPCC Fourth Assessment Report, Working Group III, ch 7.1.3 and 10.3.
6. Coke is 'degassed' coal, produced by heating coal in the absence of air, so that it loses its volatile compounds (methane, hydrogen, carbohydrates, and tar). Coke has a porous structure that gives it ideal properties for binding oxygen from iron ore in blast furnaces.
7. IPCC Fourth Assessment Report, Working Group III, ch 7.4.1.
8. IEA Energy Technology Perspectives, 2008, ch 16 and IPCC Fourth Assessment Report, Working Group III, ch 7.3.7.

9. IEA Energy Technology Perspectives, 2008, ch 16.
10. See note 8.
11. Plants take up CO_2 during growth from the atmosphere. When they are incinerated, CO_2 returns to the atmosphere. This process is thus a closed loop and net emissions are zero.
12. IPCC Fourth Assessment Report, Working Group III, ch 10.4.7.
13. IPCC Fourth Assessment Report, Working Group III, ch 10.4.6.
14. See IPCC Special Report on Methodological and Technological Aspects of Technology Transfer, 2005.
15. http://www.carbontax.org/blogarchives/2008/03/.
16. IPCC Fourth Assessment Report, Working Group III, ch 7.9.3.
17. http://ec.europa.eu/environment/climat/emission/ets_post2012_en.htm.
18. See for more detail Hepburn C et al. Climate Policy, vol 6, 2006, 137–160.
19. See note 17.
20. IEA, Issues behind competitiveness and carbon leakage: focus on heavy industry, 2008.
21. http://ec.europa.eu/environment/air/pollutants/stationary/ippc/index.htm.
22. IEA, Sectoral approaches to greenhouse gas mitigation: exploring issues for heavy industry, 2007; see also note 2.
23. IPCC Fourth Assessment Report, Working Group III, ch 10.5.

9 Land use, agriculture, and forestry

What is covered in this chapter?

Agriculture and forestry together are responsible for about 30% of greenhouse gas emissions, partly from loss of carbon from soils and vegetation and partly from agricultural activities producing methane and nitrous oxides. Demand for food is the dominant driver of developments in agriculture and deforestation. Food security has always been high on the political agenda as is visible in the strong reactions to recent increases in food prices. There is a large potential to reduce emissions. Increasing carbon in agricultural soils, livestock manure management and conserving carbon in forests by reducing deforestation, planting new forests, and better forest management can halve emissions by 2030 at reasonable costs. Policy actions to realize this potential can best be focused on reforming the many existing policies and create financial incentives for farmers and forest owners to change their practices. International climate policy instruments created by the Kyoto Protocol can contribute.

Land use trends

About one-third of global land is used for agriculture. Two-thirds of that land is grassland, one-third cropland. Forests cover about 25%. The rest (about 40%) is desert, tundra, ice, wetland, or other natural area, except for a small amount covered by urban areas (less than 0.5%). Over time shifts have occurred from forested land to agricultural land (cropland and grassland), consistent with the increase in the world population and the need for food. Over the last 40 years agricultural land has increased by about 500 Million hectare (Mha) or 10%. About half of this increase came from deforested land. Due to erosion, salt accumulation (often due to bad irrigation practices), and other processes about 20% of cropland and 10% of grassland is degraded[1]. For global land use for agriculture and forestry over the last four decades see Table 9.1.

Growing populations and improving incomes will increase demand for food. Increasing meat consumption[2] will further increase land requirements, because land use for a meat diet is much larger than that for a vegetarian diet. For 1 kg of meat 2–7 kg of grain is needed (7 for beef, 3.5 for pork, 2 for poultry, and about 1.2 for fish)[3]. It is

Table 9.1. Global land use for agriculture and forestry over the last four decades

Land use	Area 2001–2002 (Mha)	Area 1961–1970 (Mha)	Change (%)	Current rate of change (Mha/year)
Cropland (incl. permanent crops)	1535	1379	+9	
Grassland	3488	3182	+10	
Forest	3952	4126	−5	−7.3 (12.9 loss; 5.6 increase)
Desert, tundra, ice, wetlands	5850			

Source: IPCC Fourth Assessment Report, Working Group III, chapter 8 and 9; FAO Global Forest Resources Assessment 2005.

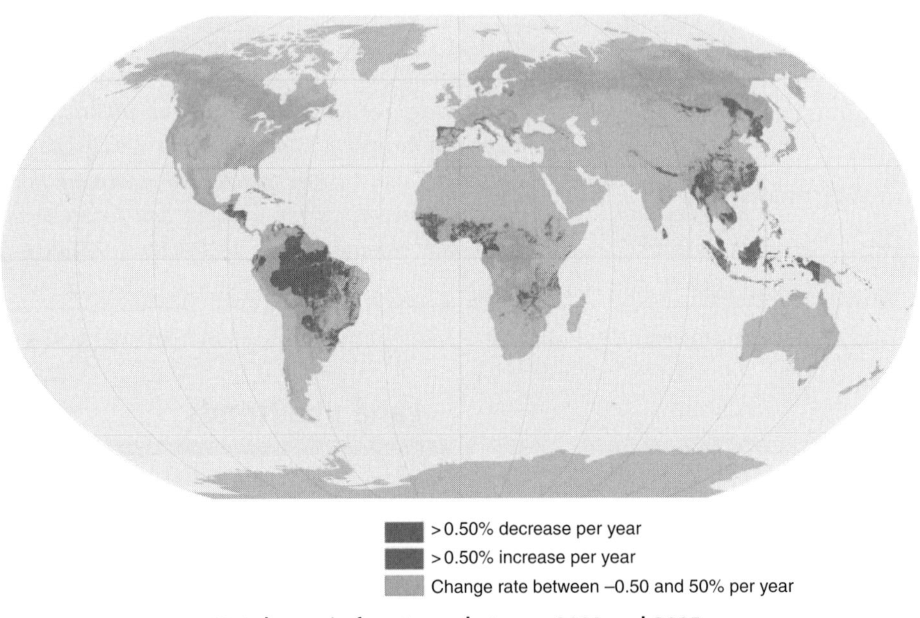

Figure 9.1 **Net change in forest area between 2000 and 2005.**
Source: FAO, Global Forest Resource Assessment 2005. See Plate 14 for colour version.

expected that another 400–500 Mha additional agricultural land will be needed between 2002 and 2020, even if crop productivity were to improve further[4].

Net loss of forest area (7.3 Mha/year) is the result of the difference between deforestation (on average about 12.9 Mha/year between 2000 and 2005) and the increase in newly forested areas (about 5.7 Mha/year)[5]. The largest losses are found in South America, Africa, and South-East Asia (see Figure 9.1). Most of the increase in forestation is in Europe and East Asia. Only part of the forests in the world is managed. Although in Europe 90% is managed, in developing countries more than 90% is unmanaged. Forest plantations only cover about 3% of the total forested area, but are growing by almost 3 Mha/year (more than half of the total forest area increase). About 30% of all forest land is degraded.

Land use and greenhouse gas emissions

Agriculture and forestry are very different from other economic sectors when it comes to greenhouse gas emissions. The reason is that agricultural soils and crops and forest represent enormous reservoirs of CO_2, in the form of organic matter and wood[6]. The amount of carbon stored in forest biomass and soils is larger than what is contained in the atmosphere. And much of that carbon is underground (see Figure 9.2). So emissions are not only determined by activities that generate emissions, but also by the loss or gain in these carbon reservoirs (absorbing CO_2 in vegetation and soils is called 'sequestration'). It is important to consider agriculture and forestry together, because of the interactions (more demand for food drives deforestation) and the coverage of lands that can be grouped under agriculture or under forestry (agro forestry and peat lands).

Figure 9.3 shows the man-made carbon fluxes together with the emissions of CH_4 and N_2O from agricultural practices and the amounts of carbon stored in reservoirs. The respective contributions are discussed below.

Agriculture

Emissions from agriculture consist predominantly of methane (CH_4) from animals, manure and rice production, and of nitrous oxide (N_2O) from nitrogen fertilizer application (see Figure 9.4). N_2O emissions from fertilized soils is the largest source (38%), followed by

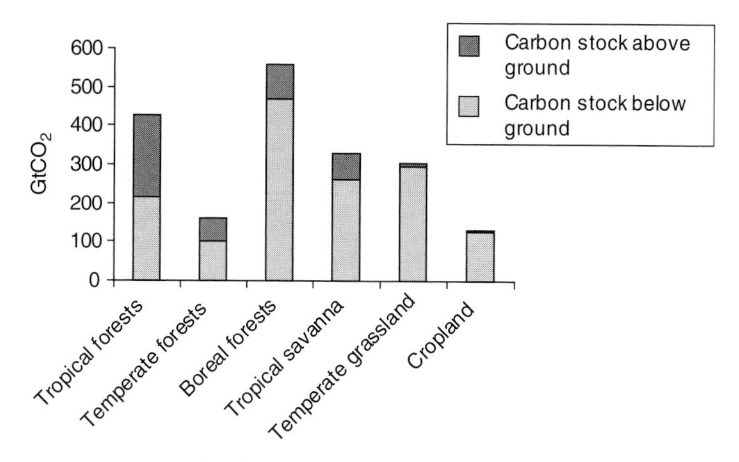

Figure 9.2 **Amount of carbon stored in agricultural and forest land.**
Source: IPCC Special Report on Land Use, Land Use Change and Forestry, 2000.

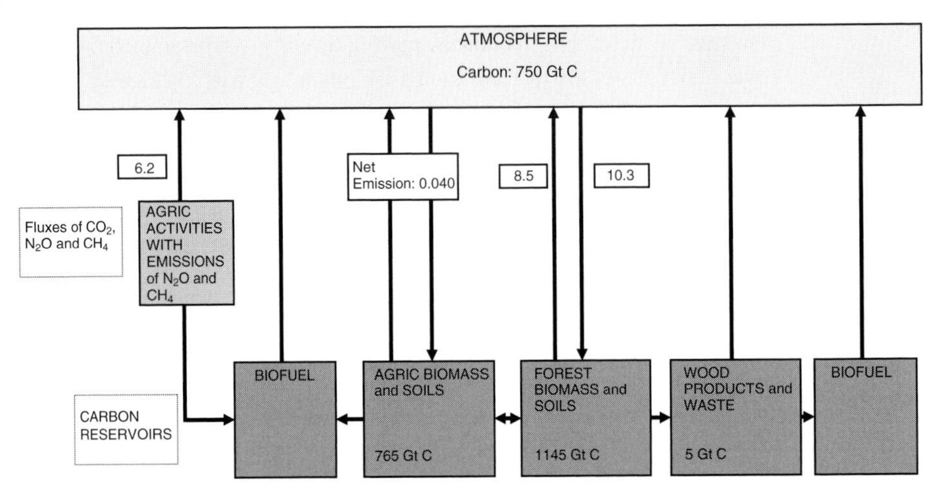

Figure 9.3 Schematic diagram of carbon reservoirs and emissions of greenhouse gases in agriculture and forestry. Reservoirs are expressed in Gt C. Fluxes are expressed in Gt CO_2-eq/year.

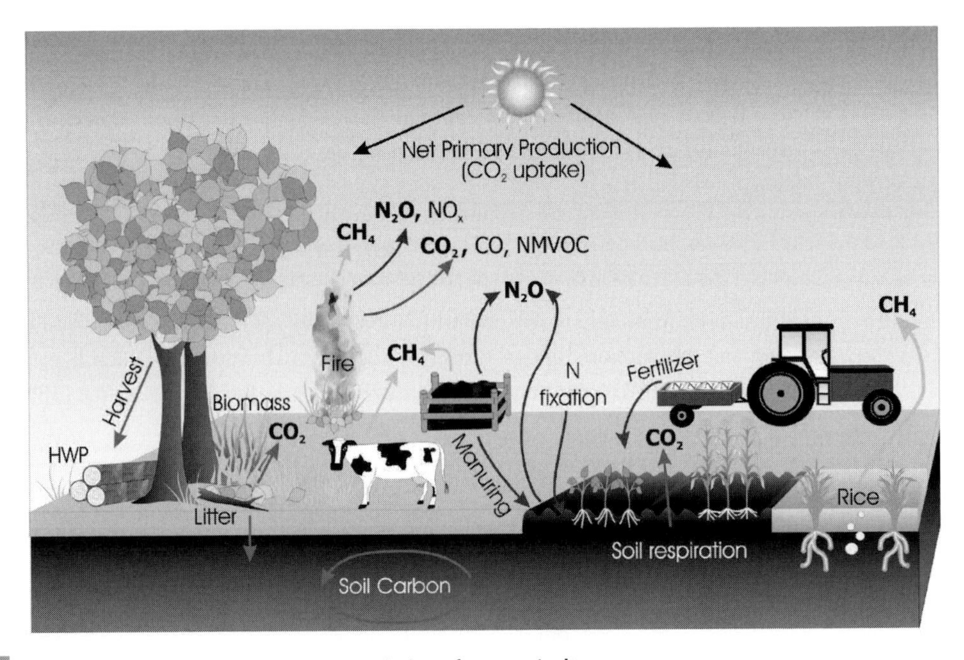

Figure 9.4 Emissions from agriculture.

Source: 2006 IPCC Guidelines for National Greenhouse Gas Inventories, Volume 4, Agriculture, Forestry and Other Land Use, chapter 1.

methane production in animals (32%), burning of crop residues (12%), rice fields (11%), and manure (7%). Although there are large amounts (fluxes) of CO_2 going into agricultural crops and soils, there are equally large fluxes going out (digestion and decomposition of agricultural crops and crop residues). The net flux is therefore small. Total CH_4 and N_2O emissions are about 6.2 $GtCO_2$-eq per year. Net CO_2 emissions due to the slowly decreasing carbon content of agricultural soils are less than 1% of that amount[7].

Regional differences in the magnitude and relative importance of CH_4 and N_2O emissions are large. Because of the importance of agriculture in developing countries and the large population, these countries are responsible for about 75% of all emissions. For rice production and crop residue burning the share is close to 100%. Emissions from manure are biggest in developed countries. Large livestock populations in Latin America, Eastern Europe, and Australia and New Zealand make this the dominant source in those regions.

Forestry

Emissions from the forestry sector are predominantly caused by loss from the large carbon reservoirs through deforestation and forest degradation (loss of trees due to selective logging or other disturbance), decomposition of wood residues, and some emissions of CH_4 from burning and N_2O from fertilized managed forests or forest plantations (about 5.8 $GtCO_2$/year), and dewatering and oxidation or burning of (deforested) peat lands (about 2.7 $GtCO_2$/year; see Box 9.1).

Wood products

Wood products are a temporary storage of carbon. Wooden houses, other structures and furniture, but also books, form a carbon reservoir of the order of 5 GtC. This is a very small amount compared to what is stored in vegetation and soils. Since wood products, including paper, have an average lifetime of about 30 years, the accumulation of carbon in wood products is limited. Wood products therefore have a very small contribution to emissions.

Biofuel

Biofuel or bioenergy is obtained from crop residues, crops, wood, or wood waste. If harvesting is done sustainably, biofuel does not contribute to emissions, since CO_2 is taken up again in the vegetation. In reality this sustainability assumption is not met, because of disturbance or fossil fuel use for harvesting and processing. Biofuel use therefore contributes to emissions. The amount depends on the specific situation.

Box 9.1 Peat lands

Peat lands are water logged, high organic soils produced by accumulation of rotting vegetation. In many countries a significant part of peatlands has been dewatered and is used for agriculture (see table below) or forest plantations. Together, agriculture and forestry are responsible for 80% of peat land loss; peat harvesting as fuel or soil supplement, urbanization and infrastructure, and flooding are responsible for the rest. Dewatered peat land produces CO_2 emissions through oxidation of organic material and through fires that keep burning underground (see picture) The biggest losses are now happening in Indonesia and Malaysia. Fires are responsible for about 2 Gt CO_2/year.

Peatland used for agriculture in selected countries

	Peatland used for agriculture (km^2)	% of total peatland
Europe	124490	14
Russia	70400	12
Germany	12000	85
Poland	7620	70
Belarus	9631	40
Hungary	975	98
Netherlands	2000	85
USA	21000	10
Indonesia	60000	25
Malaysia	11000	45

Source: Wetlands International et al. Assessment on Peatlands, Biodiversity and Climate Change, ch 3.

Peat fires.

Source: Wetlands International et al. Global Assessment on Peatlands, Biodiversity and Climate Change, 2007 and ScienceDaily.com, credit: Kim Worm Sorensen.

(Source: Wetlands International et al. Global Assessment on Peat lands, Biodiversity and Climate Change, 2007)

The total greenhouse gas emissions from agriculture and forestry are about 14.7 $GtCO_2$-eq/year, approximately 30% of the global total. The uncertainty of these numbers is high. Many of the emissions are not easily measured, such as N_2O from grasslands, CH_4 from rice production or savannah burning, and CO_2 from peat land and forest degradation. The real number could easily be several Gtonnes higher or lower.

Estimating future emissions is difficult. Agricultural emissions are going to increase, because of increasing food demand. Global grain demand is projected to increase by 75% between 2000 and 2050 and global meat demand is expected to double. More than three-quarters of growth in demand in both grains and meat is projected to be in developing countries[8]. The estimate is that emissions will go up from 6.2 to 8.3 $GtCO_2$-eq/year by 2030. How deforestation is going to develop is much more difficult to estimate. The best guess is that it will remain roughly at current levels until 2030 under a 'no climate policy' situation[9].

How can emissions be reduced and carbon reservoirs increased?

There are three broad categories of action that can be taken:

- Reducing emissions of CO_2, CH_4, and N_2O
- Increasing carbon reservoirs by increasing carbon in agricultural soils, agroforestry, and new and existing forests
- Using crops, crop residues, animal waste, wood cuttings, and wood waste as biofuel, replacing fossil fuel

In agriculture there are many specific actions that can deliver emission reductions. In many cases however there are complex relations between emissions of CO_2 and N_2O. In some circumstances emissions of N_2O could increase when CO_2 emissions are decreased, making the net effect uncertain.

The most important reduction measures are summarized in Table 9.2.

In addition to all the technical reduction options there is an important lifestyle option: change to a vegetarian diet. Vegetarian food requires less grains, land, and energy (for growing, transport, processing) than meat (see above). So changing to a vegetarian diet can avoid N_2O emissions from grasslands, CH_4 emissions from livestock and manure, CO_2 emissions from fossil fuel use, and free land for other purposes (forest, bioenergy crops).

In the forestry sector the most important actions that can be taken are summarized in Table 9.3.

Many of these actions take time to deliver results, since forest growth is slow. A disadvantage is that costs often have to be made up front and benefits come much later.

Table 9.2. Measures to reduce greenhouse gas emissions from agriculture

Category	Measure	CO_2	CH_4	N_2O	Net effect
Cropland management	Reducing ploughing, minimizing soil carbon loss	+		+/−	**
	Practices that increase returning crop residues to the soil, by leaving residues on the land and avoiding burning	+		+/−	**
	Keeping soils covered between crops and using legume crops to enhance nitrogen content of soils	+			***
	Better nutrient management, minimizing N_2O emissions	+		+	***
	Reducing CH_4 emissions from wetland rice cultivation by draining water from the field intermittently and addition of fertilizer in the dry phase	+/−	+	+/−	**
	Growing trees on farmland, in combination with livestock or food crops (agroforestry)	+		+/−	**
	Increasing the water table in drained cropland to reduce the conversion of organic soil matter into CO_2	+/−		+	*
	Set aside of part of the land for nature protection or environmental purposes, allowing increase of soil carbon	+	+	+	***
Grazing land management	Reducing fires	+	+	+/−	*
	Different grass varieties with deeper roots adding to soil carbon	+		+/−	*
	Reducing fertilizer use			+/−	*
	Increasing productivity by better water and nutrient application	+		+/−	**
Organic and peaty soils containing high carbon concentrations	Avoid draining these soils or re-establishing a high water table	+	−	+/−	**
	Minimizing ploughing of drained soils	+		+/−	**
	Keeping soils covered and avoiding tuber crops	+		+/−	*
Restoration of degraded land	Re-vegetation of eroded land	+		+/−	***

Table 9.2. (*cont.*)

Category	Measure	CO_2	CH_4	N_2O	Net effect
	Nutrient and organic matter application	+		+/−	**
Livestock management	Changing feed composition (more concentrated feed, less forage and feed supplements, such as certain oils); this reduces CH_4 emissions. Research is being done on other feed supplements		+	+	***
	Covering manure storage or composting solid manure (only for intensively managed livestock farming where herds are kept in a feedlot at least part of the time)		+	+/−	***
	(Longer-term) selective breeding of low CH_4 animals		+	+	**
Lifestyle options	Change to a vegetarian diet	+		+	***

+ indicates reduction of emissions; – indicates increase of emissions.
Net effect: more asterisks indicate higher net mitigation effect.
Source: IPCC Fourth Assessment Report, Working group III, table 8.3.

Table 9.3. Measures to reduce greenhouse gas emissions from forestry

Type	Measure	CO_2	CH_4	N_2O	Net effect
Maintain forest area	Reducing deforestation. This is by far the biggest contribution, given the big emissions from deforestation. Per hectare of forest maintained, 350–900 tonne of CO_2 emission is avoided	+	+		***
Increase forest area	The annual accumulation of carbon varies greatly between locations, tree species, and stage of the forest: it ranges from 1 to 35tCO_2 per hectare. Initially, when soils are disturbed prior to planting trees, soil carbon can be lost	+/−			***
Maintain forest density	Avoid forest degradation (preventing fires, managed	+			***

Table 9.3. (*cont.*)

Type	Measure	CO_2	CH_4	N_2O	Net effect
	logging, avoiding or reducing drainage of plantation soils)				
Increase forest density	Intensive management and nutrient application	+		–	**
	Increase the rotation period of forests	+			*
Wood products	Increase stocks of wood products and substitution of energy intensive materials by wood. This is a temporary gain in carbon reservoirs, because wood products eventually end up as waste. Some wood products however have a long life time (e.g. wooden houses, furniture) that does help to delay emissions	+			*

+ indicates reduction of emissions; – indicates increase of emissions.
Net effect: more asterisks indicate higher net mitigation effect.
Source: IPCC Fourth Assessment Report, Working group III, chapter 9.

How much can agriculture and forestry contribute to controlling climate change?

Agriculture

The (net) reduction of greenhouse gas emissions from agricultural options (see list in Table 9.2) depends on a lot of variables: climatic zone, existing practices, type of action, costs, time, etc. This means that effectiveness of mitigation strategies have to be determined locally. What works well in one place may not be effective elsewhere. In addition, information about reduction potential is limited for some regions and practices.

The relative contribution of measures till 2030 at the global level is shown in Figure 9.5. The economic potential is given[10] at three different cost levels (20, 50, and 100US$/t$CO_2$-eq avoided). In total this means that by 2030 about 4.3GtCO_2-eq/year can be reduced at costs up to US$100/t$CO_2$-eq. The figures are about 1.6 and 2.7 for costs up to US$20/t$CO_2$-eq and US$50/tCO_2-eq, respectively. The regional contributions to this global total vary. The relative contribution is shown in Figure 9.6. Most of the potential can be found in developing countries (about 70%).

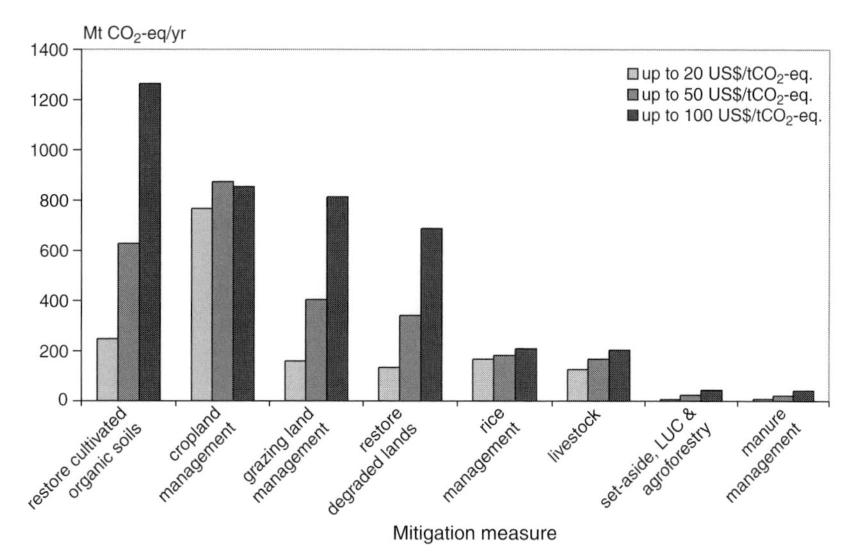

Figure 9.5 **Economic potential for greenhouse gas mitigation in agriculture at a range of carbon prices; based on SRES B2 sceanario.**
Source: IPCC Fourth Assessment Report, Working group III, figure 8.9.

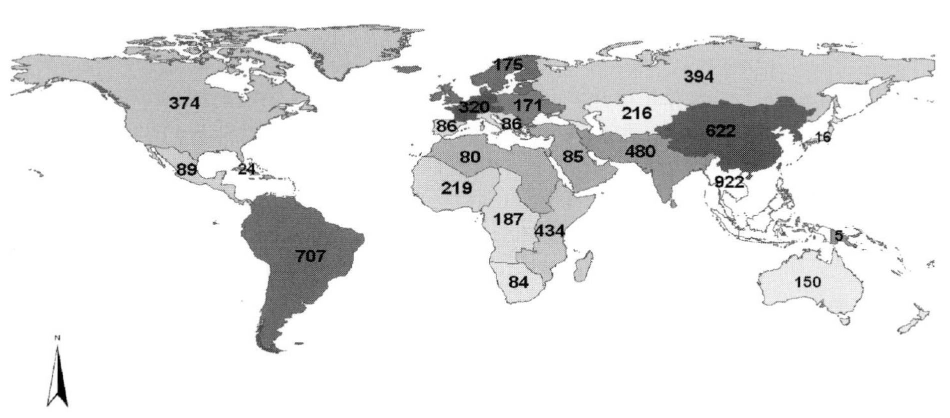

Figure 9.6 **Spread of mitigation potential for greenhouse gas mitigation in agriculture over different regions (numbers indicate relative importance of potential in regions); based on SRES B2 scenario.**
Source: IPCC Fourth Assessment Report, Working group III, figure 8.5.

About 90% of the total potential comes from increasing soil carbon reservoirs. This is completely the opposite picture to that for emissions, where CO_2 does not play a big role.

The potential of changing to a vegetarian diet is not included in these numbers. Data for the average per person emissions due to food consumption show a significant decrease of 1–2 tonnes of CO_2-eq per year when shifting towards a vegetarian or vegan diet (see Table 9.4). Numbers will differ from country to country, depending on food consumption patterns and the amount of energy used in the system.

Table 9.4. US data on emissions from food consumption

Type of diet	Annual emission per person from food consumption (tCO_2-eq/year)
Omnivorous diet	3.8
Mostly vegetarian diet	3.0
Vegetarian diet	2.7
Vegan diet	2.0

Source: http://www.conservation.org/act/live_green/carboncalc/pages/methodology.aspx

Forestry measures

Estimating the mitigation potential of forestry measures is difficult. One particular problem is that there is no scientific consensus on developments in the forest sector in the absence of climate policies. In particular future rates of deforestation are very hard to predict. So we do not have a very good idea of what difference specific measures will make.

The other big problem is that there are different methods to estimate a global mitigation potential for forestry: (1) from global forest sector models; and (2) from adding up regional bottom-up assessments. Global models assume population growth, income growth, changes in food consumption, and agricultural productivity increase, leading to a certain need for agricultural land. Then they calculate how much carbon release can be avoided compared to an assumed baseline by using land that is not needed for food production for forests, assuming a certain cost of forest management, afforestation, and avoiding deforestation for different regions. Regional bottom-up studies however start with existing forested areas and estimate what would happen to those areas with changing demand for food and with incentives for forest conservation in the form of carbon prices[11].

For the year 2030 global forest models calculate a 3 to 10 times higher reduction potential than bottom-up regional studies: 13.7 $GtCO_2$-eq/year (top-down) versus 1.3–4.2 $GtCO_2$-eq/year (bottom-up) for carbon prices of up to US$100/$tCO_2$-eq. These are large numbers when compared to the baseline estimate of 8.5 $GtCO_2$-eq/year by 2030. If the higher top-down numbers were correct this could lead to 'negative emissions' from the forest sector. Figure 9.7 shows the differences per region. Tropical countries represent about 65% of the total potential.

Among forestry experts there is widespread scepticism about the global forest sector model calculations. Assumptions are seen as too optimistic and specific regional circumstances are not adequately covered in those models. They generally have more trust in bottom-up methods. It is likely however that the bottom-up estimates incorporate barriers to realizing certain measures. Only a small percentage of what would be technically feasible is then incorporated in the estimate. In other words they probably are not giving pure economic potential estimates, but something that may be closer to a market potential (see Box 6.6). Another factor explaining the difference is that bottom-up studies are not covering all options in all regions. The true magnitude of the forestry

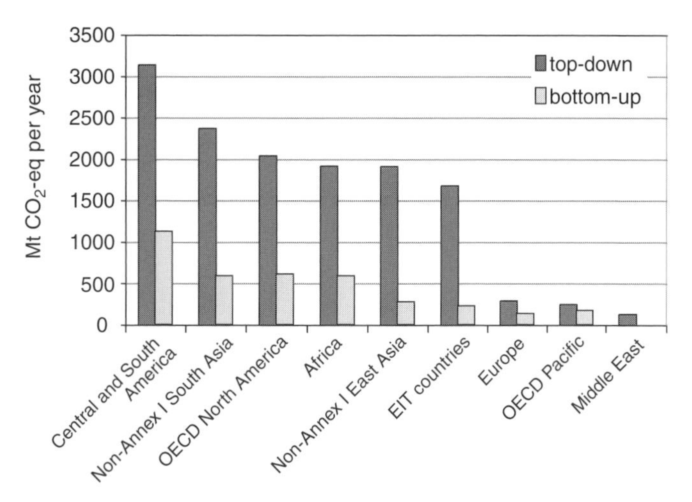

Figure 9.7 **Comparison of estimates for the 2030 economic potential of forest mitigation measures; 'top-down' estimates are from global forest models; 'bottom-up' estimates are from regional assessments.**
Source: IPCC Fourth Assessment Report, Working group III, figure 9.13.

sector mitigation potential is thus probably somewhere in the middle between the bottom-up and the top-down numbers.

Avoiding deforestation

Avoidance of deforestation represents a large share of the total mitigation potential. In South America and Africa it is by far the most important measure. Depending on the specific study, the cost level considered, and the timeframe, the contribution of reduced deforestation ranges from 30% to more than 50%. Studies for the Amazon region show that in the period up to 2050 about 40% of the Amazon forest would be lost without action and that this could be halved by an active forest protection programme, supported by financial incentives. This would avoid 60 GtCO_2 (i.e. more than 1 Gt per year on average)[12]. Costs are estimated to be relatively low compared to mitigation in other sectors: according to some global modelling studies for a carbon price of about US$30/t$CO_2$-eq almost 300 Gt$CO_2$ could be avoided in the period till 2050 (i.e. more than 5 Gt per year on average).

New forests

The biggest potential for planting of new forests can be found in East Asia, the former Soviet republics, Central and South America, and Africa, with North America and Europe also providing substantial potential. It is a matter of availability of land and competition with the economic value of other land use (as affected for instance by agricultural subsidies), as well as land ownership and legal conditions. Estimates for the share of afforestation in the forest sector mitigation potential vary considerably,

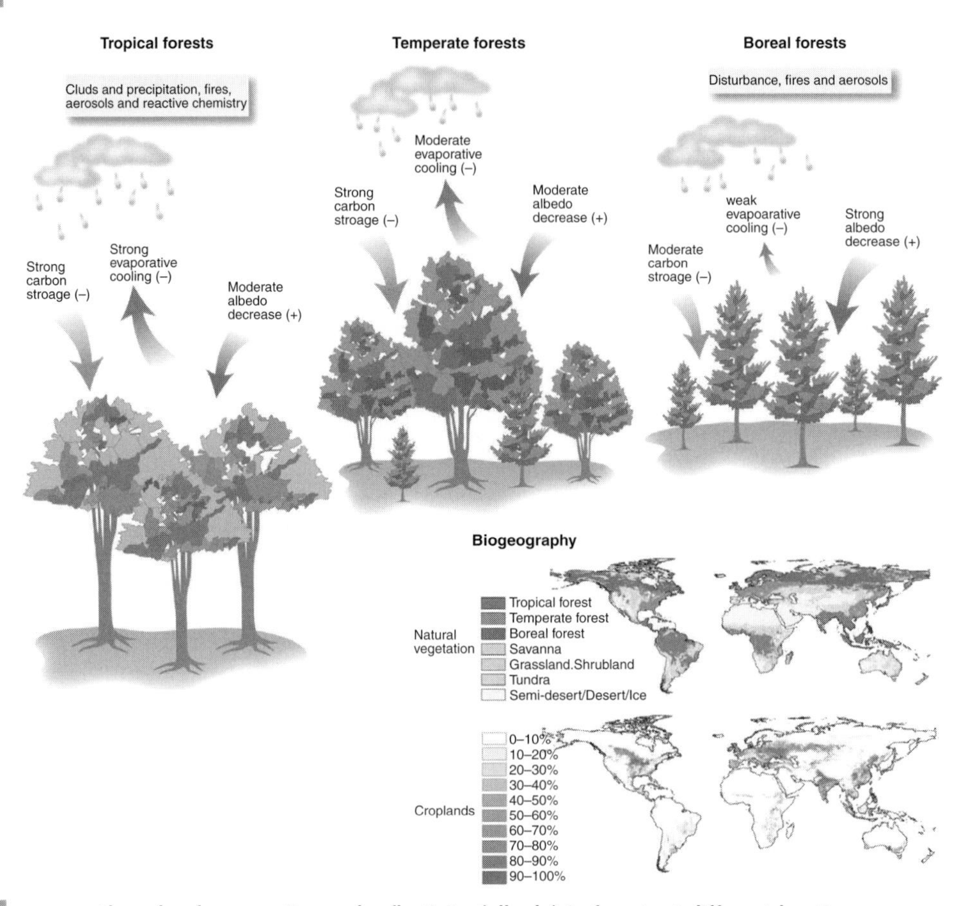

Figure 9.8 **The role of evaporation and reflectivity (albedo) in forests at different locations.**
Source: Bonan,et al. Forests and climate change: Forcings, feedbacks and the climate benefits of forests, Science, vol. 320, June 2008, p.1444. See Plate 15 for colour version.

because studies often combine the potential of new forest with that of better forest management. Russian forests are good for an economic potential in 2030 of about 0.2–0.5 $GtCO_2$/year at costs of up to US\$100/tCO_2 sequestered. In the USA 0.4–0.5 $GtCO_2$/year can be economically sequestered through new forests at costs of up to US\$100/tCO_2.

There are additional effects of planting new forests that can strengthen or weaken the effect of CO_2 sequestration: (1) evaporation of water that cools the air and forms clouds above a forest; and (2) change in reflectivity (albedo) of the land. In tropical forests evaporation is strong and clouds are formed (this is why large tropical forests generate their own rain). This has a cooling effect. In boreal forests the evaporation effect is small. A forest is darker so there is a reduced reflectivity of the land, leading to some additional warming in tropical and temperate climates. In northern areas however this decrease in albedo is much bigger, because trees do not have snow cover in winter, while grasslands do. The net effect of boreal forest planting is therefore much lower than would be expected based only on the carbon fixed (see Figure 9.8).

Forest management

Forest management consists of a range of measures, such as avoiding forest degradation by preventing and controlling fires, pest control, managed logging, avoiding or reducing drainage of plantation soils, thinning to enhance growth rates, nutrient application (partly offset by N_2O emissions), and increased rotation periods of forests. Effectiveness of measures is fully determined by local circumstances. Estimates of the total mitigation potential are therefore aggregates with a limited accuracy. In general the highest potential exists in North America, East Asia, and Russia, where management capacity exists and forest management is still underdeveloped. Global top-down models estimate forest management to be the biggest contributor to the mitigation potential in 2030, followed by afforestation and avoided deforestation.

Wood products

Substitution of steel or concrete by wood in construction can save up to 0.5 tonne of CO_2 per square meter of building floor space over the lifetime of the building. Wooden furniture and houses can keep carbon out of the atmosphere for periods of up to a century or more. Using the wood waste to generate energy does add to the mitigation effect. Every cubic metre of wood stored in the form of wood products keeps about 0.9 tonnes of CO_2 out of the atmosphere. But that is only temporary. When that wood is used in the waste stage to replace fossil fuel it can save 1.1 tonne CO_2 per tonne of wood used. Compared to the huge potential of forest conservation and forest expansion, the mitigation potential of these wood product measures is relatively small.

Overall potential

The overall mitigation potential is summarized in Table 9.5. There is a large uncertainty in the forestry mitigation potentials. Taking the lowest numbers however we can say that

Table 9.5. Total economic mitigation potential for land use and forestry

Contribution	Economic mitigation potential in 2030 ($GtCO_2$-eq/year)			Projected baseline emissions 2030 ($GtCO_2$-eq/year)
	At cost US$20/t	At cost US$50/t	At cost US$100/t	
Agriculture	1.6	2.7	2.3–6.4	8.3
Forestry	1.1–5.7	1.9–9.5	2.7–13.7	8.5
Wood products	Very small	Very small	Very small	
Total	2.7–7.3	4.6–12.2	5.0–20.1	16.8

Source: IPCC Fourth Assessment Report, Working Group III, chapters 8 and 9.

in agriculture and forestry about 30% of the projected emissions by 2030 can be reduced at costs lower than US$100/t$CO_2$-eq.

What can bioenergy contribute?

The mitigation effect of modern bioenergy[13] is realized mainly in the energy supply and transport sector. That is where the replacement of fossil fuel emissions happens. Chapters 5 and 6 discuss this in detail. The supply of biomass however comes mostly from the agriculture and forestry sector, except for some waste from households and industrial processes. The big question therefore is how much biomass can be supplied in a sustainable manner, so that food security, biodiversity protection, and water supply are not threatened. The other main question is what is the net carbon gain after subtracting the energy and emissions created by planting, managing, harvesting, transporting, and processing the biomass? The supply issue will be discussed here, but the energy balance question is discussed in Chapters 5 and 6.

As Figure 5.16 shows, the main sources of biomass are crop residues, energy crops, animal waste, and wood processing and paper making waste. Data for the biomass energy that can be produced are scarce. Indicative data are available only for 2050. Table 9.6 shows a total of 125–760 EJ/year (same order of magnitude as the total global energy use in the year 2005) and gives indicative numbers for the various sources. Since agriculture can respond quickly, these supply rates could in principle also be delivered in 2030. Crop productivity increases (of the energy crops and of the food crops that determine the availability of surplus land) would however be lower, so that the total sustainable supply for 2030 is somewhat lower than for 2050.

Comparing these supply data with demand estimates for 2030, in Chapters 5 and 6 it was concluded that biomass supply is not the limiting factor for the use of bioenergy. Demand for bioenergy is the limiting factor, caused by the relatively high cost of bioenergy compared to other alternatives.

Some doubts have been raised recently about the validity of this conclusion as food prices have increased sharply (see Figure 9.9). Are bioenergy crops causing these food price increases? Let's have a closer look.About one-third of the US maize production and more than half of the EU rapeseed production is now being converted to biofuels[14]. This suggests a considerable influence on food commodity prices. However, worldwide only 5% of oilseeds and 4.5% of cereals are used for biofuel production. Estimates of the contribution of biofuels to food price increases vary enormously. A contribution from biofuels to recent grain price increases of 30% is seen by most analyses as the maximum on average, with maize contributing much more than wheat[15]. This does not mean however that there is a structural scarcity of land for food production. There is general consensus that poor harvests, export bans, and neglected agriculture in many countries contributed strongly to the recent price increases. The productivity especially can be influenced by agricultural policy. There does not seem to be good reason to doubt the potential for bioenergy supply in the future as discussed above.

Table 9.6. Biomass supply estimates for 2050

Sector	Source	Potential supply 2050 (EJ/year)[25]
Agriculture	Crop residues	15–70
	Dung	5–55
	Energy crops	20–400
	Energy crop degraded lands	60–150
Forestry	Forest residues	12–74
Waste management	Organic waste	13
Industry	Process residues	n/a
Total		125–760

Source: IPCC Fourth Assessment Report, Working Group III, table 11.2.

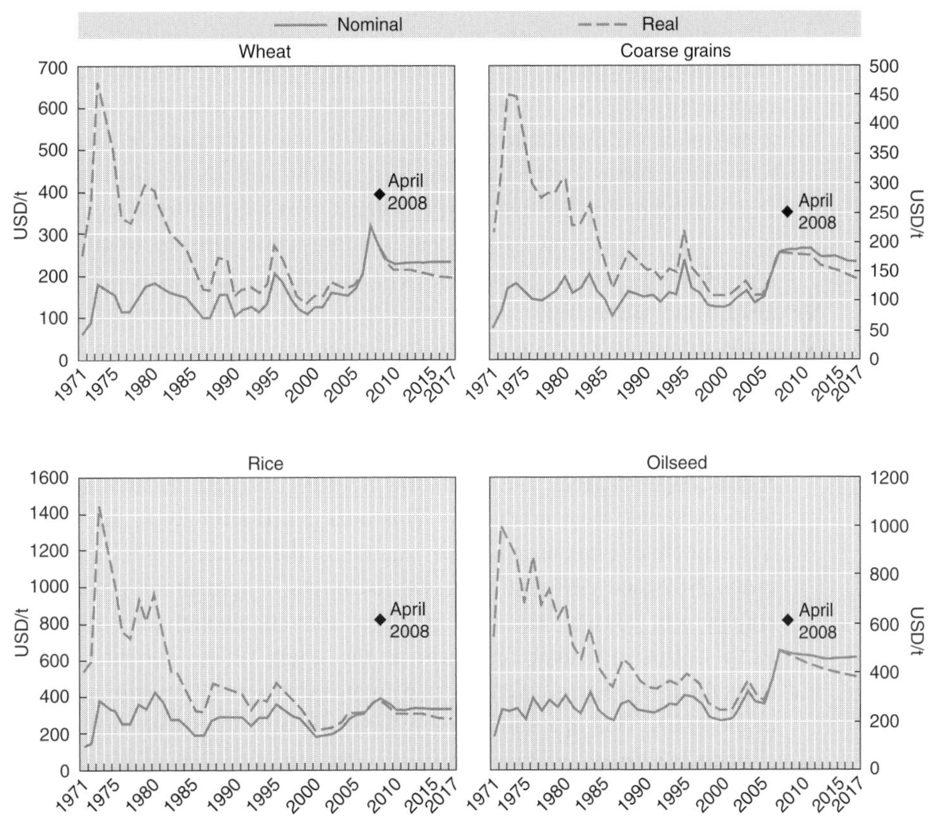

Figure 9.9 Food prices 1971–2007, April 2008, and projection till 2017. Both the nominal as well as the real (= corrected for inflation) numbers are given.
Source: OECD/FAO Agricultural Outlook, 2008–2017.

What policies are available?

Agriculture and forestry are heavily regulated: in agriculture, because food security (the guaranteed supply of adequate food) is generally seen as politically very important; and in forestry, because forests are a common good, often located on public land. This has led to a variety of regulations, price controls, subsidies, and other policy actions. This high policy density has important implications for ways of reducing greenhouse gas emissions from these sectors.

Agriculture

In agriculture, price signals are the primary factor that influences agricultural practices. And these price signals do not only come from the markets. In agriculture, subsidies play a dominant role: subsidies for production of crops or animal products, subsidies for export of agricultural products, subsidies for food processors to keep basic food affordable for poor people, and subsidies to 'set-aside' land for reasons of price control or erosion prevention (e.g. the Conservation Reserve Program (CRP) in the USA was introduced in 1985). There are non-price policies as well, such as quota systems (putting a maximum on production) and 'set-aside' rules, meaning requirements for farmers to leave a percentage of the land idle. Soil fertility policies are in place in some countries to combat and prevent soil erosion (such as China's Grain for Green programme, initiated in 1999 by the central government to address concerns about erosion, water retention, and flooding[16]), ecological policies to maintain or rebuild hedges and wooded strips or keep water tables high, and water quality policies that limit fertilizer application. Air quality policies have led to bans on burning of crop residues and grasslands in the EU and South Africa.

In addition, agriculture is very sensitive to macro-economic policy changes. When currencies were devaluated in South America in the 1970s and exports were promoted to restore trade balances, the result was a strong increase in large scale mechanized crop production and meat production. This contributed strongly to the massive deforestation. The economic restructuring of the countries of the former Soviet Union and Eastern Europe in the 1980s and 1990s led to drastic reductions in agricultural production. Oil import and employment considerations led Brazil to start its alcohol from sugar cane programmes in the 1970s[17]. Similar forces are at the origin of the current US ethanol from corn programmes (subsidy driven). Subsidy removal in agriculture in Australia and New Zealand in the 1980s[18] led to a substantial reduction in agricultural production.

Specific climate policies aiming at reduction of N_2O and CH_4 emissions are basically non-existent. And it would also be ineffective to add a set of new policies to the vast array of existing policies. By far the best approach is to change existing policies to create the right incentives to reduce greenhouse gas emissions, but with one exception: application of international Clean Development Mechanism policies that would generate funding for specific management changes leading to reduced emissions (see

Chapter 12). Use of domestic greenhouse gas emission trading programmes could also be considered[19].

What are the most promising policy changes in agriculture? Since most of the reduction potential in agriculture is in the form of soil carbon enhancement, we have to look for policies that can effectively promote that. Strong candidates are[20]:

- Banning burning of crop residues and grasslands as has already been implemented in China, South Africa, and the EU. They have benefits for air quality improvement. Since farmers do the burning in the belief that it releases nutrients more quickly, information programmes and other support may be needed to help farmers comply with such bans
- Set-aside policies as practised in the USA and EU: they have additional advantages for improving the ecological conditions in rural areas. With current high food prices there is a tendency however to abandon them (as the EU is currently considering[21])
- Soil fertility policies in the form of promoting reduced/zero tillage (practised in Brazil, Argentina, Uruguay, and Paraguay)
- Banning the dewatering of organic (peat) soils and restricting the use of such soils (no ploughing, no tuber crops)
- Subsidies for raising the water table in organic soils, to compensate for loss of productivity, as practised in parts of the Netherlands
- Mandatory land restoration of degraded lands, such as through China's Land Reclamation regulation of 1988
- Acquisition by State or private organizations of agricultural lands for nature conservation purposes and managing those lands as protected areas (as done in China and many other countries for wildlife or water quality management)
- Agricultural research and outreach to inform farmers about better farming practices

Promising policies to limit CH_4 and N_2O emissions include:

- Regulations on mandatory storage of manure (in feedlot operations) and subsidies for biogas installations (The Netherlands)
- Information programmes on vegetarian diets and alternatives for meat
- Subsidizing or regulating reduced fertilizer application in ecologically sensitive areas
- Air quality regulations controlling nitrogen oxides and ammonia from agriculture for reasons of air quality improvement (UN Convention on Long-Range Transport of Air Pollution)

Forestry

The role of price signals in forestry is even stronger than in agriculture. It is very profitable to convert forest into crop or grazing land, because the financial returns on the land can increase more than a hundred times when turning a forest into an oil palm plantation (see Figure 9.10). Implementation and enforcement of regulations on deforestation have been weak, in some countries because of corruption amongst officials. Controlling deforestation on private lands is difficult in many countries.

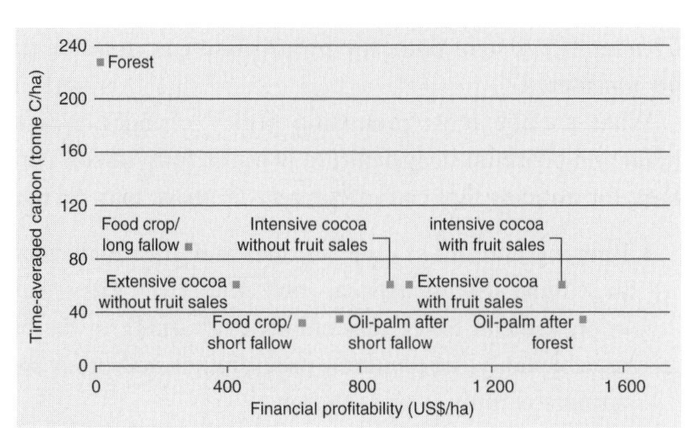

Figure 9.10 **Profitability of changing forest into agricultural use and loss of carbon (tonne C/ha); case of Cameroon.**

Source: FAO, The State of Food and Agriculture: paying farmers for environmental services, 2007.

International certification schemes for sustainably produced wood are still fragmented, strictly voluntary, and only affect a small percentage of the trade in wood. That is why most policies to reduce deforestation so far have been ineffective.

The general consensus is that stronger financial incentives than currently available will be able to reduce deforestation. The idea is that payment for maintaining a forest is justifiable because that forest provides environmental services in the form of acting as a carbon sink, keeping an amount of carbon out of the atmosphere, and preserving biological diversity as well as providing clean water. Costa Rica for instance introduced an environmental service payment system in 1997 (see Box 9.2). On the other hand moral objections are raised against these payments, because forest preservation is in the self interest of societies, given the important services they provide, and because forest preservation is often mandated by national and international law. The joint government–local community programmes for forest protection in India have been built around this principle (see Boxes 9.2 and 9.3).

Box 9.2 **Forest protection in Costa Rica**

Costa Rica is one of the few countries in Latin America to promote reforestation through incentives such as tax credits, direct payments, and subsidized loans that have benefited landowners, large and small. Among the important steps Costa Rica has taken are the following:

- The Natural Resources Administration has merged the administration of forest and protected area activities into one unified organization
- It has successfully developed a National System of Protected Areas that has a minimum of infrastructure and an institutional presence in each region of the country
- The National Forest Fund was established to handle financial issues for forests and natural resources

- Important legislation has been passed to protect the nation's forests, including the Environment Law, the Biodiversity Law, and the Forest Law
- The 'polluters pay' principle was introduced through the establishment of a tax on fossil fuels to pay for environmental services
- Many efforts have been made to protect biodiversity and generate income from it
- The Costa Rican Office of Joint Implementation was established to trade carbon emissions in the international market and Carbon Tradable Offset Certificates were developed that could serve as a model for trading other environmental services
- The government instituted a national system to certify good forest management practices
- Costa Rican forest owners have strong organizations that give them technical support for reforestation, forest management, and forest conservation. In recognition of this, Costa Rica has delegated much responsibility for forest management and conservation to private landowners

(Source: http://lnweb18.worldbank.org/oed/oeddoclib.nsf/DocUNIDViewForJavaSearch/A25EFCF3220878D585256970007AC9EE)

Box 9.3 Forest protection in India

Joint Forest Management (JFM) is now a principal forest management strategy in India. In June 1990 the government issued a resolution that made it possible for state forest departments to formally involve people in forest management through JFM1. In return for providing improved forest protection, communities receive better access to non-timber subsistence forest products and a share of net commercial timber revenues. The state retains most of the control and decision making over forest management, regulation, monitoring, timber harvesting, and forest product marketing. The government views JFM as a pivotal strategy for addressing the national policy goal of achieving 33% forest cover by 2012 (22% in 2005). The main focus of JFM in India is forest protection and conservation.

(Source: http://siteresources.worldbank.org/INDIAEXTN/Resources/Reports-Publications/366387-1143196617295/Chapter-1.pdf)

Increased forest planting (afforestation) has been reasonably successful in several regions, particularly in China, Korea, and parts of Europe. In total 5.6Mha of forest were added worldwide per year in the period 2000–2005. Government policies have been the key factor. Successful policies often serve the purpose of combating erosion (e.g. the Chinese forest planting programme in desert prone areas), or producing wood for local communities (the Indian Joint Forest Management strategy). To overcome the barrier of high upfront investment in tree planting for private land owners, governments often use investment subsidies on planting or tax deductions on investments as the primary policy instrument. In areas where demand for food is high, such afforestation programmes can only work if agricultural productivity goes up. Appropriate agricultural policies therefore are a necessary condition for successful afforestation.

Forest management to increase carbon stocks is a complex issue, where policy approaches are generally very location specific. For public forest lands management is usually entrusted to a State agency and changing the management practices is then a matter of government instructions to such agencies. On private lands policy instruments have limited effect. Capacity building programmes to educate forest managers are often used in such circumstances. Some countries, such as Costa Rica, have had success with paying forest managers for improved (carbon) management (see Box 9.2). In general, better forest management has limited potential in industrialized countries, where forest management is already quite intensive. In developing countries the potential is much bigger, but there the capacity for better management is not available[22].

Kyoto Protocol policy instruments

The Kyoto Protocol contains several possibilities to create incentives for emission reductions in agriculture and forestry. The most important is the Clean Development Mechanism (CDM, see also Chapter 12). It allows countries with emission limits under the Protocol (the so-called Annex I industrialized countries) to invest in projects in developing countries that reduce emissions and contribute to sustainable development. The reductions realized can then be deducted from their own emissions. Under the current rules, projects on reducing methane from animal waste and afforestation and reforestation projects are eligible under the CDM. The effect is that there is a bonus of about US\$10/tCO_2-eq avoided for such reduction projects in developing countries, which makes some projects financially attractive. So far only limited use has been made of the CDM in the agriculture and forestry sector. The main reason is the complex procedures to get approval for CDM projects and the exclusion of many agriculture and forestry measures from the CDM. Agricultural soil carbon enhancement, the biggest mitigation option in agriculture, is for instance not accepted so far. Measurement problems for N_2O from fertilized soils make it very hard to include this type of mitigation measures in the CDM.

New international policy instruments

Intense discussions are being held on the possibilities to create a new instrument (usually called REDD = Reducing Emissions from Deforestation and Forest Degradation[23]) to stimulate forest conservation under the Kyoto Protocol or its successor agreement for the period after 2012. The main reason is that avoidance of deforestation has such a big potential to reduce CO_2 emissions. It requires overcoming a range of difficulties that have prevented the inclusion of avoided deforestation in the current Kyoto Protocol. The difficult questions are:

- *How to determine a baseline of deforestation in a country*? Is it reasonable to assume continuation of current deforestation rates and give countries credit for any slowdown of deforestation? Or are there good reasons to say that it is the responsibility of the country to reduce the deforestation rate or even to stop deforestation altogether? When applied to forest degradation, determining a baseline is even more complicated (see Box 9.4).
- *How to avoid leakage*? Leakage refers to the phenomenon that deforestation is reduced in a certain region of a country, while elsewhere in that same country deforestation increases. Even if deforestation is reduced overall in a country, then how can we avoid its increase in another country?
- *Will the forest remain in the future*? If credit is given for retaining the carbon stock in a forest, how can we guarantee that it is not disappearing in the future (with all of the carbon still ending up in the atmosphere)?
- *How to measure the carbon stocks maintained?* As discussed above, carbon stocks are to a large extent underground. In addition carbon stocks in forests vary considerably from place to place. And when we look at forest degradation, measuring carbon stocks is even more difficult.
- *How to monitor implementation?* In light of the risk of leakage do we need to accurately monitor forests country wide, or even in the whole world? Are the current forest monitoring systems, as used by the FAO for its regular forest assessments, adequate or are new and more precise methods needed?
- *How to create financial incentives?* There are in principle various ways to operationalize an REDD system. It could be coupled to the existing carbon market, i.e. credits from avoided deforestation or degradation could be sold internationally to countries or companies that are subject to emission limits and trading systems. This could be done on a project by project basis as in the CDM; it could even be integrated in the CDM. To reduce the risk of leakage it could also be done on a country wide basis.

Box 9.4 Crediting for reduced deforestation

In its simplest form a baseline is established of emissions due to deforestation in a base-period. Then any reduction over a certain commitment period compared to the baseline is credited to the country (see figure). More sophisticated approaches could take a declining baseline or an increasing domestic share of avoided deforestation.

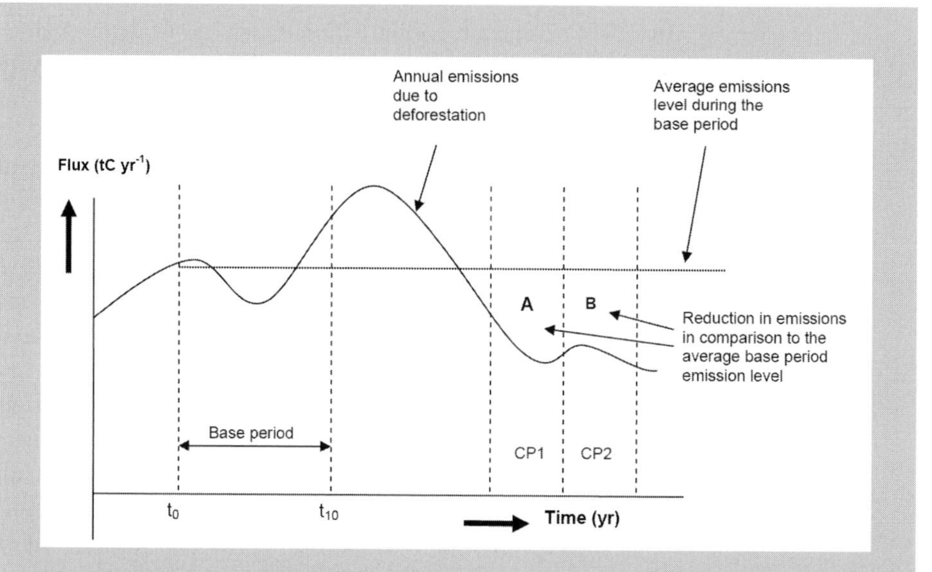

Solid line indicates annual emission levels due to deforestation. The dotted horizontal line is the average emissions during the base period. Area A is the reduction in emissions during the first commitment period below the base period's emission level. Area B is the same but in the second commitment period, if there was to be one.

Source: Trines E et al. Integrating agriculture, forestry and other land use in future climate regimes: Methodological issues and policy options. Report 500102 002, Netherlands Environment Assessment Agency, October 2006.

Interaction with adaptation and sustainable development

Mitigation action in the agriculture and forestry sector can help reduce the vulnerability to climate change. Increasing soil carbon in agricultural soils will make them more drought resistant. The same applies to forest conservation. Large forest areas, especially in the tropics, create their own climate and rainfall through evaporation and cloud formation. Forest conservation and expansion also helps protect biodiversity, which will be under stress when the climate changes. Forest fires and insect plagues, which are likely to increase in a changing climate, can be countered with better forest management. Bioenergy crops can provide farmers with an additional source of income that can help them compensate reduced incomes from other crops when yields go down due to higher temperatures and more irregular rainfall.

It depends however on the way these mitigation actions are performed. Forest plantations that replace a primary forest always lead to loss of biodiversity[24]. More intensive management of forests will also harm biodiversity by disturbance and the effect of fertilizer. Bioenergy crops could harm biodiversity if grown on former grasslands or marginal unused lands.

So what does this mean for the role of agriculture and forestry?

A few key points stand out. First, in agriculture and forestry the stocks of carbon in soil and vegetation are more important than the emissions from activities. This plays a role in emissions caused by deforestation, but also in measures to absorb CO_2 from the atmosphere in soils and vegetation.

There is uncertainty about emissions from agriculture and forestry, particularly from the forest sector. You can find quite different numbers in various publications. The contribution by peatlands is not always included and is very uncertain. Agriculture emissions of CH_4 and N_2O are also much more uncertain than CO_2 emissions from energy use. Emissions from agriculture and forestry are large however.

The uncertainty of the economic mitigation potential is also large. Estimates for the forestry sector vary by almost a factor of 10. Again, they are very large, and if the optimistic estimates are correct, this could create large negative emissions in the forestry sector. For agriculture the biggest contribution comes from enhancing the soil carbon content.

In terms of policies needed to capture the mitigation potential, this sector is quite different from others. Given the presence of extensive existing regulations, it is more important to adjust existing (non-climate) regulations than to invent new (climate) policies. There is one important exception though: additional financial incentives in forestry can make a difference in terms of reducing deforestation, forest planting, and better forest management. And these financial incentives will have to come from new international climate change mechanisms. Actions to reduce emissions from agriculture and forestry go hand in hand with adapting to a changed climate.

Notes

1. FAO, http://www.fao.org/newsroom/en/news/2008/1000874/index.html.
2. In developed countries 56% of protein is obtained from animal products; in developing countries this is about 30%. With increasing incomes in developing countries it is expected they will move towards the dietary pattern of developed countries; see IPCC WG III, table 8.2.
3. See Brown LR. Outgrowing the Earth: The Food Security Challenge in an Age of Falling Water Tables and Rising Temperatures, W.W. Norton & Co., NY, 2005.
4. IPCC Fourth Assessment Report, Working Group III, ch 8.2.
5. IPCC Fourth Assessment Report, Working Group III, ch 9.2.1.
6. $1 m^3$ of wood contains on average 0.92 tCO_2.
7. Emissions from farm machinery and trucking are covered under the transportation sector.
8. International Assessment of Agricultural Knowledge, Science and Technology for Development (IAASTD), Global Summary for Decision Makers, 2007.
9. IPCC Fourth Assessment Report, Working Group III, ch 9.3.

10. See Box 6.6 for definition of economic potential.
11. If there is a carbon price as a result of policy to limit greenhouse gas emissions, forest sector managers have an incentive to manage their forests better, plant new forests, or to avoid deforestation. This works through a system of tradable emission permits. Industrial installations can then opt to buy permits from foresters that avoid emissions or fix CO_2 in trees, instead of taking their own measures (if forest measures are cheaper).
12. IPCC Fourth Assessment Report, Working Group III, box 9.1.
13. Traditional biomass is not considered to be a mitigation option, because it is often unsustainably harvested. Its use is declining as incomes of people in developing countries grow. Its use is still large however providing about 7.5% of total primary energy (see Chapter 5). Improving the efficiency of wood stoves however is a mitigation option discussed in Chapter 7 on the buildings sector.
14. See USDA Foreign Agricultural Service, EU-27 Oilseeds and Products, Annual Report 2008, http://www.ebb-eu.org/stats.php, and http://www.esru.strath.ac.uk/EandE/Web_sites/02–03/biofuels/quant_biodiesel.htm.
15. See Banse M, Nowicki P, van Meijl H. Why are current world food prices so high? LEI Wageningen UR, Report 2008–040, Wageningen 2008.
16. FAO, State of Food and Agriculture, 2007.
17. See Chapter 4, Box 4.2.
18. http://www.newfarm.org/features/0303/newzealand_subsidies.shtml.
19. New Zealand is planning to include agriculture in its emission trading system, see Kerr S, Ward M. 'Emissions Trading in New Zealand: Introduction and Context, 2007, see http://www.ecoclimate.org.nz/ETS.htm.
20. IPCC Fourth Assessment Report, Working Group III, ch 8.6.
21. See http://ec.europa.eu/news/agriculture/080520_2_En.htm.
22. IPCC Fourth Assessment Report, Working Group III, ch 9.6.3.
23. http://unfccc.int/methods_science/redd/items/4531.php.
24. See Stokstad E. ScienceNews, Vol 320, 2008, pp 1436–1438.
25. 1 Exajoule (EJ) = 10^{12} Joule; see chapter 5, Box 5.1.

10 How does it fit together?

What is covered in this chapter?

This chapter discusses the overall mitigation potential at various cost levels, the contributions of the sectors, and the question of where the potential is located. It concludes that the mitigation potential is big enough to bring global emissions back to current levels by 2030. With this potential stabilization of atmospheric concentrations of greenhouse gases at levels of 450ppm CO_2-eq is still within reach, provided that the potential in developing countries is also tapped. Geo-engineering is not needed, and that is comforting, because the risks and uncertainties of such planetary experiments are huge. In all countries substantial potential exists with so-called negative costs (i.e. where investment is profitable), but costs differ as a result of strong differences in national circumstances. Cost for the economy as a whole is limited when cheap options are implemented first. On average annual economic growth rates will not be reduced by more than a few tenths of a percentage point, and that is without taking co-benefits for energy security, health, and employment into account. Investments required for implementing the reduction options will have to shift strongly to efficiency of energy use and low carbon energy, and are bigger than those without stringent climate policy, but are compensated by much lower energy costs. Implementing low carbon technology rapidly in developing countries is crucial to controlling climate change. National priorities in developing countries for modernization, energy security, and trade are the main drivers. Governments in the North and South should remove obstacles and create the right conditions.

Adding up the sector reduction potentials

After looking into the economic reduction potentials for the various sectors in Chapters 5, 6, 7, 8, and 9, it is time to discuss the total reduction potential for the world as a whole. This then should be compared with the reduction needs for the various stabilization levels identified in Chapter 3 to see if low level stabilization is possible.

Adding up sector potentials sounds simple. There are some complications however. When evaluating the sector potentials, reductions in electricity use and heat from power plants and district heating installations were included. This affects the demand for electricity and heat that was assumed in Chapter 5 in estimating the reduction potential for the energy supply sector. To avoid double counting, this needs to be reconciled. The easiest way to do that is to recalculate the energy supply reduction potential for the reduced demand after subtracting the demand reductions from the various energy end-use sectors. In doing so, the reduction potential in the energy supply sector becomes 2.4–4.7 $GtCO_2$-eq/year by 2030, for costs up to US\$100/ tCO_2-eq. Without this correction the numbers were almost twice as high: 4.0–7.2 $GtCO_2$-eq/year[1].

There are other complications. Baseline assumptions in the various sectors are not exactly the same, because available studies differ. And since reduction potentials are sensitive to the baseline assumed, adding up sector potentials introduces additional uncertainty.

The other major problem is the lack of numbers on the potential of some reduction opportunities[2]:

- Fluorinated gases from energy supply, transport, and buildings. There are only a few numbers for 2015: about 0.4 $GtCO_2$-eq/yr for HFCs at costs ranging from negative to above US\$100/tCO_2-eq[3]
- The potential of Combined Heat and Power in the energy supply sector is uncertain and probably about 0.2–0.4 $GtCO_2$-eq
- Methane from gas pipelines and coal mining in the energy supply sector. Estimates for methane reduction from coal mining for 2020 are 0.2–0.4 $GtCO_2$-eq/year
- Freight transport
- Public transport, urban planning, change of transport mode, and speed limits
- More advanced opportunities in buildings
- Energy efficiency in the non-energy intensive industries
- Reduced use and replacement of energy intensive materials

This means the numbers given are underestimating the reduction potential by at least 10–15%.

Finally, energy prices do have an impact on the calculation of economic reduction potentials. Available data on reduction potentials are usually calculated with oil and gas prices much lower than today. The transport sector is the most sensitive because of the importance of oil. Generally speaking economic reduction potential would be higher if oil prices remain high for a long time. In other sectors, where mostly coal and gas are used, the influence is smaller.

Further it is important to remember that the calculation of economic reduction potential uses 'social costs', i.e. longer payback times as used in public sector investments (5–30 years). They calculate what is economically rational for societies as a whole. That is different from the way private sector decision makers look at profitability of investments. They use much shorter payback times. See Box 6.6 for definitions of mitigation potential.

Figure 10.1 shows the results as they emerge from Chapters 5, 6, 7, 8, and 9. Note the large uncertainty ranges. Economic reduction potentials are given for different cost

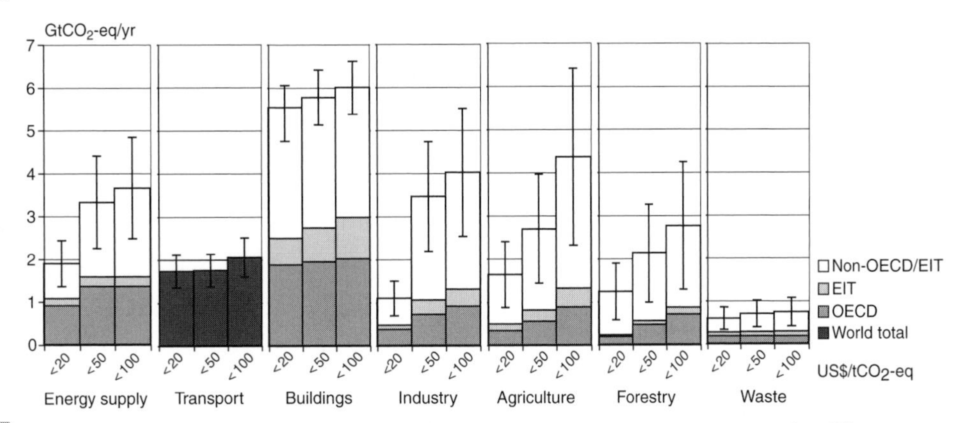

Figure 10.1 **Global economic mitigation potential for the most important economic sectors, for different cost categories and geographical regions. Note: industry and waste have been grouped together in Chapter 8 and are shown separately here; the numbers shown here for forestry are at the low end of the range listed in Chapter 9.**

Source: IPCC Fourth Assessment Report Working group III, fig SPM.4.

levels and for different categories of countries. What is striking is the large share of developing countries in the reduction potential. This is consistent with the general knowledge about low efficiencies of energy use and lack of capital to invest in modern installations. Overall more than 50% of the potential is found in developing countries[4].

Since potentials for measures with negative costs are not available for all sectors, there is just one category of costs up to US\$20/tCO_2-eq in Figure 10.1. However, about 6GtCO_2-eq in total is available at zero or negative costs in 2030.

Reliable quantitative estimates of the potential from behavioural change are not available. They are of course real, but small compared to the potential from technical options.

A global cost curve?

Ideally all sectoral reduction options are grouped into one integrated abatement cost curve. In its latest assessment report the IPCC did not produce such a cost curve, because published data did not allow it. However, McKinsey and Company, in collaboration with Vattenfall and others, making use of their extensive set of private industrial data, did produce such a global abatement cost curve recently[5]. Figure 10.2 shows a simplified version, where only a limited set of reduction options is highlighted (see Box 10.1 for an explanation of how to read such an abatement cost curve). The total reduction potential at costs $<$ € 60/tCO_2-eq (roughly equal to US\$100/t) is about 37GtCO_2-eq/year in 2030. This is considerably higher than the range found in the latest IPCC report (16–31GtCO_2-eq for costs $<$ US\$100/tCO_2-eq).

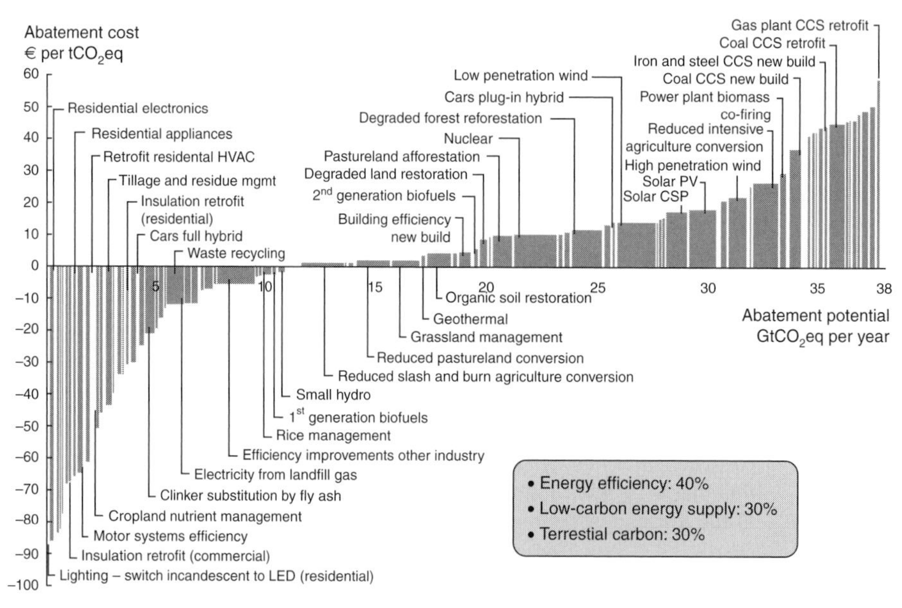

Figure 10.2 **Global greenhouse gas abatement cost curve.**

Source: McKinsey and Company, 2009.

Box 10.1 How to read an abatement cost curve?

The abatement cost curve describes two numbers:

1. The potential to reduce CO_2 equivalent (CO_2-eq) emissions. For example, a global infrastructure to use cellulose ethanol as a fuel would reduce CO_2-eq emissions by almost a billion metric tonnes per year in 2030, compared with continuing to use fossil fuels.
2. How much that measure costs for every tonne of CO_2-eq emissions it saves. For example, the abatement cost for cellulose ethanol is calculated by dividing the costs of building and operating a cellulose ethanol infrastructure by the number of tonnes of CO_2-eq it saves compared with the current fuel mix.

The first number, the abatement potential, is plotted on the horizontal axis, and the second number, the cost, on the vertical. The measures have been arranged in order of cost, with the cheapest on the left, and the most expensive on the right. Only measures with an estimated cost of less than € 60/tCO_2-eq are included in the analysis. This is not to make any forecasts about what a potential future carbon price should be, but rather a reflection that a cut has to be made at some price and that it is increasingly difficult to calculate the costs of technologies the further they are from being commercial today.

How does this compare with global top-down studies?

Complex integrated models, describing the whole economy and the climate system, are being used as well to estimate economic mitigation potential (we will call them 'top-down' models; see Chapter 3). They have of course much less detail about the specific elements of sector activities and about mitigation technologies. On the other hand they usually have something that bottom-up analyses lack: an integration of all activities into the overall economy. The advantage is that supply and demand of energy are by

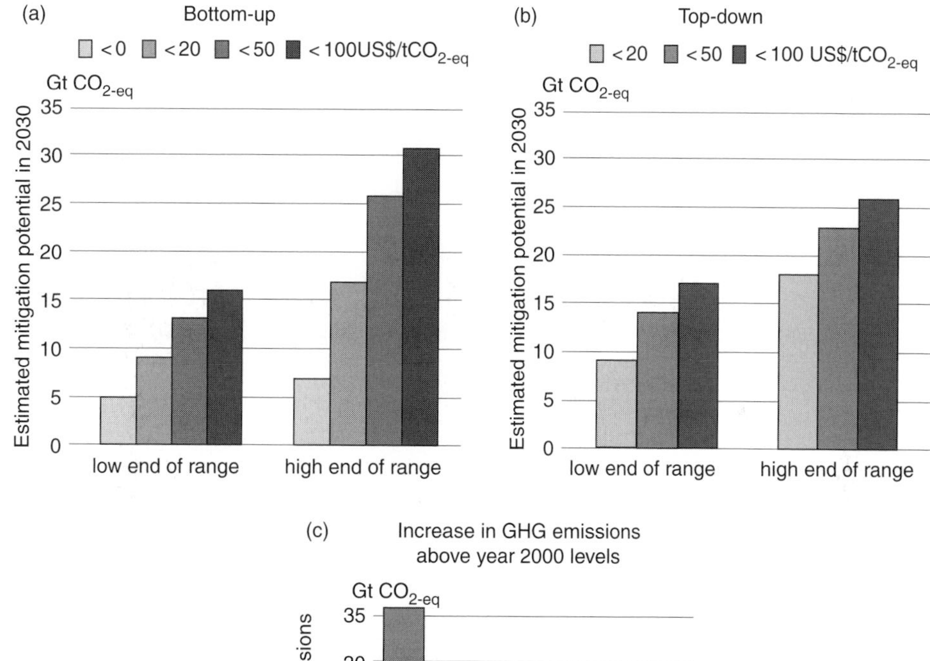

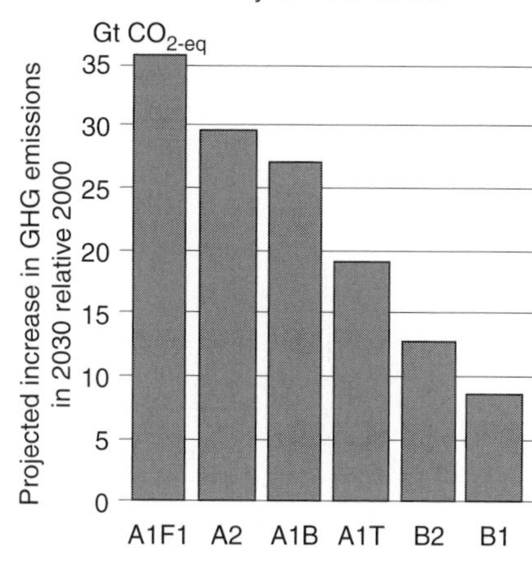

Figure 10.3 **Bottom-up and top-down estimates for the global reduction potential in relation to the estimated emissions increases in the baselines.**

Source: IPCC, Fourth Assessment Report, Synthesis Report, figure SPM.9.

definition the same (in bottom-up studies corrections are needed; see above). And energy prices in these models are the result of demand and supply and automatically adjusted over time (bottom-up studies normally have to assume a certain energy price).

So it is interesting to compare the economic reduction potential estimates from these two different approaches. Figure 10.3 shows they are roughly of the same order of magnitude. That is somewhat of a surprise, because for a long time top-down models used to give much lower estimates than bottom-up assessments. One important explanation was that top-down models assume that no negative cost reduction options exist.

Figure 10.3 also shows that for cost levels up to US\$100/tCO_2-eq the reduction potential is enough to have a good chance of fully compensating the projected growth in the baseline (the bars on the right). In case baseline growth is not that strong (the B1, B2, or A1T scenarios), emissions could even be brought back to below 2000 levels by 2030 at costs up to US\$100/t CO_2-eq.

The consistency of the top-down and bottom-up results only holds for economy wide estimates. At sector level there are large discrepancies. A major reason is the difference in sector definitions between the top-down and bottom-up calculations. Other explanations are the partial coverage of the energy supply sector in bottom-up estimates and their better coverage of the buildings and agriculture sectors[6].

How far do we get with these reduction potentials?

The big question is of course how far we get towards stabilization of concentrations in the atmosphere with these reduction potentials. As discussed in Chapter 3, emissions

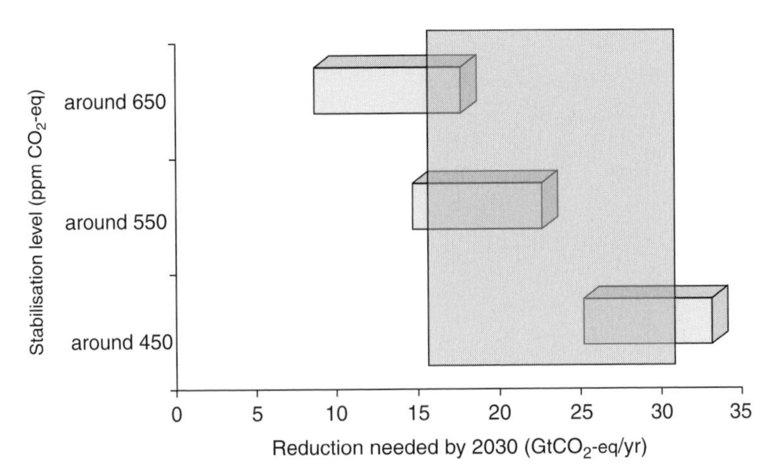

Figure 10.4 Comparison of available CO_2-eq reduction potential for 2030 (shown as transparent box for costs up to US\$100/tCO_2-eq with minimum and maximum) and CO_2-eq emission reductions needed by 2030 (compared to 2000) for stabilization at various levels expressed in ppm CO_2-eq (range due to different possible baseline emissions).

Source: IPCC Fourth Assessment Report, Working Group III, table 11.3 and van Vuuren et al. Climatic Change, vol 81(2), March 2007, pp 119–159.

Table 10.1. Comparison of reductions needed for different stabilization levels and economic mitigation potential available

Stabilization level (ppm CO_2-eq)	Global Mean temp. increase at equilibrium (¤C)	Estimated CO_2-eq reduction needed by 2030 compared to baseline[a]	Annex I 2030 CO_2-eq mitigation potential (bottom-up)		Global 2030 CO_2-eq mitigation potential (bottom-up)	
			<US$50/t	<US$100/t	<US$50/t	<US$100/t
445–490 (around 450)	2.0–2.4	25–33	6–9	7–11	12–25	16–31
535–590 (around 550)	2.8–3.2	14–22	6–9	7–11	12–25	16–31
590–710 (around 650)	3.2–4.0	8–17	6–9	7–11	12–25	16–31

[a] Taken from van Vuuren et al for two different baselines (IPCC SRES B2 and A1B).

Source: IPCC Fourth Assessment Report, Working Group III, table 11.3 and van Vuuren et al. Climatic Change, vol 81(2), March 2007, pp 119–159.

trajectories towards stabilization have to peak and then go down steeply. The lower the stabilization level, the earlier this peaking has to occur and the earlier deep reductions need to be reached. Table 3.1 shows the time frames for different stabilization levels. We can combine those data with the reduction potentials for 2030 for costs up to US\$100/tCO_2-eq, shown above. The available reduction potential is underestimated as a result of lack of information and the effect of behavioural change. It can be concluded that emission reduction potentials at costs <US\$100/t are probably sufficient to reach the lowest stabilization level, except where baseline emission growth is very strong. In that case reduction options with costs higher than US\$100/tCO_2-eq need to be added.

What is also relevant is to look at the regional contributions. Table 10.1 shows that industrialized countries alone (OECD countries and countries with economies in transition) cannot deliver enough reductions to stabilize at any level below about 700 ppm CO_2-eq. To achieve the lowest stabilization level, tapping the whole global reduction potential up to US\$100/t is essential.

Do we need to look at geo-engineering options as well?

In discussions about the need for deep emission reductions suggestions have been made that the regular reduction measures, as discussed in Chapters 5, 6, 7, 8, and 9 and summarized above, will not be enough. The reason given for these claims is that the economic potential simply will not be tapped due to lack of political will and resistance from vested interests. This then leads people to propose large scale interventions in the solar radiation that reaches the earth or in the functioning of the planetary carbon cycle. These proposals are usually referred to as 'geo-engineering'[7].

The *first category*, reducing the net solar radiation that reaches the earth, covers for instance:

- Distributing large amounts of fine particles (such as soot or sulphur), metal strips, or other reflecting materials in the upper atmosphere of the earth. This would reduce incoming solar radiation. Such particles would have to be replenished because they would only have a lifetime of several years.
- Installing a kind of mirror in space, at a point that is staying between the sun and the earth, so that incoming solar radiation is reduced. Preliminary calculations say this mirror needs to have a surface of about 100 km^2, which means it would have to be fabricated in space.
- Spraying finely dispersed sea water into low level clouds above the oceans in order to make them[3] 'whiter' and reflect more solar radiation. Early calculations say an amount of water of about 10 m^3 per second would be needed.
- Putting large amounts of floating reflecting strips in the oceans that increase the reflection of solar radiation.

The only experience we have with these proposed planetary engineering methods is what happens when there is a major volcanic eruption where large amounts of sulphate particles are thrown into the upper atmosphere. The effect can be measured. The eruption of Mount

Pinatubo in the Philippines in 1991 threw more than 17 million tonnes of SO_2 into the atmosphere and ash reached heights of more than 30km. It led to a drop in average global temperature of about 0.4 °C and stratospheric ozone depletion increased. The effects lasted for 2–3 years. This phenomenon has led to proposals to dump sulphur particles from high flying airplanes. Other materials also have been proposed. Risks in terms of stratospheric ozone depletion, air pollution, and regional impacts and costs are not well understood.

Other methods mentioned above are purely theoretical at the moment and their side effects not understood. The proposal to seed clouds with finely dispersed sea water could have significant impacts on weather and precipitation patterns.

The *second category*, changing the global carbon cycle, covers proposals to 'fertilize' the oceans with large amounts of iron compounds or nitrogen fertilizer. The idea is to enhance the growth of plankton in areas where iron or nitrogen in ocean water is low and limiting plankton growth. This supposedly would remove carbon from the ocean surface layer through dead plankton biomass that sinks to the ocean floor.

Iron deficiency occurs in about 30% of the oceans, mostly the Southern Ocean and the Pacific Ocean near the equator and near the Arctic. A number of large scale tests have been performed with several tonnes of iron sulphate. Enhanced growth of plankton has been observed. However, the few checks that have been done on how much of the plankton sinks to the ocean floor show only a very limited effect. Less than 10% of the plankton sinks to deep waters. Most of the dead plankton is decomposed and recycled back into the ocean surface layer. There are other problems with ocean fertilization. Very little is known about the impacts of large scale application. It could lead to oxygen depletion of parts of the ocean; it could lead to changes in the plankton composition with unknown consequences for ecosystems and the food chain; it could lead to emissions of methane or nitrous oxide. Nitrogen fertilization has similar problems.

For the time being ocean fertilization has no real value as a mitigation option. The question is if it ever will have, given the huge uncertainties and the risks of doing major damage with large scale operations. Nevertheless there are some commercial operations[8] that claim they can remove CO_2 in this way at attractive costs and they suggest this option to be politically viable. These claims have no chance of being internationally accepted. A much more robust option of dissolving captured CO_2 in ocean waters (see Chapter 5) is widely seen as too risky to be considered as an acceptable mitigation option. Ocean fertilization is an order of magnitude more risky.

In general all geo-engineering proposals have one important deficiency (on top of the uncertainties and lack of understanding of their potential side effects). They do nothing about the direct effects of higher CO_2 concentrations in the atmosphere. The most important consequence of that, acidification of the oceans (see Chapter 1), is therefore not addressed. Serious disruptions of oceanic ecosystems and the food chain can happen as a result of ocean acidification. A second general issue is that geo-engineering proposals are promoted by interest groups that would be losing out as a result of major shifts away from fossil fuel. It draws attention away from using all the existing technologies to drastically reduce CO_2 emissions by gambling on an unproven technology. The third major problem with geo-engineering is the fact that it proposes large scale experiments with the earth. While our first experiment, drawing large amounts of fossil fuel from the earth and burning it to drive human development, is about to lead to disaster, it is proposed to carry out another experiment to counter the impacts of the first. Shouldn't we think twice about this?

How is the overall mitigation picture for individual countries?

For individual countries or groups of countries (such as the European Union) many studies have been done of the national (or group) mitigation potential. Objectives of such studies differ. Finding the optimal implementation of a policy target is one. The EU performed such studies to find the lowest cost implementation of its Kyoto target of −8% compared to 1990. Figure 10.5 shows the sector distribution of the reductions that would give the lowest overall costs. It illustrates the general finding that applying an equal reduction percentage to all sectors is more costly than allowing different percentages in accordance with the relative costs of measures.

Another objective of country studies is to find out if and at what costs deep emission reductions are possible. Japan has studied for instance a 70% reduction of GHG emissions by 2050, compared to 1990[9]. Conclusions were that this is feasible at annual abatement costs of about 1% of GDP in 2050. Economic growth would continue at an average rate of 1–2% per year till 2050, while the population would be shrinking. It showed a strong contribution of energy efficiency, leading to a 40–45% reduction in energy demand. Emission reduction for the respective sectors was estimated at 20–40% for industry, about 70% for transportation, 40–50% for buildings, and a strong transition to low carbon energy supply, based on nuclear, gas with CCS, renewables, and hydrogen.

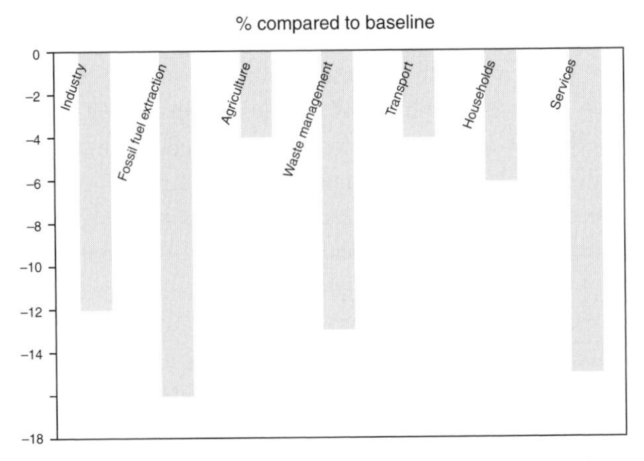

Figure 10.5 **EU implementation of −8% Kyoto target to achieve lowest possible costs: relative contributions of sectors, expressed as reduction percentage compared to the baseline.**
Source: EU DG Environment, http://europa.eu.int/comm/environment/enveco/climate_change/summary_report_policy_makers.pdf.

(a)

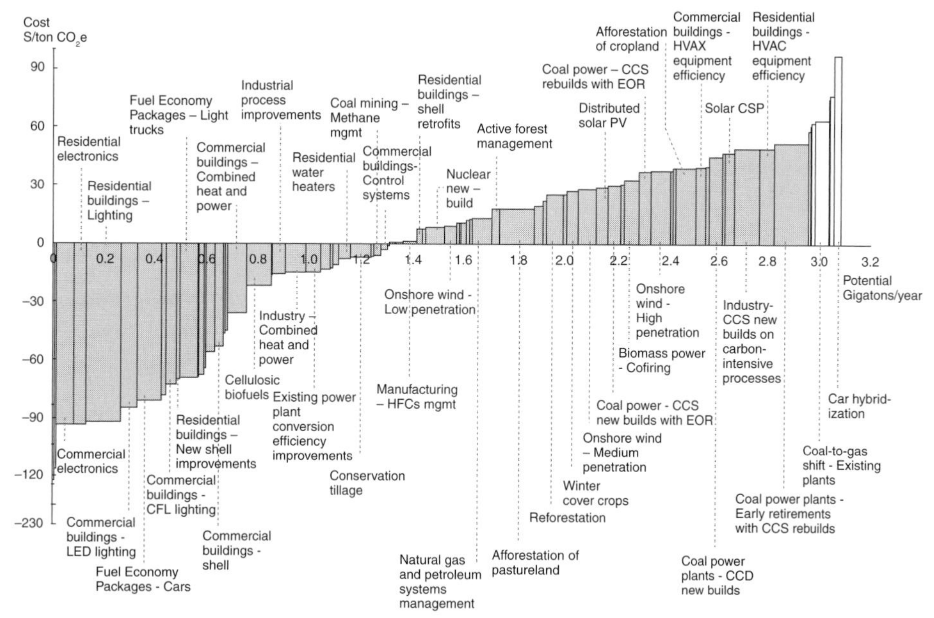

(b) Australian carbon abatement cost curve for 2020

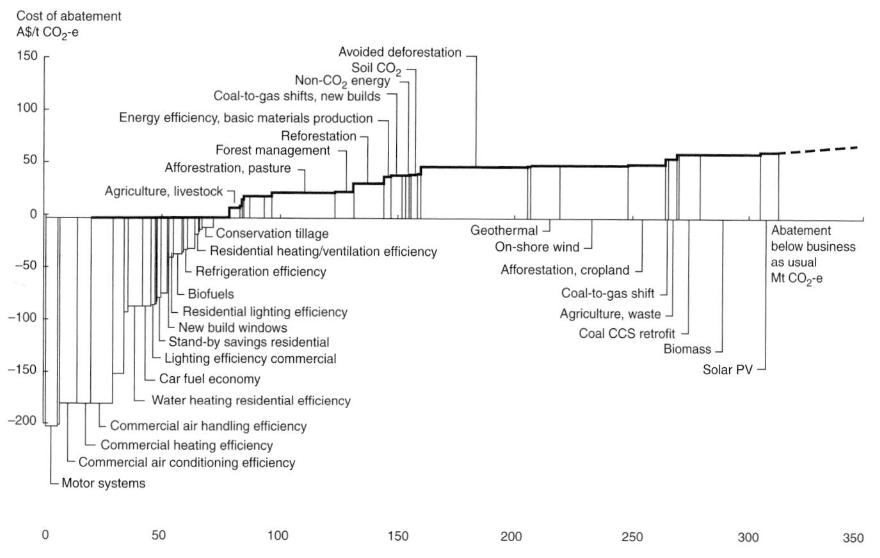

Figure 10.6 **Mitigation cost curves for the USA (a) and Australia (b).**
Source: Reducing US greenhouse gas emissions: how much at what costs?, McKinsey and Company, 2007; An Australian cost curve for greenhouse gas reductions, McKinsey and Company, 2008.

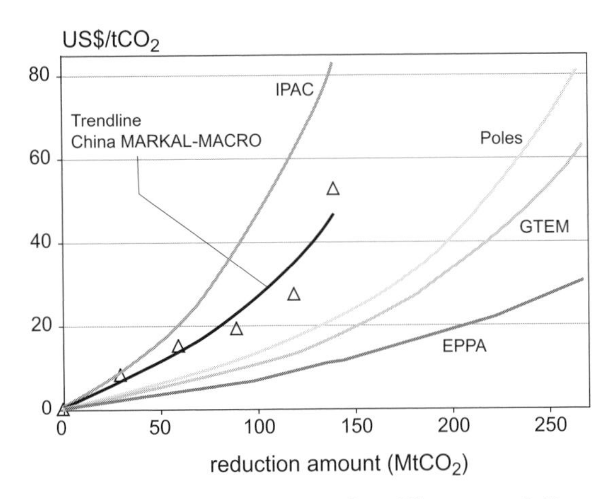

Figure 10.7 **Cost curves for China in 2010, as used in different modelling studies.**
Source: IPCC Fourth Assessment Report, Working Group III, fig 11.6.

National circumstances differ between countries. Some countries have large coal deposits, others have large hydropower resources, while yet others have an economy heavily reliant on agriculture and forestry. Some countries have already implemented policies to promote energy efficiency, others have not. That means the economic potential for mitigation and the type of reduction measures at a particular cost level also vary. In other words: for one specific cost level, reductions compared to the baseline will be different from country to country. The McKinsey-Vattenfall cost curves, as introduced above, illustrate that clearly (see Figure 10.6 for some country examples). These differences in national abatement cost curves are one of the main reasons for the differentiated reduction targets for industrialized countries under the Kyoto Protocol (see further in Chapter 12).

There is one big problem with country cost curves: they are very uncertain. The above examples from the McKinsey studies are not the only ones for the respective countries. Different cost curves are available for any country and available modelling studies are not using the same cost curve. Figure 10.7 gives an example for China. At a cost level of US\$20/tCO_2 avoided, reduction potentials vary by a factor of 4 (between 0.05 and 0.2GtCO_2/year).

A closer look at the cost of mitigation actions

Costs were referred to when economic mitigation potentials were discussed in Chapters 5, 6, 7, 8, and 9. Those were the costs of emission reductions. They depend on the reductions to be achieved. For a trajectory towards stabilization of atmospheric concentrations of around 450ppm CO_2-eq, it was shown that measures need to be taken with costs of the most expensive ones going up to about US\$100/tCO_2-eq avoided.

Table 10.2. **Cost of mitigation by 2030**

<table>
<tr><th>Stabilization level aimed at (ppm CO_2-eq)</th><th>Global abatement costs (% of GDP)</th><th>Macro-economic costs (% of GDP loss in 2030)</th><th>Reduction in average annual growth rate until 2030 (percentage points)</th></tr>
<tr><td>Around 650</td><td>0.1</td><td>−0.6–1.2</td><td><0.05</td></tr>
<tr><td>Around 550</td><td>0.3–0.5</td><td>0.2–2.5</td><td><0.1</td></tr>
<tr><td>Around 500</td><td></td><td rowspan="2"><3</td><td rowspan="2"><0.12</td></tr>
<tr><td>Around 450</td><td>1–1.5</td></tr>
</table>

Source: IPCC Fourth Assessment Report, Working Group III, Chapter 3.3.5.3 and van Vuuren et al. Netherlands Environmental Assessment Agency, 2006.

In Chapter 3 a treatment of economic costs is presented. It discusses the concept of total mitigation costs (the expenditures for realizing reductions). One important notion is that the total costs are obtained by multiplying the tonnes to be reduced by the average cost per tonne (i.e. not with the cost of the last tonne reduced, the so-called 'marginal cost'). For the various ambition levels of global emission reduction, abatement costs are given in Table 10.2. For the most ambitious one, aiming at stabilization at 450ppm CO_2-eq, global abatement costs in 2030 will be about 1–1.5% of global GDP. This is comparable to the amount spent globally on beverages or on the military today[10].

What about the costs for the economy as a whole?

There is another way of expressing costs and that is the GDP increase 'foregone' by taking emission reduction measures. As discussed in Chapter 3, 'foregone' means that other economic activities, along the lines of what societies have been doing in the past, would have resulted in a larger increase of GDP. Many economic models make the simple assumption that current economies are functioning in an optimal way (i.e. are 'in equilibrium'). In other words, the markets work perfectly and taxes are giving the revenue at the lowest possible economic loss. In such a situation doing something different (i.e. mitigating climate change) would always reduce economic output (GDP). Since this assumption of ideal economies is not the reality, some of the models have implemented ways to simulate suboptimal economies. In such models introducing taxes on emissions can sometimes lead to an improvement of economic output, i.e. an increase in GDP.

There is also the issue of technological change that influences cost estimates from computer models. Most models do not assume any influence of climate policy on the rate at which technological innovation takes place. However, it is plausible that such an influence exists, i.e. more rapid technological innovation with stringent climate policy.

Table 10.3. Selected country Kyoto targets and economic costs

Country/group	Kyoto Protocol target (% of emissions in 2008–2012 below 1990 level)	Estimated GDP change in 2010, compared to a baseline, after US withdrawal
EU-15	−8	−0.05
Belarus	−8	+0.4
Canada	−6	−0.1
Hungary	−6	+0.2
Japan	−6	−0.05
Poland	−6	+0.2
New Zealand	0	
Russian Federation	0	+0.4
Ukraine	0	+0.4
Norway	+1	
Australia	+8	

Note: GDP changes assume full Annex I emission trading.
Source: Boehringer C, Loeschel A. Market power and hot air in international emissions trading: the impacts of US withdrawal from the Kyoto Protocol, Applied Economics, vol 35 (2003), pp 651–663.

When these assumptions are built into the models the costs of achieving certain emission reductions go down. This explains the fairly wide cost ranges as shown in Table 10.2, with some estimates giving 'negative costs' (meaning economic benefits).

Individual countries can face costs that are higher or lower than the global average. Individual country costs depend strongly on the international arrangements about contributions of countries to the global mitigation effort. A good example is the Kyoto Protocol agreement for industrialized countries. These countries agreed to achieve a 5% reduction of their collective emissions below 1990, to be reached on average in the period 2008–2012[11]. For each country or group of countries individual reduction percentages were agreed, varying from −8% for the EU-15[12] to +10% for Iceland. The costs of achieving those targets vary between countries. Table 10.3 gives some typical results of economic studies. As can be seen, agreed targets were chosen in such a way that costs for OECD countries would be comparable, while countries with economies in transition were given opportunities to benefit, in light of the drastic economic restructuring they were facing.

The USA and Australia refused to join the Kyoto Protocol because they claimed costs to their economies were too high, although both countries had agreed with the text in Kyoto. Australia joined Kyoto recently after a change in government. Model calculations show that for a situation with USA and Australia participation (in which case there would be a greater demand for emission reduction credits from emissions trading and higher costs to the economy) costs for the USA were of the order of 0.2–0.4 % lower GDP in the year 2010[13]. These costs were not higher than for most other OECD countries and the excessive costs argument was therefore not rational.

Spill-over effects

Climate policy does change the relative value of resources and commodities. In a low-carbon economy the demand for fossil fuels and energy intensive goods will decline, if not in absolute terms, then certainly relative to a baseline (the so-called spill-over effects). Countries exporting fossil fuels and energy intensive goods will then notice the effects. OPEC (Oil Producing Exporting Countries) has made a strong political point about that since the beginning of the international negotiations on controlling climate change. The argument was simple: if actions to control climate change are having a negative impact on our oil revenues, we need to be compensated.

The question is: do they have a point? Economic modelling studies were done to investigate the effects for the implementation of the Kyoto Protocol, where industrialized countries reduce their emissions and OPEC countries have no obligations on their emissions. The results show strongly increasing oil revenues due to increased consumption, but somewhat lower than in the absence of the Kyoto Protocol (the most pessimistic study gave a 25% reduction in 2010 compared to a non-Kyoto baseline). Macro-economically though the impact is relatively small: a decline in 2010 GDP of about 0.05% compared to what it otherwise would have been. And these results did not include the positive spill-overs from enhanced availability of energy efficient technologies. Nor were the sharply risen oil prices of 2007–2008 taken into account. There is so much revenue flowing now to oil exporting countries that the case for compensation has lost steam. The debates however did lead to clauses in the Kyoto Protocol putting an obligation on industrialized countries to minimize the adverse impacts of their mitigation actions on other countries.

Investments

How much money will have to be invested to get to a low carbon economy? Is that money available? And what will be the timing of these investments? These are questions that worry many people.

Let us first look at what needs to be invested in energy supply and energy use anyway, irrespective of climate change control. According to IEA estimates between now and 2030 something like US$22 trillion (22000 billion) will have to be invested to keep up with energy demand and to renew the energy infrastructure. About 50% of this investment will have to be made in developing countries[14]. The 22 trillion is equivalent to about US$1 trillion (1000 billion) per year. Compared to the total investments in infrastructure, buildings, industrial plants, the energy systems, and other things (US$7.8 trillion/year) it is a little more than 10%.

For a low carbon economy, i.e. trajectories towards stabilization at 450–550ppm CO_2-eq, the energy system needs to be restructured. This means investments will have to shift from fossil fuel based energy supply to energy efficient end use equipment and low carbon energy supply (renewables, fossil fuel with CCS, nuclear). Estimates from IEA[15] show that for a trajectory towards 550ppm CO_2-eq it takes total *additional* investments in power supply and end use efficiency of about 4 trillion US$ or slightly less than 20% of the investment that would be needed anyway. There are however about two times as much savings in energy costs due to lower fossil fuel use and higher energy efficiency. For a 450ppm CO_2-eq scenario the additional investment costs are higher, about 9 trillion US$ till 2030 or about 35% of the investments that have to be made anyway. Savings in energy use amount to about 6 trillion US$.

Total investment requirements is one thing, investment by the private sector something else. Social needs may make major shifts in investments attractive, but does that mean individual businesses are making these investments? Energy supply has been privatized in many countries and even where governments control energy supply, criteria for investments follow a business logic. In most circumstances electricity supply companies invest in increased supply at the lowest possible costs. So unless there are regulations forbidding it or real carbon prices, coal based power plants (without CCS) come out as the most attractive option in many places. With CCS for large scale power plants still being too experimental to be mandated by governments, real carbon prices still being zero in most places, natural gas prices as high as they are, and opposition to nuclear power plants still strong in many countries, there are no strong business reasons not to build a traditional coal fired power plant, when taking a short term perspective.

What could change these investment decisions? Protests from environmental groups against coal fired plants can sometimes make a difference for companies who are sensitive to their public image[16]. Expectations about carbon price increases can also lead to different decisions. In the EU for instance there is now a carbon price of about Euro 20/tonne CO_2 (about US$ 30/t) as a result of the EU −8% Kyoto target and the EU Emission trading System. A decision on a further unilateral reduction to −20% below 1990 by 2020 has been taken, which will lead to higher carbon prices. Together with intentions to move to auctioning of emission allowances under the EU ETS, this is now beginning to have an impact on investment decisions by electric power companies.

Timing of investments is critical, since long term stabilization levels depend strongly on how fast emissions will be brought down (see Chapter 3). The most logical approach is to make use of the replacement of existing technology (the so-called capital stock turnover). In modelling studies this replacement takes place after the economic life time. That is the time in which the investments are depreciated (in other words 'is written off'). In practice however, it is very profitable for companies to keep installations going well beyond their economic lifetime. There are no more capital costs, but only operating costs. Even when new installations would have lower operating costs because they are more energy efficient, keeping the old installation going is often more profitable. Only when operating costs are drastically lower or other reasons exist, such as regulations or product specifications, will old installations be scrapped[17]. There is no guarantee that new, low carbon technologies will come in fast, unless there are clear incentives for companies to do so.

How big are the co-benefits?

Reducing GHG emissions has a number of co-benefits. The most important ones are:

- Reduced air pollution: shifts from coal to gas or renewable energy and energy efficiency improvements lead to lower emissions of fine particles and sulphur and nitrogen oxides; lower methane emissions reduce the formation of tropospheric ozone
- Increased energy security: energy efficiency and renewable alternatives for oil reduce the dependence of many countries on oil imports; foreign currency expenditures for oil can be reduced
- Employment: strengthening energy efficiency and production of renewable energy is relatively more labour intensive than large scale fossil fuel based electricity supply

These co-benefits are usually not taken into account when considering the costs of mitigation measures. When factored in, they can make a big difference however.

Reduced air pollution

Air pollution has big impacts on human health, agricultural production, and natural ecosystems. Reducing air pollution can thus have important benefits. For industrialized areas it is well established that a 10–20% CO_2 reduction typically leads to a 10–20% reduction in SO_2 and NOx and a 5–10% reduction in fine particle emissions. The associated health benefits are substantial. If these health benefits are quantified they account for a reduction of mitigation costs of anywhere from US$2 to more than US $100/t$CO_2$ avoided, depending on the assumptions made and the types of air pollution included. This means the health benefits alone could compensate for all of the mitigation costs in certain cases. Agricultural and ecosystem benefits, particularly from reduced tropospheric ozone, will add to these benefits. They have not been well quantified on a global scale. A study for China however showed that a 15–20% CO_2 reduction from the baseline would lead to an agricultural productivity increase that fully compensates the costs of CO_2 reduction[18].

Energy security

Energy security is a top political concern these days. With rising oil prices and oil demand and only a handful of major oil producers, it is primarily the concern about

This chart compares the energy security and climate characteristics of different energy options. Bubble size corresponds to incremental energy provided or avoided in 2025. The reference point is the 'business as usual' mix in 2025. The horizontal axis includes sustainability as well as traditional aspects of sufficiency, reliability, and affordability. The vertical axis illustrates lifecycle greenhouse gas intensity. Bubble placements are based on quantitative analysis and WRI expert judgement.

Power Sector (this size corresponds to 20 billion kWh)

Transport Sector (this size corresponds to 100 thousand barrels of oil per day)

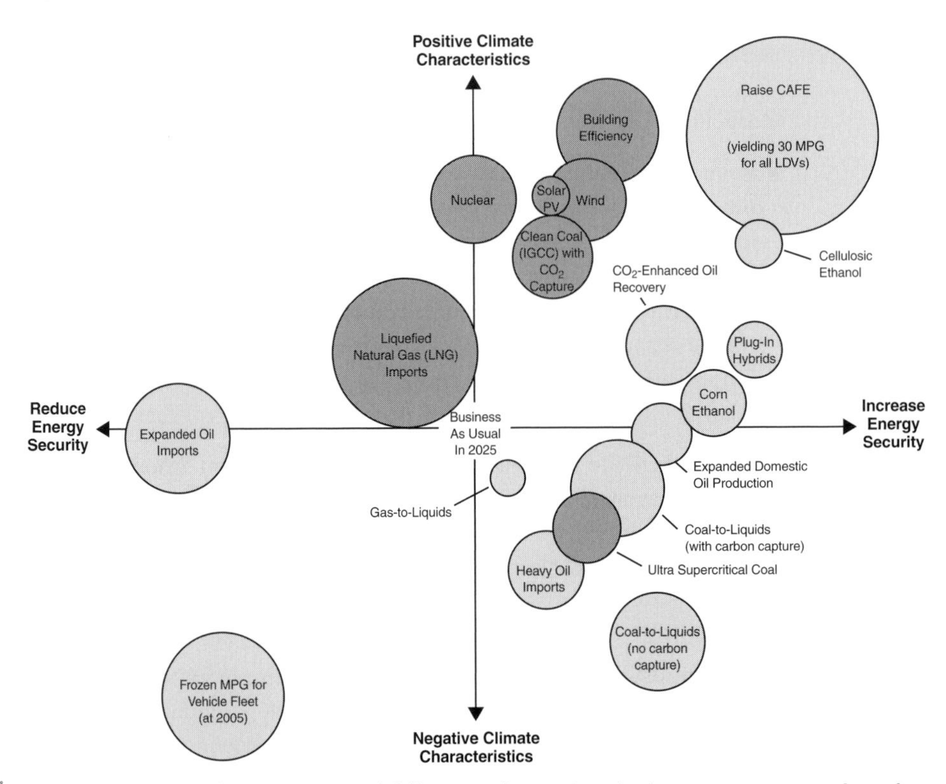

Figure 10.8 **Energy security and climate aspects of different policy options in the USA energy supply and transport sector.**
Source: Weighing US Energy Options: the WRI Bubble Chart, World Resources Institute, 2007.

interruption of oil supply that worries political decision makers. For natural gas the situation is more regionally determined, but in some areas is not much different. Improving energy efficiency and shifts to renewable forms of energy for reasons of climate control go perfectly hand in hand with improving energy security. The other way around however, i.e. taking action to increase energy security, is not always helpful for reducing GHG emissions. A shift from gas to coal for power production or moving towards gasoline production from coal or gas brings us further away from a low carbon economy. In Figure 10.8 the relations between climate control and energy security measures is shown for the USA. It clearly shows there are large win-win opportunities, particularly in energy efficiency improvements for cars and buildings, but also problematic trade-off issues, such as for import of LNG, super efficient coal fired power plants without CCS, and fuels production from coal or gas.

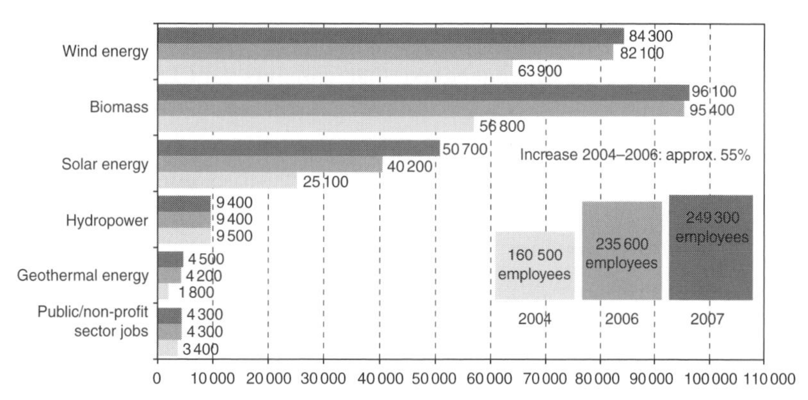

Figure 10.9 **Employees in the German renewable energy sector 2004, 2006 and 2007. Figures for 2006 and 2007 are provisional estimates.**

Source: Modified after graphic from BMU Projekt "Kurz– und langfristige Auswirkungen des Ausbaus der erneuerbaren Energien auf den deutschen Arbeitsmarkt", KI III 1; interim report March 2008.

Employment

Several studies confirm that shifting to a low carbon economy has positive effects on employment[19]. Particularly energy efficiency improvements in the building sector have a good potential to generate substantial additional employment. The European Commission estimated that a 20% energy efficiency improvement in the EU by 2020 would generate a million new jobs. In Germany the strongly growing renewable energy supply industry is now employing about 250000 people (see Figure 10.9), and Polish estimates claim that renewable energy supply is about 10 times as labour intensive as traditional fossil fuel based supply.

Technology transfer

Technology is key to a low carbon economy. As discussed above the technology is available to avoid all of the projected increase of GHG emissions between 2000 and 2030. Most of these technologies are commercially available today and some are expected to be commercialized by 2030. Table 10.4 shows the most important technologies for the various economic sectors.

It was also pointed out above that much of the mitigation potential is found in developing countries (see Figure 10.1), which means these low carbon technologies need to be applied in developing countries without delay. And that requires these technologies to be readily available in developing countries, which is currently often not the case. This is the technology transfer challenge.

The international political debate about technology transfer in the context of controlling climate change has become much polarized. Developing countries take the position that it is the responsibility of OECD countries to make modern low carbon technology available to them at low or zero costs. They normally refer to an article in the Climate Change Convention that commits OECD countries to support developing countries with finance and technology, although that article is formulated very generally. OECD countries reject these claims and point out they do not own the technologies and suggest developing countries make themselves attractive for foreign investment that brings modern technologies. This polarized debate has stood in the way of finding pragmatic solutions to speeding up the diffusion of modern low carbon technology to developing countries.

What do we know about the driving forces and the obstacles for technology transfer? There are three main mechanisms identified.

Technology transfer through foreign direct investment

Foreign Direct Investment (FDI) is often positive for low carbon technology transfer, if foreign companies bring in their own best technology. That is not automatically happening, because companies sometimes are afraid that patented technologies are stolen in countries where the protection of intellectual property rights is not actively enforced. Foreign companies may also be tempted to put second hand technology in place, while they invest in the best available technology at home. For instance, a detailed study of FDI driven technology transfer from three big US automakers to Chinese joint ventures showed that outdated pollution control technology was transferred and little was done to build local technological capabilities[20]. This can of course be prevented if developing countries have the technical and administrative capacity to demand the best technology to be used.

FDI contributes most to capital flows to developing countries. In 2006 it was US$380 billion[21]. For comparison, energy related Official Development Assistance (ODA) was about US$7.5 billion in 2005[22]. It is very hard to identify what the implied low carbon technology transfer in FDI is, but what is known is that about 30% of FDI goes to manufacturing. Most FDI goes to a limited number of developing countries however (see Box 10.2).

Box 10.2 Ten biggest FDI recipient developing countries (billions of US$ in 2006, accounting for about 60% of all FDI to developing countries)

China, incl. Hong Kong	113
Brazil	19
Saudi Arabia	18
India	17
Mexico	15
Egypt	10
Thailand	10

UAE	8
Chile	8
Malaysia	6

(Source: UNCTAD World Investment Report 2007)

Investment risk is an important consideration for private sector investors and that risk is determined by a large number of local circumstances in the respective country (so-called 'enabling conditions'). The most important are[23]:

- Political stability
- Transparent legal and regulatory system
- Skilled labour
- Available financial services and banking provisions

Governments in developing countries have the power, but not always the capacity, to create favourable conditions for foreign investment in low-carbon technology.

Technology transfer through export from developing countries

International trade is another strong driver for low carbon technology transfer, particularly for internationally traded energy intensive industrial products. As discussed in Chapter 8, the latest production plant for chemicals, fertilizer, aluminium, or steel is usually the most energy efficient, whatever its location. The reason is the very competitive international market for these products that make cost reduction through energy efficient production a necessity. It also applies to manufactured products (household appliances, motors, etc.) meant for export to industrialized countries. These products often have to comply with standards on energy efficiency or absence of containment of fluorinated GHGs. This can be a strong driver for low carbon technology transfer.

Technology transfer through domestic innovation

The third – and most interesting – driver for low carbon technology transfer is domestic technology innovation to serve national priorities. Brazil's sugar cane alcohol development programme (see Chapter 6) is a good example. A national priority to become less dependent on imported oil spurred the development of a modern alcohol industry that is among the most efficient in the world.

The political priority for developing renewable energy in India has led to a strong wind turbine industry, with the biggest company Suzlon now being the 5th largest global wind turbine supplier[24]. The company acquired the best available technology, for instance by buying one of the leading European companies producing gearboxes (see Box 10.3).

China is another good example of this 'technology transfer through domestic innovation' approach[25]. Driven by its priority for energy security and its huge domestic market, it has become the world's cheapest supplier of supercritical coal fired plants (with much higher efficiency than traditional coal fired plants). It has also become the biggest producer of solar water heaters (market of US$2 billion per year and 600000 people employed), biogas digesters, wind turbines, electric bicycles, and scooters, and the second biggest supplier of solar PV cells. While electric bicycles and scooters are just a niche market in industrialized countries, China has almost 2000 production facilities and more than 20 million units were sold in 2007, a US$6 billion market (see also Chapter 6). China is already the third largest biofuel producer. Acquiring new technologies from abroad has become an integral part of innovation and does not depend on foreign investment.

Box 10.3 Suzlon wind energy

The Suzlon story began in 1995 with just 20 people, and in a little over a decade has become a company of over 13000 people, with operations across the USA, Asia, Australia, and Europe, fully integrated manufacturing units on three continents, sophisticated R&D capabilities, market leadership in Asia, and ranked 5th in terms of global market share.

Faced with soaring power costs, and with infrequent availability of power hitting his business hard, Mr. Tanti looked to wind energy as an alternative. His first encounter with wind energy was as a customer, having secured two small-capacity wind turbine generators to power his textile business. Moving quickly, he set out to acquire the basic technology and expertise to set up Suzlon Energy Limited – India's first home-grown wind technology company.

Suzlon began with a wind farm project in the Gujarat state of India in 1995 with a capacity of just 3 MW and has, at the end of 2007, supplied over 6000 MW world over. Suzlon has grown more than 100% annually and registered a 108% growth in the financial year ended 2007 – over twice the industry average – in a supply restricted environment. Today Suzlon is ranked as the 5th largest wind power equipment manufacturer with a global market share of 10.5%. The company seized market leadership in India over 8 years ago, and has consistently maintained over 50% market share, installing over 3000 MW of wind turbine capacity in the country.

The company adopted innovation at the very core of its thinking and ethos. This led to full backward integration of the supply chain. Suzlon by this approach has developed comprehensive manufacturing capabilities for all critical components – bringing into play economies of scale, quality control, and assurance of supplies. Taking this focus forward, Suzlon acquired Hansen Transmissions of Belgium in 2006. The acquisition of the world's second leading gearbox maker gives Suzlon manufacturing and technology development capability for wind gearboxes, enabling an integrated R&D approach to design ever more efficient wind turbines.

Suzlon's R&D strategy emphasizes the need to lower the cost per kilowatt-hour, in order to create ever more competitive technology and products. Making technology development a central objective, Suzlon has leveraged Europe's leadership, talent, and experience in wind energy technology, setting up R&D centres in the Netherlands and Germany. Combined with a strong engineering backbone in India, the approach brings together the expertise of different centres of excellence to build 'best of all worlds' products.

Looking for growth not just in India, but across the world, Suzlon looked past traditional markets for wind energy, and entered new and emerging high growth markets. This step has success in the rapid global expansion of Suzlon's business with orders from Australia, Brazil, China, Italy, Portugal, Turkey, and the USA.

Suzlon, with its internationalized business model, fully integrated supply chain, and R&D focus on cost per kWh reduction, is today an agile, fast moving organization that is well equipped to take on a dynamic, changing market place with innovative products and solutions. (Source: http://www.suzlon.com)

The three models described above are of course complementary. Essentially they are business oriented with governments creating the right conditions. This model can transform the international political debate on low carbon technology transfer from an 'if you want me to do something you have to give me the technology' approach to an approach driven by national self-interest.

Irrespective of the conceptual model, there are many obstacles to effective low carbon technology transfer, where 'effective' means the best low carbon technology and widespread application, not just some individual projects. For that to happen a large number of things need to be in place. Technical knowledge, including capacity to assess technologies and organizational capacity, is one. Financing, such as availability of capital, understanding of low-carbon technologies by banks, removal of subsidies for fossil fuel based technologies in developing countries that compete with low carbon technologies, and shift of export subsidies in industrialized countries from traditional fossil fuel based technologies to low carbon ones, is another. Lack of standards and transparent regulation, including the presence of corruption, and inadequate enforcement of contracts and property rights create a bottleneck. Incentives to invest in new low carbon technologies are often missing, because of existing tax laws, import restrictions, or other constraints. Lack of business networks and ways to communicate positive results of innovation to other companies hamper the spread of low carbon technologies. If one of the essential components is lacking, the whole process of technology diffusion comes to a halt. The chain is as strong as its weakest link!

These multiple barriers can be overcome by targeted policies however. In developed countries they relate particularly to reforming the system of export subsidies by issuing specific environmental guidelines for export crediting agencies that are active in many countries. Reducing tied aid (mandatory use of finance for equipment from the donor country), actively pursuing low carbon technology introduction in development assistance programmes, and discouraging the misuse of patents by manufacturers of low carbon technologies in developed countries are other important elements. In developing countries policies need to be aimed at education and training, reforming legal, regulatory and financial systems, proper assessment of the technology needs of the country for achieving its development goals, introduction of low carbon standards for technologies, enforcement of intellectual property rights, and stimulating markets for low carbon technologies[26].

Technology development

Although many low carbon technologies are commercially available right now, additional technologies need to be brought from the R&D stage to commercialization in order to have an adequate toolbox to control climate change. Table 3.2 shows some of these technologies that are expected to be commercially available by 2030. Beyond 2030, technologies like biomass based chemical processes and biomass fuelled power plants with CCS need to become commercial, while low carbon technologies already being applied are further improved and made cheaper. This requires a vibrant R&D infrastructure and adequate funding. A sobering fact is that government funding for energy research has gone down since the early 1980s and is now at about half the 1980 level in dollar terms (see also Chapter 11). This can be explained by the massive privatization of energy supply in many countries, but private R&D investments have not compensated much of this loss. Current trends are thus completely opposite to what would be needed to control climate change in the longer term.

Commercialization of technologies is done by the business community, not by governments. So it is important to understand the way companies are handling R&D investments. The objective of a company is to create future profits through new products for which it has to carry out R&D. The market prospects for such new products are therefore critical. In the case of low carbon technologies these market prospects depend heavily on government policy. The clearer governments are about future policies and regulations the better companies can anticipate. Return on R&D investment is also an important consideration. It has been well established that for a society as a whole the return on R&D investments is very good. For an individual company however it is quite uncertain. The reason is that competitors may be more successful with comparable new products or patent protection is not effective. Companies are therefore sometimes hesitant to invest in R&D. Governments can address these risks by providing support in the form of tax deductions or R&D subsidies, something that is fully justified by the high social return on R&D investment[27].

The relation between mitigation and adaptation

In Chapter 3 it was concluded that mitigation and adaptation are both needed to control the risks of climate change. It is a matter of 'and-and', and not 'or'. For that reason it is wise to look for synergies with adaptation, when deciding on a package of mitigation actions. In any case mitigation actions that make societies more vulnerable to climate change ought to be avoided. In some sectors there are strong interactions between mitigation and adaptation. In agriculture and forestry in particular, mitigation measures to enhance carbon sinks in soils can make these soils less vulnerable to drought, which is a good adaptation action. On the other hand, forest and biomass plantations that replace natural forests can reduce biodiversity and food security that is already under stress from climate change. This is therefore a measure that is not good for adaptation. Low carbon

Table 10.4. **Synergies and trade-offs between mitigation and adaptation measures**

MITIGATION OPTIONS	SYNERGY with adaptation	TRADEOFFS with adaptation
Energy: low carbon supply options, efficiency	• Energy required for adaptation does not increase emissions	• Unsustainable biomass production may reduce resilience and damage biodiversity
Forestry: avoid deforestation, plant trees, soil management	• Reduces vulnerability through water management	• Biodiversity (plantations)
	• Protects biodiversity	• Competition with food production
Agriculture: soil management	• Reduces vulnerability to drought and erosion	

Source: IPCC Fourth Assessment report, Working Group III, ch 11.

energy supply is good for adaptation in the sense that many adaptation measures require energy (water pumping, air conditioning, water desalinization). For synergies and trade-offs between mitigation and adaptation measures see Table 10.4.

Notes

1. IPCC Fourth Assessment Report, Working Group III, ch 11.3.1.3.
2. IPCC Fourth Assessment Report, Working Group III, ch 11.3.1.5.
3. IPCC Special Report on Safeguarding the ozone layer and the global climate system: Issues related to hydrofluorocarbons and perfluorocarbons, 2005.
4. IPCC Fourth Assessment Report, Working Group III, ch 11.3.1.5, table 13.1 and figure SPM.4.
5. Pathways to a low-carbon economy, Version 2 of the global greenhouse gas abatement cost curve, McKinsey and Company, 2009.
6. IPCC Fourth Assessment Report WG III, chapter 11.3.2.
7. IPCC Fourth Assessment Report, Working Group III, ch 11.2.2.
8. See http://www.abc.net.au/science/articles/2008/06/05/2265635.htm, http://www.businessgreen.com/business-green/news/2226689/climos-defends-ocean and www.acecrc.org.au/uploaded/117/797514_18fin_iron_6sept07.pdf.
9. 2050 Japan Low-Carbon Society' scenario team, Japan scenarios and actions towards Low Carbon Societies, 2008, http://2050.nies.go.jp.
10. For GDP numbers: http://web.worldbank.org/WBSITE/EXTERNAL/DATASTATISTICS/0,,contentMDK:21298138~pagePK:64133150~piPK:64133175~theSitePK:239419,00.html]; for military expenditures: http://www.globalsecurity.org/military/world/spending.htm.
11. The USA was part of this agreement, but never ratified the Protocol.

12. At the time of the Kyoto Protocol agreement in 1997, the EU had 15 member states; the agreed reduction percentages therefore apply to the so-called EU-15.
13. IPCC Third Assessment Report, Working Group III, table TS.5.
14. International Energy Agency, World Energy Outlook 2007, p.94.
15. International Energy Agency, World Energy Outlook 2008, ch 18.
16. For a list of plans and public protests on new coal fired power plants in Germany see http://in.reuters.com/article/oilRpt/idINNLO15053920081001.
17. See Lempert RL et al. Capital cycles and the timing of climate change policy, Pew Center on Global Climate Change, Arlington, USA, 2002 for an in-depth discussion.
18. O'Connor et al, Technical Paper 206, OECD Development Centre, Paris, 2003.
19. IPCC Fourth Assessment Report, Working Group III, ch 11.8.2.
20. See Sims Gallagher K. Foreign Direct Investment as a vehicle for deploying cleaner technologies: technology transfer and the big three automakers in China, PhD thesis, Fletcher School of Diplomacy, Tufts University, Boston, USA, 2003.
21. UNCTAD, World Investment Report, 2007.
22. This is for bilateral and multilateral assistance, see Tirpak D, Adams H. Climate Policy, vol 8(2), 2008, pp 135–151.
23. See for an elaborate discussion IPCC Special Report on Methodological and Technological Issues in Technology Transfer, 2005.
24. http://www.suzlon.com/.
25. See China's clean revolution, Climate Group, London, 2008.
26. See note 24.
27. IPCC Fourth Assessment Report, Working Group III, ch 2.7.2.

11 Policies and measures

What is covered in this chapter?

Without policies that provide incentives to act, the enormous potential for reduction of greenhouse gas emissions will not be realized. This chapter is therefore devoted to discussing the various policies that can be used to influence behaviour of consumers and companies, the circumstances under which they are effective, and the administrative capacities required. With emission reduction objectives becoming more stringent, there is a shift from 'soft policy instruments' such as funding for research and development, information, voluntary agreements, and green government procurement to 'hard instruments' like regulations, taxes, and tradable permit systems. Particularly trading systems are becoming popular. Climate policies are just half the story. It is equally important to use non-climate policies, such a general tax, macro-economic, trade, and other environmental policies to change behaviour in a more climate friendly direction. In practice combinations of policies are always needed to achieve optimal results. Lessons have been learned from implementation of policies about what works best in what sector. Ultimately it is the total package of policies in a country that will determine greenhouse gas emissions and some examples of that will be discussed.

Realizing mitigation potential requires government policies

The point has been made over and over again in the previous chapters: without specific policy action by governments the potential to bring greenhouse gas emissions substantially down will not be realized. There are just too many incentives to continue business as usual practices and too many barriers to capture the reductions that would pay for themselves. So the question then becomes what are the most effective policies?

There is a substantial collection of studies available on this question, drawing on a range of environmental, energy, and transportation policy approaches. This material can be used to draw conclusions regarding climate change policies. There is now also a growing experience of implementation of climate policies in many countries that have started to address climate change. Both sources can be used to try and answer our primary question.

The focus will be on national and local policies, because those are the ones that have a direct influence on decisions that affect greenhouse gas emissions. International policies, as they emerge from international agreements between countries or from international institutions, will be discussed in Chapter 12.

Governments cannot implement effective policies on their own. Social scientists have introduced the term 'governance' (as opposed to 'government') to capture the changing complexity of modern societies. Business, non-governmental organizations, and civil society[1] all play an important role in shaping social change[2]. With increasing globalization of the economy and the acceptance of market mechanisms in many countries the idea of governments regulating desired social change has lost its appeal. Implementing new policies does require the support and involvement of these groups. And business, NGOs, and civil society often are the instigators of change that then is captured in new policies. Some companies for instance have found that performing in a socially responsible manner and pursuing a sustainable development strategy is in fact good business. NGOss are often able to mobilize public support for environmental causes that governments can build upon.

As argued in Chapter 4, controlling climate change cannot be realized with climate change policies alone. Creating incentives to move towards a low carbon economy have to be embedded in policies that directly address economic activities themselves. Tax policies can make a huge difference in investment preferences. Trade policies determine the market for low carbon technologies. Energy security policies can steer development of the energy system in a low carbon or a high carbon direction. A discussion on the most effective policy instruments therefore needs to be put in a broader context.

Types of policy instruments

There is a range of policy instruments relevant to controlling climate change available to governments. A list of the main instruments is given in Box 11.1. They can be applied at local, sub-national, national, or supra-national (as in the case of the European Union for instance) level. The IEA Policies and Measures database has a large number of records of existing applications of various policy instruments[3].

Box 11.1 **Definitions of the main policy instruments relevant to controlling climate change**

Regulations and standards: These specify the abatement technologies (technology standard) or minimum requirements for pollution output (performance standard) that are necessary for reducing emissions.

Taxes and charges: A levy imposed on each unit of undesirable activity by a source.

Tradable permits: These are also known as marketable permits or cap-and-trade systems. This instrument establishes a limit on aggregate emissions by specified sources, requires each source to hold permits equal to its actual emissions, and allows permits to be traded among sources.

Voluntary agreements (VAs): An agreement between a government authority and one or more private parties with the aim of achieving environmental objectives or improving environmental performance beyond compliance to regulated obligations. Not all VAs are truly voluntary; some include rewards and/or penalties associated with participating in the agreement or achieving the commitments.
Subsidies and incentives: Direct payments, tax reductions, price supports, or the equivalent thereof from a government to an entity for implementing a practice or performing a specified action.
Information instruments: Required public disclosure of environmentally related information, generally by industry to consumers. These include labelling programmes and rating and certification systems.
Research and development (R&D): Activities that involve direct government funding and investment aimed at generating innovative approaches to mitigation and/or the physical and social infrastructure to reduce emissions. Examples of these are prizes and incentives for technological advances.
Non-climate policies: Other policies not specifically directed at emissions reduction but which may have significant climate related effects.

Note: The instruments defined above directly control greenhouse gas emissions; instruments may also be used to manage activities that indirectly lead to greenhouse gas emissions, such as energy consumption.

(Source: IPCC Fourth Assessment Report, Working Group III, box 13.1)

Regulations

Regulations are widely used in environmental protection. They come either as generally applicable standards or site specific operating permits. Generally applicable standards can be divided into two separate classes: technology standards, prescribing the means to be used to control emissions; or performance standards, requiring a maximum energy use or emission per unit of product.

An example of a technology standard would be the requirement to install a specific CO_2 capture system at a coal fired power plant, the installation of an incinerator on an HCFC plant to destroy HFC-23 in the plant's exhaust gas, or the requirement to install solar water heaters in certain types of buildings. In many cases the specificity of the situation is such that tailored permitting conditions are being used to prescribe the required action. However, this requires well trained and adequately staffed regulatory agencies that do not exist everywhere.

Examples of performance standards are building codes that require a maximum amount of energy use per unit of floor space or automobile fuel efficiency standards, mandating a maximum fuel use or CO_2 emission per kilometre. Standards can be used to get rid of the most inefficient products or processes by following the best available products on the market. This is often the case for energy efficiency standards for household appliances. They can also be used to 'force' technological improvement by setting standards for a future date that are more stringent than the best available products on the market. A good example of the latter is the EU decision to set a maximum CO_2 emission of 130 gCO_2/km for new passenger cars on average to be reached by 2015, while the current average is still around 160 gCO_2/km. The EU is also decided on even more stringent standards for 2020. Performance standards give more flexibility to companies, architects, and builders to reach the goal in the most efficient way.

In general, regulations and standards provide no incentives to companies to move to technologies that go below the current standards. There is often even the fear among companies that doing so would trigger more stringent regulations. One way of addressing that problem is to regularly revise standards according to technological development or to set 'technology forcing' standards for a future date. This poses big challenges for regulatory agencies however.

Regulatory approaches have proven to work well when dealing with mass products, such as automobiles or household appliances. For buildings they have also worked well. Many countries have building codes in place. See Box 11.2 on the application of building codes in China. Another area where regulations have performed well is the banning of ozone depleting and powerful greenhouse forcing fluorinated chemicals.

Box 11.2 **Building codes in China**

Approximately 2 billion m^2 of floor space is being built annually in China, or one-half of the world's total. Based on the growing pace of its needs, China will see another 20–30 billion m^2 of floor space built between the present and 2020. Buildings consume more than one-third of all final energy in China, including biomass fuels (IEA, 2006). China's recognition of the need for energy efficiency in the building sector started as early as the 1980s but was impeded due to the lack of feasible technology and funding. Boosted by a nationwide real estate boom, huge investment has flowed into the building construction sector in recent years.

On 1 January, 2006, China introduced a new building construction statute that includes clauses on a mandatory energy efficiency standard for buildings. The Designing Standard for Energy Conservation in Civil Building requires construction contractors to use energy efficient building materials and to adopt energy saving technology in heating, air conditioning, ventilation, and lighting systems in civil buildings. Energy efficiency in building construction has also been written into China's 11th Five-Year National Development Programme (2006–2010), which aims for a 50% reduction in energy use (compared with the current level) and a 65% decrease for municipalities such as Beijing, Shanghai, Tianjin, and Chongqing as well as other major cities in the northern parts of the country. Whether future buildings will be able to comply with the requirements in the new statute will be a significant factor in determining whether the country will be able to realise the ambitious energy conservation target of a 20% reduction in energy per gross domestic product (GDP) intensity during the 11th Five-Year Plan of 2005–2010.

(Source: IPCC Fourth Assessment Report, Working Group III, box 13.3)

Box 11.3 The Carbon Emissions Reduction Target obligation in the UK

The Carbon Emissions Reduction Target (CERT) – which came into effect on 1 April 2008 and will run until 2011 – is a regulatory obligation on energy suppliers to achieve targets for promoting reductions in carbon emissions in the household sector. It is the principal driver of energy efficiency improvements in existing homes in Great Britain. It marks a significant strengthening of efforts to reduce household carbon emissions – with a doubling of the level of activity of its predecessor Energy Efficiency Commitment (EEC).

CERT will deliver overall lifetime CO_2 savings of 154 $MtCO_2$, equivalent to annual net savings of 4.2 $MtCO_2$ by 2010, and equivalent to the emissions from 700 000 homes each year, and will stimulate about GBP 2.8 billion of investment by energy suppliers in carbon reduction measures.

In addition to the energy efficiency measures of the current EEC, suppliers will be able to promote microgeneration measures; biomass community heating and CHP; and other measures for reducing the consumption of supplied energy. CERT will maintain a focus on vulnerable consumers and will include new approaches to innovation and flexibility. Suppliers must direct at least 40% of carbon savings to a priority group of low income and elderly consumers. Extending the priority group to include the over 70s seeks to ensure that a large number of fuel poor households, who are not eligible under the current criteria, become eligible for support.

In addition, the newly launched ACT ON CO_2 advice line will help customers take advantage of suppliers' offers under CERT.

(Source: www.defra.gov.uk/environment/climatechange/uk/household/eec/index.htm)

The reasons that regulations work better for these types of situations than other instruments are varied: for consumer products and automobiles for instance the complexity of comparing products, the difficulty of considering purchase price and lifetime operating costs, and the multitude of other non-energy considerations in individual purchase decisions make financial incentives ineffective. For buildings there is an additional problem that the user of the building (the one that pays the energy bills) is often different from the one deciding on the construction or the refurbishment.

For existing buildings one of the most effective policy approaches has been 'demand side management (DSM)'. This means giving electricity companies the task or the opportunity to reduce the demand for electricity in existing buildings in exchange for a possibility to earn money by selling less. Incentives can for instance be created by allowing the companies to include the cost of the DSM programmes in the price they charge for electricity. Since energy efficiency improvement is usually cheaper than building a new generating plant, consumer prices for electricity are lower than without the DSM programmes. DSM programmes can be voluntary or required by the regulations that affect electricity generators. The approach has been particularly popular and successful in the USA[4]. Recently it has been introduced in the UK (see Box 11.3). Application in other parts of the world is a matter of making the necessary changes in the way electricity generators are regulated.

Financial incentives for users, like a tax on energy, are not very effective in such situations. The same holds for smaller companies that often do not have the expertise or the capacity to do rigorous cost minimization and therefore often do not make use of profitable low carbon technologies.

A very different argument in favour of regulatory approaches is the limited administrative capacity in many developing countries. This makes technology standards and performance standards, which can often be copied from other countries, the easier way to control greenhouse gas emissions. When administrative capacity becomes bigger and more sophisticated, taxes, subsidies, and tradable permit approaches may become more attractive.

Taxes and levies

The principle of a tax or levy is simple: increase the price of energy use or greenhouse gas emissions so that less energy is used and measures to reduce energy use or emissions become profitable. A uniform tax or levy has the advantage that all energy users or greenhouse gas emitters face the same carbon price and in theory all measures up to a certain cost level (depending on the level of the tax) are taken, provided they do take the measures that are profitable.

Taxes and levies are widely used on energy products (often as excise duties, but increasingly as CO_2 charges), on motor vehicles (mostly as purchase, registration or road tax, increasingly differentiated according to the CO_2 emissions of the vehicle), and on waste. In a few countries (Norway, Sweden, Denmark, UK) a tax or levy is charged directly on CO_2 emissions[5].

Taxes or levies on energy or emissions can have negative impacts on poor people, since they normally have few possibilities to reduce the tax burden by investing in energy efficiency improvement and emission reduction. Also their expenditures on energy often form a substantial part of their income. The main drawback of taxes and levies however is that they are generally very unpopular amongst businesses and voters. So politically it is very hard to raise taxes to a level where they are really effective in influencing decisions or to adjust the tax over time to get the desired effects. As a consequence many taxes have lots of exemptions, usually to accommodate concerns of influential lobby groups. Or there are ways to avoid the tax by taking alternative actions (see for instance the UK Climate Change Levy in Box 11.4).

Box 11.4 The UK Climate Change levy

The UK has a tradition of action on climate change that dates from the early acceptance of the problem by the Conservative Prime Minister Margaret Thatcher in 1988. The Labour government in 1997 reaffirmed the commitment to act and to use market-based instruments wherever possible; however, it voiced concerns on two aspects of this commitment: Firstly, that such measures might have a disproportionate effect on the poor which, in turn, might affect the coal mining communities (an important constituency) and, secondly, that this commitment might perpetuate a perception that the Labour government was committed to high taxes. A key element of the UK's climate policy is a climate levy. The levy is paid by energy users (not extractors or generators), is levied on industry only, and aims to encourage renewable energy. An 80% discount can be secured if the industry in question participates in a negotiated 'climate change agreement' to reduce emissions relative to an established baseline. Any one company over-complying with its agreement can trade the resulting credits in the UK emissions trading scheme, along with renewable energy certificates under a separate renewable energy constraint on generators. However, a number of industrial emitters wanted a heavier discount and, through lobbying, they managed to have a voluntary emissions trading scheme established that enables companies with annual emissions above 10000 tCO_2-eq to bid for allocation of subsidies. The 'auction' offered payments of 360 million and yielded a de-facto payment of 27€ per tonne of CO_2. Thus, the trading part of the scheme has design elements that strongly reflect the interest groups involved. The levy itself has limited coverage and, consequently, households and energy extractors and generators have no incentive to switch to low carbon fuels. However, its design does take household vulnerability, competitiveness concerns, and the sensitivity of some sectoral interests into account. Thus, while the levy has contributed to emission reduction, it has not been as effective as a pure tax; a pure tax may not have been institutionally feasible.

(Source: IPCC Fourth Assessment Report, Working Group III, box 13.2)

As discussed above in the section on regulation, the price signal established through a tax or levy is not always leading to the desired response (i.e. lower use of the commodity or taking measures to reduce emissions). That is particularly the case for decisions by individuals where cost minimization is not the most important factor, for instance when buying household appliances or a car and when choosing a house or apartment. In larger companies where cost minimization is a priority, the effect of taxes is much better, but business is often exempted in order not to undermine their competitiveness internationally. In general the effectiveness of taxes and levies is modest. The UK Climate Change Levy for instance has resulted in about 2% reduction of CO_2 emissions so far. There is one success story of a CO_2 tax: the Norwegian CO_2 tax played a big role in the establishment of the Sleipner CO_2 capture and storage project at a natural gas production platform off the Norwegian coast. Paying for the CCS installation was more attractive than paying the tax.

Tradable permits

Another way to give emissions of CO_2 and other greenhouse gases a price (other than through taxation) is to issue allowances (or permits) for a limited amount of emissions and to allow trading of these permits. This is also called a 'cap and trade system'. Scarcity is created by limiting allowances to less than what is going to be emitted. Then buying and selling of these allowances will create a price. Individual companies that are likely to emit more than their permits entitle them to can decide to invest in emission reductions or to buy permits from other companies. If investments bring emissions below the allowance, they can sell excess permits or keep them for later use (so-called 'banking'). The first large scale application of a tradable permit system happened in the US in the 1980s for SO_2 under the Clean Air Act. For CO_2 the EU Emission Trading System (see Box 11.5) is the largest tradable permit system in operation.

There are a number of important design issues for a tradable permit scheme: the coverage (which sources, which gases?), the way permits (allowances) are issued, and enforcement issues. Economic theory is clear about those issues: the broader the coverage, the more the permits are auctioned (i.e. sold to the highest bidder) and the stricter the penalties for non-compliance, the more effective and efficient the system will be. In practice however, this ideal is not met.

Coverage is often partial because of difficulties administering large numbers of small sources (such as cars and households). Emission sources that are hard to measure accurately (non-CO_2 emissions from agriculture for instance) are another reason to keep certain emission sources out of the emissions trading system. The EU ETS for instance covers only CO_2 and only about 40% of the total EU greenhouse gas emissions.

Allocation of permits is a sensitive issue. Coming from a situation where greenhouse gas emissions to the atmosphere from companies were free, governments generally give emission allowances away for free to companies (called 'grandfathering'). The step to auctioning is generally too big for getting sufficient political support for introducing a tradable permit system. There is a tendency however to gradually shift to auctioning. Under the EU ETS for instance EU Member States can auction up to 10% of the allowances in the period 2008–2012 and by 2020 70% of allowances to industries not subject to international competition will be auctioned. This shift was made easier when it was discovered that freely allocated permits to electricity generators in the EU nevertheless led to increasing the price of electricity on the basis of the value of these permits. Electricity companies were accused of making 'windfall profits'.

Auctioning permits creates a new problem: what to do with the (substantial) revenue from auctioning? Ministries of Finance usually demand these to be part of general revenue. Others propose to use part of these revenues to stimulate development and deployment of low carbon technologies. Yet others suggest that part of these revenues could be used to help developing countries to make a rapid transition to a low carbon economy. The debate is still ongoing.

The amount of permits received is another very crucial thing for a company. It determines to a large extent how much a company should reduce its emissions or how many permits it should buy. No surprise therefore that there is normally heavy lobbying to get more permits. Under the EU ETS allocation to individual companies for the period 2008–2012 was left to EU Member States. This led to strong differences in allocation between comparable companies in different Member States, generating competitiveness concerns. As a result, in the third phase of the EU ETS (after 2012) there will be centralized allocation of permits by the European Commission, based on a commonly agreed system. One particularly important point in allocation is how to reward past emission reductions by companies. It would be not be fair to ignore past actions. Using performance standards (emissions per unit of product) is a good way to solve that problem: above average performers would receive somewhat more and below average performers somewhat less permits.

A related issue is to whom are the permits issued. In principle there is a choice: for instance, issuing permits to the users of electricity (downstream) or to the producers of it (upstream). The general trend is to use upstream permitting in order to reduce the administrative burden of dealing with large numbers of small users/emitters. The disadvantage is that smaller consumers only notice a higher price for electricity that may not trigger the desired reductions in electricity use.

The position of newcomers, i.e. new companies that enter the market, and of companies that strongly expand production often leads to heated debates. In a system where permits are given for free, normally governments keep some permits in reserve for newcomers, but those that expand production will have to buy the additional permits on the market. In an auctioning system these problems disappear, because every company would have to buy the permits.

Compliance with the system is a very important issue. It should be very unattractive for companies to violate the system by emitting more than the permits it possesses dictate. One important element of a good compliance system is to have accurate monitoring of emissions. The other crucial element is to set a penalty that is substantially higher than the price of permits in the market. There is a complication though, because the CO_2 permit price cannot be predicted. Large fluctuations of the price do happen, although normally during the earlier phases of introducing an emissions trading system. These fluctuations create uncertainty for companies in estimating the costs of the permits and in deciding upon investments in emission reduction projects. As a reaction to this phenomenon proposals about setting up a tradable permit system in the USA do contain elements of 'price caps' (setting a price level above which free permits are issued by the government). These proposals are very controversial however[6].

Tradable permit systems are only used so far in industrialized countries. That certainly has to do with the administrative and enforcement capabilities that are needed to run such a system. But applying tradable permit systems for climate change control in developing countries would also be politically difficult because of the need of the economy to grow and improve the living conditions of people, which leads to a strong increase in greenhouse gas emissions. Technically and politically that raises problems.

Box 11.5 The EU Emissions Trading System

The EU Emissions Trading System (EU ETS) is the world's largest tradable permits programme. The programme was initiated on 1 January, 2005, and it applies to approximately 11500 installations across the EU's 25 Member States. The system covers about 45% of the EU's total CO_2 emissions and includes facilities from the electric power sector and other major industrial sectors. The first phase of the EU ETS runs from 2005 until 2007. The second phase will begin in 2008 and continue through to 2012, coinciding with the 5-year Kyoto compliance period.

Member States develop National Allocation Plans, which describe in detail how allowances will be distributed to different sectors and installations. During the first phase, Member States may auction off up to 5% of their allowances; during the second phase, up to 10% of allowances may be auctioned off.

Market development and prices: A number of factors affect allowance prices in the EU ETS, including the overall size of the allocation, relative fuel prices, weather, and the availability of certified emission reductions (CERs) from the Clean Development Mechanism (CDM). The EU ETS experienced significant price volatility during its start-up period, and for a brief period in April 2006 prices rose to nearly 30€ per tonne; however, prices subsequently dropped dramatically when the first plant-level emissions data from Member States were released. The sharp decline in prices focused attention on the size of the initial Phase I allocation. Analysts have concluded that this initial allocation was a small reduction from business as usual emissions.

Consistency in national allocation plans: Several studies have documented differences in the allocation plans and methodologies of Member States. Researchers have looked at the impact on innovation and investment incentives of different aspects of allocation rules and have found that these rules can affect technology choices and investment decisions. When Member States' policies require the confiscation of allowances following the closure of facilities, this creates a subsidy for continued operation of older facilities and a disincentive to build new facilities. They further find that different formulas for new entrants can impact on the market.

Implications of free allocation on electricity prices: A significant percentage of the value of allowances allocated to the power sector was passed on to consumers in the price of electricity and that this pass-through of costs could result in substantially increased profits by some companies. The authors suggest that auctioning a larger share of allowances could address these distributional issues. In a report for the UK government, a similar cost pass-through for the UK and other EU Member States was found.

(Source: IPCC, Fourth Assessment Report, Working Group III, box 13.4)

Voluntary agreements

Voluntary agreements (VAs) are agreements that are negotiated between a government and a group of private companies or other entities. They are therefore also called 'negotiated agreements'. VAs are different from 'voluntary actions': unilateral commitments of one or more companies without government involvement (discussed below). VAs have become quite popular: amongst private companies, because it gives them a lot of influence over what needs to be done and how it is done and helps them to establish a leadership image, but also amongst governments, because it avoids difficult battles about legal policy decisions. See Box 11.6 for some examples.

VAs come in many different forms, in terms of goals, stringency, role of government, and 'penalties' for non-compliance. They range from agreements on 'best efforts' to reduce energy efficiency and minimize emissions to agreements to meet very specific quantitative performance standards at a specific point in time (such as the European Automobile

Agreement referred to in Box 11.6). VA goals are generally not very stringent, which is caused by the voluntary nature. Not all companies normally join a VA and the VA often applies to domestic companies only. Competitiveness considerations make companies reluctant to commit to very stringent goals. The commitment of governments in VAs also varies. It ranges from communicating the results of the VAs to financial support with data collection or research and development. Many VAs do not have any form of 'penalty' for non-compliance, but some do, mostly in the form of legislation that governments will introduce if the goals of the agreement are not met.

Box 11.6 **Examples of national voluntary agreements**

- **The Netherlands Voluntary Agreement on Energy Efficiency:** A series of legally binding long term agreements based on annual improvement targets and benchmarking covenants between 30 industrial sectors and the government with the objective to improve energy efficiency.
- **Australia 'Greenhouse Challenge Plus' programme:** An agreement between the government and an enterprise/industry association to reduce GHG emissions, accelerate the uptake of energy efficiency, integrate GHG issues into business decision making, and provide consistent reporting. See http://www.greenhouse.gov.au/challenge.
- **European Automobile Agreement:** An agreement between the European Commission and European, Korean, and Japanese car manufacturing associations to reduce average emissions from new cars to 140 gCO_2/km by 2008–2009. See http://ec.europa.eu/environment/CO2/CO2_agreements.htm.
- **Canadian Automobile Agreement:** An agreement between the Canadian government and representatives of the domestic automobile industry to reduce emissions from cars and light-duty trucks by 5.3 $MtCO_2$-eq by 2010. The agreement also contains provisions relating to research and development and interim reduction goals.
- **Climate Leaders:** An agreement between US companies and the government to develop GHG inventories, set corporate emission reduction targets, and report emissions annually to the US EPA. See: http://www.epa.gov/climateleaders/.
- **Keidaren Voluntary Action Plan:** An agreement between the Japanese government and 34 industrial and energy converting sectors to reduce GHG emissions. A third party evaluation committee reviews the results annually and makes recommendations for adjustments. See http://www.keidanren.or.jp.

(Source: IPCC, Fourth Assessment Report, Working Group III, box 13.4)

Environmental effectiveness of VAs has been the subject of many studies. The findings are mixed. The majority of agreements have not achieved significant emission reductions beyond what would have happened under a business as usual scenario. However, some more recent agreements, in a few countries, have led to faster implementation of best available technology and to measurable emission reductions. The most successful VAs have clear and quantitative targets, a defined baseline situation to compare with, independent third party monitoring and review, and a credible threat of legislative action when goals are not met. VAs fit better in some cultures than in others. In Japan for instance there is a long tradition of close cooperation between industry and government and compliance of VAs is taken very seriously. These mixed findings are often ignored by fierce proponents of VAs that do not like to see a move towards more stringent policy instruments. In the run up to the introduction of the EU Emissions Trading System for instance there was strong resistance by German and Dutch industry associations who argued that their VAs were more effective than the envisioned emissions trading system. Now the system is in place, industry has adjusted very well to it.

Introducing other policy instruments does not mean that VAs no longer have a role to play. They can often be supplementary to other policies as a way of raising awareness and mobilizing the innovation capacity of industry. VAs can also be used to promote actions of non-commercial entities, such as Social Housing Corporations[7], local governments, and water management authorities. They then become a tool to coordinate policy at different levels of government.

Subsidies and other financial incentives

Subsidies are popular because they are politically attractive. And that not only holds for direct subsidies, but also for price support (guaranteed prices for renewable electricity for instance) and tax deductions or exemptions. They are in fact indirect subsidies. Subsidies are widespread, but not always helping a low carbon economy. Many countries for instance provide subsidies on fossil fuel products or fossil fuel based electricity. In OECD countries these fossil fuel subsidies are 20–80 billion US dollars per year and the amounts in developing countries and countries with economies in transition are even higher. These subsidies are often justified to assist poor people, but in practice most of the subsidies end up in the hands of people who do not really need them. The result is increased consumption, lack of incentives to use energy efficiently and unfair competition with renewable energy. Removal of such subsidies is politically very difficult, which explains the pervasiveness of existing fossil fuel subsidies.

Subsidies can be effective however to help the market development of low carbon technologies. They are widely used for that purpose and have been successful. For renewable electricity feed-in tariffs (a guaranteed price at which utilities have to buy the electricity from suppliers, see Chapter 5) and producer subsidies (a certain amount per kWh produced) are used in more than 50 (developed and developing) countries[8]. Important in using subsidies as a policy instrument is the need to reduce them over time to reflect the cost reduction of the technology due to the fact that the market is expanding and to reflect cost increases of fossil fuel alternatives. What is also important is to focus the subsidy as precisely as possible on those that need it and not where the low carbon technology would be used anyway.

Subsidies are relatively expensive policy instruments, because not all the money gets to the right place and because subsidies are often continued at too high a level for political reasons[9].

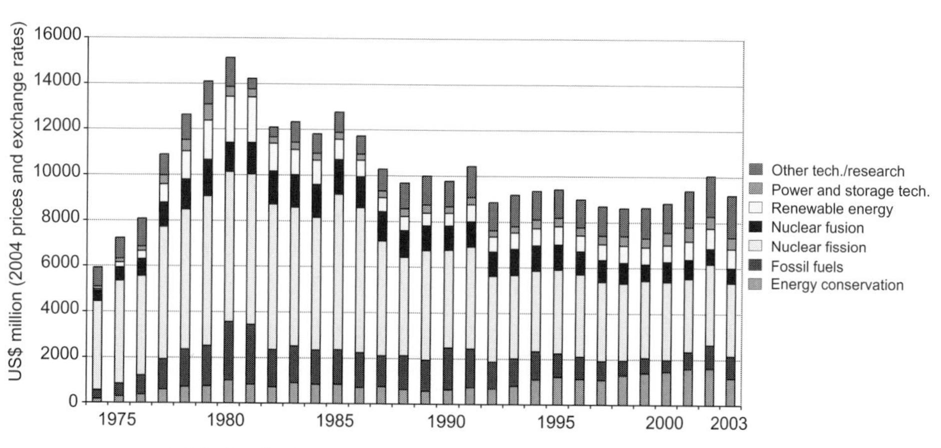

Figure 11.1 **Public funded energy R&D expenditures from IEA Member States.**
Source: IPCC, Fourth Assessment Report, Working Group III, figure 13.1.a.

Research and development

Funding and focussing research and development (R&D) is an important policy instrument. For controlling climate change in the longer term it is essential that new, improved and cheaper low carbon technologies become commercially available and R&D is crucial for that. R&D is however sometimes used as a substitute for direct policy action in cases where the political will is lacking or strong opposition against climate change policy is present.

Energy related R&D funding by governments has declined substantially after the oil crisis of the 1970s. It is now almost half the 1980 level and there is no systematic increase, not even after the Climate Change Convention came into force in 1994. Figure 11.1 shows the trend and the share of the various topics. Private R&D funding has also declined[10].

R&D alone has only a limited effect on changing greenhouse gas emissions. The reason is that cost reduction of new technologies is driven more by the learning effect of actually building and implementing them (see Chapter 10). But a successful long term transition to a low carbon economy cannot be achieved without a much strengthened R&D effort[11].

Information instruments

Awareness about the impacts of climate change and the opportunities to control it is vital to effective action. It is important for taking individual action in households and companies, but also to build public support for local and national policies to seriously reduce greenhouse gas emissions. Civil society (the broad array of non-governmental and business organizations in society) plays a big role in this area. Governments cannot do this on their own.

Product labelling, either mandated by law or on a voluntary basis, is widely used to inform consumers about energy use of household appliances, cars, and even houses[12]. There are many different systems in use, mostly focussing on energy use. Some attempts have been made to start labelling of food products regarding the indirect CO_2 emissions due to transport. This has led to heated debates if this is a proper reflection of the carbon content of food items, because only transport is covered and because it ignores important social issues. See the example about labelling of air freighted perishable goods in Box 11.7.

Box 11.7 **How green are your beans?**

In 2006 the UK supermarket giant Tesco announced its plan to introduce carbon labelling. They are therefore working on developing a universally accepted and commonly understood measure of the carbon footprint of every product they sell – looking at its complete lifecycle from production, through distribution to consumption. The issue that has been the focus of much attention is that of 'food miles': the carbon cost of transporting food from around the world and domestically between centralized distribution points and stores. Air freighted fruit and vegetables are often highlighted in this debate, and both Tesco and Marks & Spencer have recently introduced 'air freighted' labels to enable consumers to make informed choices.

Whilst imported produce is easy to single out for its climate impact, any significant move away from these products would have negative impacts on producers in Kenya, Thailand, and other countries that have built up perishable exports industries. As Hilary Benn, UK Minister for International Development, notes: *'The food miles debate poses a real dilemma. People say I want to do my bit to stop climate change. So, should I only buy local and boycott produce from abroad, especially things flown in – or should I support poor farmers to improve their income, to take care of their families, to work and trade their way out of poverty?'*

Research by IIED has found that if consumers were to boycott fresh produce air freighted from Africa, the UK's total emissions would be reduced by less than 0.1%, but impacts on workers, communities, and economies in countries that have invested in developing a niche in perishable goods would be much more significant. This danger is certainly seen as important by industry players in exporting countries. As Jane Ngige, Chief Executive of the Kenya Flower Council said: *'We consider ourselves as partners with UK supermarkets . . . One minute we are talking about fair trade and market compliance, the next this is less of an issue and the issue is lessening the carbon footprint of the developed world possibly by cutting markets in Africa'.* Kenya's High Commissioner in London, Joseph Muchemi, has also criticized the labelling scheme which he says may lead to a boycott of such products. However, according to Tesco, *'our customers love Kenyan produce. There has been no reduction of sales but instead they seem to have gone up'*.
(Source: What assures consumers on Climate Change, Consumers International, 2007)

In addition, many governments run or commission information campaigns to inform consumers of opportunities to reduce energy use, reduce CO_2 emissions, and save money. Companies are also positioning themselves more and more as 'green' in the hope to appeal to consumer awareness. Lessons on what is needed to be trustworthy for consumers so that they do buy 'green products' have been formulated (see Table 11.1)

Table 11.1. Lessons on green marketing

What works in assuring consumers	Application to climate change
Consistency. Companies that consumers trust give out the same message in everything they do – through their products, labels and promotions, customer service staff, corporate communications, and through partnerships with trusted experts	**Yahoo,** for example, in developing a strategy for going carbon 'neutral' has sought to do this in a way that is not only rigorous and transparent but embedded in its corporate culture. Company founder David Filo announced the plan by posting a question on the company's 'Yahoo Answers' site asking customers how the company should go about achieving carbon 'neutral' status
Seeing is believing. A picture may be worth a thousand words but a demonstration beats everything	Given the invisible and global nature of greenhouse gas emissions this is difficult to do, and there is a danger of focusing on highly but tokenistic actions visible. **The Co-operative Group's** Solar Tower in Manchester UK and **Wal-Mart**'s move to put solar panels on the roof of some of its stores are key examples where companies have used highly visible demonstrations alongside less visible operational changes and policies
Serious intent. Consumers accept that companies are profit-motivated, but they object when there is a discrepancy between what they say and what they do, or where they appear to be 'greenwashing' in their approach	This is likely to be a major challenge. **Few companies** are yet able to claim that they have done everything possible to reduce their own impacts, and that consumption of their products is consistent with a 'one planet lifestyle' in which emissions levels could be cut by 90%
Trust in the messenger. Companies need to take a broad view of assurance and develop both formal and informal mechanisms to get their message across, not forgetting their own employees as ambassadors of the company. And while it is fact that consumers say they do not trust celebrities, reality proves some however respond to them, though there is increasingly a call for more accountability	**News Corporation** uses both the power of individual relationships and mass media to get the climate change message across to consumers. While in the UK BSkyB engineers have been dropping off low energy light bulbs when they install equipment in people's homes, the company has set up a MySpace channel dedicated to climate change and Chairman Rupert Murdoch has committed to engaging its millions of readers, viewers, and web users around the world on this issue
Layers of information. Companies that people trust provide an ethics built-in guarantee within their brand and back this up with the right information when and	**Marks & Spencer's** 'Plan A' and **Whole Foods Market** both tie the company's commitments to climate change into its overall brand offering and back it up with further information and endorsements including store ambassadors – in the case of Marks & Spencer – for those who want to find out more

Table 11.1. (*cont.*)

What works in assuring consumers	Application to climate change
where it is needed to help consumers make decisions without having to always 'read the small print'	
Linking responsibility, quality, service, and value for money	**Toyota** succeeded in marketing its hybrid, Toyota Prius, as a mass market vehicle. Buyers are attracted not only to the car's fuel efficiency but to its iconic status as an environmentally friendly vehicle popularized by many celebrities
Helping customers select choices. The majority of consumers do not want to have to take into account too many extra factors in their everyday decision making	**The Co-operative Group** is employing a choice reduction strategy to only stocking energy efficient compact fluorescent light bulbs and high energy efficiency rated kitchen appliances

Source: What assures consumers on Climate Change, Consumers International, 2007.

Information instruments alone are unable to realize significant emission reduction. Their importance lies in supporting other instruments to be more effective.

Voluntary actions

Although not a policy instrument per se, voluntary actions do play a role in shaping public policy and mobilizing society to tackle climate change. As indicated above, voluntary actions are different from voluntary agreement in the sense that governments do not play a role.

There are many examples of business, NGO, and joint initiatives aiming to make a difference[13]. Public disclosure of greenhouse gas emissions for instance can help raise awareness amongst private companies about their contribution to emissions that can trigger reduction measures. The Carbon Disclosure Project is one of the biggest private efforts in place today (see Box 11.8). It was initiated to assist institutional investors in assessing the risks of investing in companies. The Global Reporting Initiative[14] is another private initiative involving thousands of companies and institutions, focussing on promoting sustainability reporting by companies.

Box 11.8 Carbon Disclosure Project

The Carbon Disclosure Project (CDP) is an independent not-for-profit organization aiming to create a lasting relationship between shareholders and corporations regarding the implications for shareholder value and commercial operations presented by climate change. Its goal is to facilitate a dialogue, supported by quality information, from which a rational response to climate change will emerge.

CDP provides a coordinating secretariat for institutional investors with combined assets of over $57 trillion under management. On their behalf it seeks information on the business risks and opportunities presented by climate change and greenhouse gas emissions data from the world's largest companies: 3000 in 2008. Over 8 years CDP has become the gold standard for carbon disclosure methodology and process. The CDP website is the largest repository of corporate greenhouse gas emissions data in the world.

CDP leverages its data and process by making its information requests and responses from corporations publicly available, helping catalyze the activities of policymakers, consultants, accountants, and marketers.

(Source: http://www.cdproject.net/)

Other voluntary initiatives are the World Business Council on Sustainable Development 'Cement Sustainability Initiative'. A number of the biggest cement manufacturers from across the globe report information about energy use and CO_2 emissions and develop standards to promote lower emissions. The World Steel Association[15] has comparable voluntary programmes for its member companies. Joint NGO–private sector initiatives are the WWF Climate Savers[16] and the Pew Business Environment Leadership Council[17] where companies are supported by positive publicity of NGO's if they pledge to undertake emission reductions actions as part of their green marketing strategies.

A special form of voluntary action is green government procurement. In this category fall (national, regional, or local) government purchases that are screened for low carbon products, government buildings that are made more energy efficient on a voluntary basis, and also governments purchasing renewable electricity for their government buildings or installing solar PV cells to generate their own. Of course these measures are normally subject to some form of budgetary approval by elected councils or parliament, and in that sense they are different from voluntary action by private entities. They nevertheless can create good examples, can be moved quickly without the need for legislation, and can help to build markets for low carbon products. Requiring local and national governments to implement green procurement is a next step that has now been initiated in the EU, which brings this approach into the regulatory category[18].

Non-climate policies

As was extensively discussed in Chapter 4, integrating climate change into other socio-economic and development policies is one of the most effective ways to change investment patterns, behaviour, energy use, and greenhouse gas emissions. It can influence the drivers of social and economic development and realize a transition to a prosperous low carbon economy. It also engages a whole new range of stakeholders, which in many countries are more influential than those that shape environmental or climate change policies. By combining the goal of reducing greenhouse gas emissions with promoting social and economic progress, resistance against climate change action can be effectively overcome. See Chapter 4 for a more in-depth discussion.

What are the strengths and weaknesses of the various policy instruments?

When discussing the main types of policy instruments above, it became clear that each instrument has its strengths and weaknesses, often even dependent on the national circumstances in which it is applied. To help assessment of what works best under what conditions, the different policy instruments can be checked against four criteria[19]:

- *Environmental effectiveness*: how effective is the instrument in realizing emission reductions? This is not only affected by the type of instrument of course (information versus regulation for instance), but also to a large extent by the stringency of the goals set and the enforcement of the policy.

- *Cost-effectiveness:* what are the social costs of achieving a specific environmental effect by the respective instrument? The most cost-effective policy is the one that achieves a desired goal at the lowest costs. In comparing policy instruments this becomes somewhat problematic, because different instruments cannot all achieve the same goal in terms of emission reduction.
- *Distributional considerations:* how does the instrument affect different groups in society? Is there a fair distribution of who pays the costs and who reaps the benefits? Policies that are perceived as being unfair to specific groups normally have a hard time getting through the political decision making process, although sometimes lobbying power is more important.
- *Institutional feasibility:* can the instrument get through political decision making and can it be implemented and enforced given the institutional infrastructure? Institutional capacity varies a lot between countries, so this is an issue that often leads to different scores for advanced industrialized countries versus poor developing countries.

Table 11.2 gives a concise overview of how the various policy instruments score against these criteria. As discussed above for each of the policy instruments, they all have their strengths and weaknesses. Environmental effectiveness of regulation and tradable permits is higher than for the other instruments, with information, R&D, and voluntary agreements being on the soft end of the scale. From a cost-effectiveness point of view market approaches through taxes or tradable permits generally score better than others, but the specific design of the instrument can make a big difference. Distributional and equity considerations are important for all instruments and all of them require careful design or compensation to create a level playing field. Institutional feasibility can be a real problem for some instruments in countries with limited administrative capabilities. If there is not a well functioning tax system or no experience with regulated markets, the use of taxes and tradable permits can be problematic. As said above, policy instruments have to be tailored to the specific circumstances in a country and a sector and should be used in combination to be effective.

What are the lessons from practical experience?

Since the entry into force of the United Nations Framework Convention on Climate Change (UNFCCC) in 1994 and more specifically after the agreement on the Kyoto Protocol in 1997, countries have begun to implement policies to reduce emissions. So there is now a reasonable experience with climate policies from which conclusions can be drawn. Table 11.3 summarizes these conclusions for the criterion of environmental effectiveness, for each of the main economic sectors.

The differences between sectors are striking. In energy supply, financial instruments (taxes and subsidies) are the most effective, with only renewable energy obligations as an effective regulatory instrument. In the building sector, the picture is completely different. Regulatory approaches are clearly superior there. For transport, agriculture, and forestry it

Table 11.2. National climate policies and their performance against four evaluation criteria

	Criteria			
Instrument	Environmental effectiveness	Cost-effectiveness	Meets distributional considerations	Institutional feasibility
Regulations and standards	Emissions level set directly, though subject to exceptions. Depends on deferrals and compliance	Depends on design; uniform application often leads to higher overall compliance costs	Depends on level playing field. Small/new actors may be disadvantaged	Depends on technical capacity; popular with regulators in countries with weakly functioning markets
Taxes and charges	Depends on ability to set tax at a level that induces behavioural change	Better with broad application; higher administrative costs where institutions are weak	Regressive; can be ameliorated with revenue recycling	Often politically unpopular; may be difficult to enforce with underdeveloped institutions
Tradable permits	Depends on emissions cap, participation, and compliance	Decreases with limited participation and fewer sectors	Depends on initial permit allocation. May pose difficulties for small emitters	Requires well functioning markets and complementary institutions
Voluntary agreements	Depends on programme design, including clear targets, a baseline scenario, third party involvement in design and review and monitoring provisions	Depends on flexibility and extent of government incentives, rewards, and penalties	Benefits accrue only to participants	Often politically popular; requires significant number of administrative staff
Subsidies and other incentives	Depends on programme design; less certain than regulations/standards	Depends on level and programme design; can be market distorting	Benefits selected participants, possibly some that do not need it	Popular with recipients; potential resistance from vested interests. Can be difficult to phase out

Table 11.2. (*cont.*)

	Criteria			
Instrument	Environmental effectiveness	Cost-effectiveness	Meets distributional considerations	Institutional feasibility
Research and development	Depends on consistent fundings when technologies are developed and polices for diffusion. May have high benefits in the long term	Depends on programme design and the degree of risk	Benefits initially selected participants; potentially easy for funds to be misallocated	Requires many separate decisions. Depends on research capacity and long term funding
Information policies	Depends on how consumers use the information; most effective in combination with other policies	Potentially low cost, but depends on programme design	May be less effective for groups (e.g. low-income) that lack access to information	Depends on cooperation from special interest groups

Note: Evaluations are predicated on assumptions that instruments are representative of best practice rather than theoretically perfect. This assessment is based primarily on experiences and published reports from developed countries, as the number of peer reviewed articles on the effectiveness of instruments in other countries is limited. Applicability in specific countries, sectors, and circumstances – particularly developing countries and economies in transition – may differ greatly. Environmental and cost effectiveness may be enhanced when instruments are strategically combined and adapted to local circumstances.

Source: IPCC, Fourth Assessment Report, Working Group III, table 13.1.

Table 11.3. Selected sectoral policy instruments that have been shown to be environmentally effective in the respective sector in at least a number of national cases

Sector	Policies[a], measures and instruments shown to be environmentally effective	Key constraints or opportunities
Energy supply	Reduction of fossil fuel subsidies Taxes or carbon charges on fossil fuels	Resistance by vested interests may make them difficult to implement
	Feed-in tariffs for renewable energy technologies Renewable energy obligations Producer subsidies	May be appropriate to create markets for low emissions technologies
Transport	Mandatory fuel economy, biofuel blending, and CO_2 standards for road transport	Partial coverage of vehicle fleet may limit effectiveness
	Taxes on vehicle purchase, registration, use and motor fuels, road and parking pricing	Effectiveness may drop with higher incomes
	Influence mobility needs through land use regulations, and infrastructure planning Investment in attractive public transport facilities and non-motorized forms of transport	Particularly appropriate for countries that are building up their transportation systems
Buildings	Appliance standards and labelling	Periodic revision of standards needed
	Building codes and certification	Attractive for new buildings Enforcement can be difficult
	Demand-side management programmes Public sector leadership programmes, including procurement	Need for regulations so that utilities may profit Government purchasing can expand demand for energy-efficient products
	Incentives for energy service companies (ESCOs)	Success factor: Access to third party financing
Industry	Provision of benchmark information	May be appropriate to stimulate technology uptake
	Performance standards Subsidies, tax credits	Stability of national policy important in view of international competitiveness

Table 11.3. (*cont.*)

Sector	Policies[a], measures and instruments shown to be environmentally effective	Key constraints or opportunities
	Tradable permits	Predictable allocation mechanisms and stable price signals important for investments
	Voluntary agreements	Success factors include: clear targets, a baseline scenario, third party involvement in design and review and formal provisions of monitoring, close cooperation between government and industry
Agriculture	Financial incentives and regulations for improved land management, maintaining soil carbon content, efficient use of fertilizers, and irrigation	May encourage synergy with sustainable development and with reducing vulnerability to climate change, thereby overcoming barriers to implementation
Forestry/forests	Financial incentives (national and international) to increase forest area, to reduce deforestation, and to maintain and manage forests	Constraints include lack of investment capital and land tenure issues. Can help poverty alleviation
	Land use regulation and enforcement	
Waste management	Financial incentives for improved waste and wastewater management	May stimulate technology diffusion
	Renewable energy incentives or obligations	Local availability of low cost fuel
	Waste management regulations	Most effectively applied at national level with enforcement strategies

[a] Public R&D investment in low emissions technologies have proven to be effective in all sectors.
Source: IPCC, Fourth Assessment Report, Working Group III, table SPM.7.

is a mixture of financial and regulatory instruments that have shown to be the most environmentally effective. Industry is a special case: tradable permits, voluntary approaches, and information instruments have played a strong role there.

As part of the UNFCCC obligations industrialized countries report annually on the implementation of climate policy. Box 11.9 gives a summary of the most recent trends in the use of policy instruments.

Box 11.9 Summary of policies and measures used by Annex I countries

Annex I Parties, with few exceptions, are increasingly relying on harder (economic and regulatory) instruments over softer (voluntary) instruments to elicit emission reductions. In addition, new and innovative policy approaches have gained prominence and share in overall policy portfolios such as market-based mechanisms, including tradable certificate schemes.

Carbon taxes have played a key role in some countries for some time, but newer quotas and tradable certificates systems (i.e. regulations with an element of economic flexibility) are growing more quickly and are already more widely used. In countries where both carbon taxes and emissions trading are implemented, governments are seeking synergy between the two instruments to ensure comprehensive coverage of emission sources: in most cases, emissions trading targets a fixed number of mostly large sources and installations, while carbon tax remains in sectors that are not easily incorporated under emissions trading. Emissions trading is the largest and most visible form of tradable certificate systems, but green certificates (renewable energy sources), white certificates (energy efficiency), and landfill allowance trading schemes are growing as well. Moreover, regulatory approaches are widely used to mitigate emissions from industrial processes, for example emissions of PFC, HFC, and SF_6.

(Source: UNFCCC secretariat report FCCC/SBI/2007/INF.6 19 November 2007)

'Lean and mean'

An effective national policy is thus always a matter of putting together a portfolio of policy instruments. However, the leaner such a portfolio is, while covering all important sectors and activities, the better it is. Packages of policies can easily become overlapping, creating unnecessary burdens for and confusion amongst different actors and putting pressure on administrative and regulatory institutions. Design of an effective and efficient policy portfolio is crucial.

All policy instruments have their limitations and that is a strong incentive to go for combinations of policy instruments. Tradable permit systems could in theory cover the whole economy. In practice however they become very complex and labour intensive when dealing with large numbers of smaller emitters and emission source with a high degree of uncertainty (see the section on tradable permits above). That is the reason that such systems are usually restricted to large emitters, creating the need for other instruments to cover the other emitters. Another weak point of tradable permit systems is that they do not easily provide incentives for development of future low carbon technologies. Additional policy instruments to promote R&D and demonstration plants then need to be introduced[20].

Similarly, carbon taxes could theoretically cover the whole economy as well, because actors, particularly individual consumers, often do not react to the financial incentives created by a tax. In those circumstances regulatory instruments can be much more effective. Political problems with taxes also put limits on what a tax policy can do.

National policy packages

What ultimately counts is how the overall national policy package fits together. Many countries have by now put together such packages, both countries that have emission caps under the Kyoto Protocol and developing countries that are addressing greenhouse gas emissions as part of their national sustainable development plans. China for instance has a sustainable development plan in place for the period till 2010 that will lead to significantly lower CO_2 emissions than otherwise would have occurred. See Chapter 4 for a detailed description. India has recently published its National Action Plan on Climate Change (see Box 11.10). Of course, these national policies are not meant to lead to absolute reduction of emissions. Given the huge development challenge, that is not yet possible. But these plans will be able to keep emissions below what they otherwise would have been. What is fundamental to developing country policies is that they are driven by non-climate change considerations, such as energy security, modernization of industry, improving air quality, or combating erosion. Climate change benefits almost always come as a co-benefit.

Box 11.10 Indian National Action Plan on Climate Change

Emphasizing the overriding priority of maintaining high economic growth rates to raise living standards, the plan 'identifies measures that promote our development objectives while also yielding co-benefits for addressing climate change effectively'. It says these national measures would be more successful with assistance from developed countries, and pledges that India's per capita greenhouse gas emissions 'will at no point exceed that of developed countries even as we pursue our development objectives.'

National Missions

National Solar Mission: The NAPCC aims to promote the development and use of solar energy for power generation and other uses with the ultimate objective of making solar competitive with fossil-based energy options. The plan includes:

- Specific goals for increasing use of solar thermal technologies in urban areas, industry, and commercial establishments
- A goal of increasing production of photovoltaics to 1000 MW/year

- A goal of deploying at least 1000 MW of solar thermal power generation. Other objectives include the establishment of a solar research centre, increased international collaboration on technology development, strengthening of domestic manufacturing capacity, and increased government funding and international support

National Mission for Enhanced Energy Efficiency: Current initiatives are expected to yield savings of 10 000 MW by 2012. Building on the Energy Conservation Act 2001, the plan recommends:

- Mandating specific energy consumption decreases in large energy-consuming industries, with a system for companies to trade energy savings certificates
- Energy incentives, including reduced taxes on energy efficient appliances
- Financing for public–private partnerships to reduce energy consumption through demand-side management programmes in the municipal, buildings, and agricultural sectors

National Mission on Sustainable Habitat: To promote energy efficiency as a core component of urban planning, the plan calls for:

- Extending the existing Energy Conservation Building Code
- A greater emphasis on urban waste management and recycling, including power production from waste
- Strengthening the enforcement of automotive fuel economy standards and using pricing measures to encourage the purchase of efficient vehicles
- Incentives for the use of public transportation

National Water Mission: With water scarcity projected to worsen as a result of climate change, the plan sets a goal of a 20% improvement in water use efficiency through pricing and other measures.

National Mission for Sustaining the Himalayan Ecosystem: The plan aims to conserve biodiversity, forest cover, and other ecological values in the Himalayan region, where glaciers that are a major source of India's water supply are projected to recede as a result of global warming.

National Mission for a 'Green India': Goals include the afforestation of 6 million hectares of degraded forest lands and expanding forest cover from 23% to 33% of India's territory.

National Mission for Sustainable Agriculture: The plan aims to support climate adaptation in agriculture through the development of climate-resilient crops, expansion of weather insurance mechanisms, and agricultural practices.

National Mission on Strategic Knowledge for Climate Change: To gain a better understanding of climate science, impacts, and challenges, the plan envisions a new Climate Science Research Fund, improved climate modelling, and increased international collaboration. It also encourages private sector initiatives to develop adaptation and mitigation technologies through venture capital funds.

Other Programmes

The NAPCC also describes other ongoing initiatives, including:

- Power Generation: The government is mandating the retirement of inefficient coal-fired power plants and supporting the research and development of IGCC and supercritical technologies
- Renewable Energy: Under the Electricity Act 2003 and the National Tariff Policy 2006, the central and the state electricity regulatory commissions must purchase a certain percentage of grid-based power from renewable sources
- Energy Efficiency: Under the Energy Conservation Act 2001, large energy-consuming industries are required to undertake energy audits and an energy labelling programme for appliances has been introduced

Implementation

Ministries with lead responsibility for each of the missions are directed to develop objectives, implementation strategies, timelines, and monitoring and evaluation criteria, to be submitted to the Prime Minister's Council on Climate Change. The Council will also be responsible for periodically reviewing and reporting on each mission's progress. To be able to quantify progress, appropriate indicators and methodologies will be developed to assess both avoided emissions and adaptation benefits.

(Source: Pew Center Summary, http://www.pewclimate.org/international/country-policies/india-climate-plan-summary/06-2008)

Policy programmes in industrialized countries are often more directly focussed on emission reductions and co-benefits do not play such an important role. They are often very broad with large numbers of policy instruments complementing each other. A good example is the climate policy of the European Union. The 27 Member States have put together a comprehensive set of policies to reach the target of reducing greenhouse gas emissions collectively to 8% below the 1990 level over the period 2008–2012. An important part is formulated at EU level, but that is supplemented with extensive policy packages at national level. The EU has put together such a package for reaching its unilateral objective of reducing greenhouse gas emissions further to 20% below the 1990 level by the year 2020[21]. A summary of that policy package is presented in Box 11.11. Box 11.12 shows the complementary national policy programme for Germany, covering the actions in addition to implementation of EU policy. As part of the internal effort sharing within the EU Germany is supposed to deliver a reduction of 21% compared to 1990.

Japan, which has a reduction obligation of 6% below the 1990 level under the Kyoto Protocol, is following a very different approach than the EU and other industrialized countries in terms of its policy package[22]. It has not introduced tradable permit systems (a limited voluntary version is being introduced), nor has it requirements or feed-in tariffs for renewable energy. On the other hand, it has a strong energy efficiency standards programme, with automatic strengthening[23]. It also has extensively used voluntary agreements between government and industry and has invested heavily in research and development[24]. It is also one of the few countries to use policies aiming at lifestyle changes, such as guidelines for minimum temperatures in air conditioned buildings or 'lights out at night' in offices.

Going against the (federal) tide on climate change in the US, the State of California has been one of the forerunners in developing climate change policies. It built this on a long history of active environmental policy and electricity regulatory actions. The latter for instance led to implementing extensive so-called 'Demand Side Management' programmes that make it attractive for electricity generators to invest in end-use efficiency improvement, while being able to make a profit. This was realized by regulations tying the investments in end-use efficiency to the electricity prices that companies can charge. There are building codes and appliance standards in place, and there are many policies to stimulate the generation of renewable energy. As a result of all efficiency policies it is estimated that about 20 power plants of 500MW have been avoided since the beginning of these programmes in the 1970s. Building of coal fired power plants has been effectively banned, although coal based electricity is imported into the State from elsewhere. As a result the average emissions of CO_2 per capita in California are about half that of the rest of the USA[25]. New car standards have been introduced for 2016 and 2020, bringing emissions down to levels comparable to what is now being discussed in the EU[26]. These are challenged in court however by the US federal government, which considers car emission standards to be the prerogative of the federal government. Strong overall emission targets have also been set for the State: reduction of greenhouse gas emissions to 1990 levels by 2020 and 80% below 1990 levels by 2050. The share of non-hydropower renewables in electricity has been set at 20% in 2020 and 33% in 2020 (it was 11% in 2006)[27].

Box 11.11 EU integrated climate, energy, and transport policies for the period till 2020

The package of policies and measures that is currently proposed for implementation in the period till 2020 by the 27 EU Member States is as follows:

- EU Emissions Trading System (EU ETS):
  - emission cap to be tightened to −21% below 2005 by 2020 for covered sectors including air transport sector and parts of chemical industry in ETS
  - harmonized allocation of allowances to avoid competitiveness problems
  - increased auctioning of allowances: 70% auctioning of allowances to industries not subject to international auctioning by 2020 and 100% by 2027
  - linking of EU ETS to other emission trading systems and (in a limited way) to CDM
- CCS:
  - acceptance of CCS in ETS
  - regulations regarding liability and safety
- Non-ETS sectors (60% of total GHG emissions):
  - emissions cap −10% below 2006 by 2020
  - differentiated (according to GDP per capita) individual caps for Member States from −0% to +20% compared to 2005 by 2020

- Renewable energy:
  - 20% mandatory minimum share of renewable energy in final energy use by 2020 for EU as a whole
  - differentiated individual minimum shares of renewable energy for each Member State, varying from 10% to 49%
  - freedom for Member States to chose policies to realize this mandatory minimum
- Transport:
  - minimum use of 10% biofuel in transport, with minimum standards for carbon reduction and sustainability
  - average maximum vehicle emissions standard for new cars of 130 gCO_2/km, to be achieved in 2015
  - additional measures to reach 10 g/km further reduction on average
- Buildings:
  - more stringent minimum standards for building codes
- Energy efficiency:
  - energy efficiency standards for consumer goods
  - enhanced energy labelling for goods without standards

(Source: http://ec.europa.eu/environment/climat/home_En.htm; http://ec.europa.eu/energy/index_En.html)

Box 11.12 National climate policy Germany

The most important elements of the 2008 integrated energy and climate programme of Germany are:

- *General*: promotion and rapid implementation of EU legislation
- *Combined heat and power*: modification of subsidies to increase CHP share of electricity to 25% by 2020
- *Renewable energy*:
  - modification of feed-in tariffs, improvement of the electricity grid to handle fluctuating supply and zoning regulations for off-shore wind power; should lead to renewable electricity share of 5–30% by 2020
  - introducing feed-in tariffs for biogas, leading to a 6% share by 2020
- *CCS*: financing of 2–3 large scale CCS demonstration plants
- *Smart metering*: liberalizing the market for electricity meters and regulatory changes to allow variable price regimes
- *Energy efficiency:*
  - changing tax deductions for industry after 2012 to reward energy efficiency
  - subsidies for energy efficiency advice to business and households
  - market introduction subsidies for new energy efficient technologies
  - information campaigns
  - promotion of export of energy efficient technologies
  - enhanced energy labelling of consumer goods

- *Buildings*:
  - tightening building codes, including requirement for minimum use of renewable energy for heating
  - regulations to require actual energy consumption is charged to apartments
  - extension and modification of subsidies for energy renovation of existing buildings
  - energy renovation of government buildings
- *Transport*:
  - vehicle tax reform to make it CO_2 emission dependent
  - improved vehicle energy labelling
  - differentiation of road toll for trucks according to CO_2 emissions
- *Fluorinated gases:*
  - tightening of regulation on leakage from refrigeration
  - subsidies for introducing zero emission alternative technologies in refrigeration and air conditioning
- *Public procurement*: guidelines for energy efficient procurement for federal government agencies and encouragement of state and local governments to do the same
- *Research and development*: increased R&D funding
- *International assistance*: additional funding of low carbon energy and adaptation projects in developing countries, funded from the proceeds of auctioning allowances under the EU ETS

(Source: Federal Ministry for Environment, Nature Conservancy and Nuclear Safety: Key Elements of an Integrated Energy and Climate Programme, Decisions of the German Cabinet of Ministers, 2007)

Implementation and enforcement

Climate policies mean nothing without active implementation and enforcement. On paper policies may look good; however, if there is no clear, transparent, and competent implementation through qualified agencies and no enforcement through effective monitoring, inspection, verification, and issuance of penalties they are ineffective. Some of these aspects were discussed above when looking at the effectiveness of certain types of policy instruments in different circumstances. Policies that require strong administrative capabilities, such as fiscal and market instruments, could be ineffective in many developing countries. Regulations require inspection and enforcement to be effective. Unfortunately those aspects are often neglected.

There are no good overviews of compliance records of countries with their own legislation. International and EU networks of compliance and enforcement practitioners try to improve the quality of implementation[28].

Notes

1. Civil society is defined as 'the arena of uncoerced collective action around shared interests, purposes and values', see Rayner S., Malone E., Security, governance and the environment. In Lowi M., Shaw B. (eds). Environment and Security: discourses and practices, Macmillan, 2000

2. IPCC Fourth Assessment Report, Working Group III, ch. 12.2.3.
3. http://www.iea.org/textbase/pm/grindex.aspx.
4. IPCC Fourth Assessment Report, Working Group III, ch. 6.8.3.1.
5. See IEA Policies and Measures Database, http://www.iea.org/textbase/pm/index_clim.html.
6. The most important counterargument against price caps is that they will undermine the environmental effectiveness of tradable permits, its strongest property. Proposals for price caps have triggered other proposals for price floors (i.e. a minimum price at which permits are auctioned), in order to strengthen the instrument. See IPCC Fourth Assessment Report, Working Group III, ch. 13.2.1.3.
7. See http://www.ec.europa.eu/energy/demand/legislation/doc/neeap/netherlands_en.pdf.
8. REN21, Renewable Energy Status Report, 2007.
9. IPCC Fourth Assessment Report, Working Group III, ch. 13.2.1.5.
10. For US numbers see http://rael.berkeley.edu/files/2005/Kammen-Nemet-ShrinkingRD-2005.pdf.
11. IPCC Fourth Assessment Report, Working Group III, ch. 13.2.1.6.
12. See Energy labelling of housing in the Netherlands in http://www.ec.europa.eu/energy/demand/legislation/doc/neeap/netherlands_En.pdf.
13. IPCC Fourth Assessment Report, Working Group III, ch. 13.4.2.
14. http://www.globalreporting.org.
15. http://www.worldsteel.org.
16. http://www.panda.org/about_wwf/what_we_do/climate_change/our_solutions/business_industry/climate_savers/ index.cfm.
17. http://www.pewclimate.org/companies_leading_the_way_belc/.
18. http://ec.europa.eu/environment/gpp/index_en.htm.
19. IPCC Fourth Assessment Report, Working Group III, ch. 13.2.2.2.
20. IPCC Fourth Assessment Report, Working Group III, ch 13.2.2.1.
21. http://ec.europa.eu/environment/climat/home_en.htm.
22. Annual Report on the Environment and the Sound Material-Cycle Society in Japan 2008, Ministry of the Environment, Japan, 2008.
23. This is the so-called 'top-runner programme', see http://www.eccj.or.jp/top_runner/index.html.
24. WWF, G8 climate change scorecard, http://a*ssets*.panda.org/downloads/g8scorecardsjun29light.pdf*imaterds.*
25. No reason to wait: the benefits of greenhouse gas reductions in Sao Paolo and California, The Hewlett Foundation, Palo Alto, CA, 2005.
26. http://www.arb.ca.gov/cc/ccms/ccms.htm.
27. http://www.climatechange.ca.gov/ab32/index.html.
28. http://www.inece.org/index.html; http://ec.europa.eu/environment/impel/index.htm.

12 International climate change agreements

What is covered in this chapter?

Is international cooperation to address climate change a matter of sharing a common burden between countries or of maximizing the benefits of preventing climate change damages? The perspective taken determines to a large extent the shape of international agreements. This chapter discusses the current international arrangements and how they work. The Climate Change Convention and its Kyoto Protocol in particular. It tries to make the complex international negotiation process and its outcomes understandable for non-specialists. It also discussed the ongoing negotiations on arrangements for the period after 2012, when the agreed actions under the Kyoto Protocol to reduce greenhouse gas emissions expire.

Will these negotiations result in aggressive action to curb emissions worldwide, so that the worst impacts can be avoided? Or will political differences and North–South tensions lead to a delay in ambitious action and serious damages in the long term? What are the key issues and what are the prospects for agreement? These are issues discussed in this chapter.

Why are international agreements needed?

Climate change is a typical example of a 'global commons' problem. Everybody on the planet benefits from a stable climate, but that can only be achieved if everybody participates. Not participating, i.e. leaving it to others to reduce greenhouse gas emissions and keep forests intact, is tempting: the contributions from most individual countries are small and no economic sector covers more than 25% of the total emissions (see Chapter 2). Many small contributors together are responsible for a large share of course, so international collaboration is vital. In addition, dealing with the unavoidable impacts of climate change on poor countries and common resources such as ocean ecosystems requires solidarity. International agreements can provide that.

There is another perspective to this however. More and more it becomes obvious that addressing climate change through emission reduction and adaptation is in a country's self-interest. The damages caused by climate change are in most cases bigger than it would cost to avoid them (see Chapter 3). And moving towards a low carbon economy provides huge benefits to most countries in terms of lower energy costs, efficient industrial production, improved energy security, cleaner air, and job creation (see Chapter 4 for an elaborate discussion). In that perspective international cooperation is a way to do all this much more effectively and efficiently by doing the cheap things first and by creating bigger markets for low carbon energy and products. Solidarity to deal with climate change impacts and adaptation remains important even in this perspective.

The perspective countries take determines the framing of international negotiations. Is it about sharing a common burden of dealing with climate change (with the incentive to minimize the contribution) or is it about benefiting from the opportunities of joint action (with the incentive to join such an agreement)? To be honest, in today's world the former is still the dominant view. And the dominant attitude is still to minimize contributions. Investments in low-carbon technologies are still seen as costs. Business associations still speak mostly for members that have to adjust their business and much less for companies that produce the efficient products and renewable energy. Politicians still listen predominantly to the voices of those that resist change. Things are changing gradually however. The financial crisis of 2008 may be sparking a rethinking of what sustainable economic development is.

The Climate Change Convention and the Kyoto Protocol: lessons learned

There is now an established set of international agreements to deal with the problem of climate change. In the first place there is the United Nations Framework Convention on Climate Change and its Kyoto Protocol. Related to these, but completely independent, are many other international agreements between states and/or private entities: agreements on R&D in the framework of the International Energy Agency, financial arrangements of multilateral development banks to invest in emission reduction projects, programmes to promote energy efficiency, renewable energy, CO_2 capture and storage and other mitigation technologies, as well as joint regional expert centres.

Climate Change Convention

The Climate Change Convention, officially called the United Nations Framework Convention on Climate Change (UNFCCC), was agreed upon in 1992 at the World Summit of Environment and Development in Rio de Janeiro. It had been negotiated in a period of about 2 years after the concern of scientists about the changing climate and the global impacts of it had convinced political leaders that is was time to act. The first assessment report of the Intergovernmental Panel on Climate Change (IPCC), published in 1990 and the First World Climate Conference in Sundsvall, Sweden that same year galvanized those concerns. As a result the UN General Assembly in December 1990 decided to set up a negotiating committee to work out an agreement[1,2]. This led to the UNFCCC that was agreed in 1992 and entered into force in 1994 after 55 Countries (representing 55% of industrialized countries' emissions) had ratified it (i.e. officially approved through their national parliaments or other mechanisms).

The UNFCCC is, as the title says, a framework agreement. It has only limited specific obligations to reduce emissions of greenhouse gases, but formulates principles, general goals, and general actions that countries are supposed to take. It also established institutions and a reporting mechanism, as well as a system for review of the need for further action. Over time is has received almost universal subscription[3] (see Box 12.1).

Box 12.1 The United Nations Framework Convention on Climate Change (UNFCCC): key elements

Principles:

- 'common but differentiated responsibility'
- special consideration for vulnerable developing countries
- 'precautionary principle'
- 'polluter pays'
- promote sustainable development

Goals: the ultimate goal (article 2) is to 'stabilize greenhouse gas concentrations in the atmosphere at a level that would prevent dangerous anthropogenic interference with the climate system. Such a level should be achieved within a time frame sufficient to allow ecosystems to adapt naturally to climate change, to ensure that food production is not threatened and to enable economic development to proceed in a sustainable manner.'

Participation: almost universal (191 countries and the European Union, 1 September, 2008)

Actions required:

- Minimize emissions and protect and enhance biological carbon reservoirs, so-called 'sinks' (all countries); take action with the aim to stop growth of emissions before 2000 (industrialized, so-called Annex I countries)
- Promote development, application, and transfer of low carbon technologies; Annex I countries to assist developing countries
- Cooperate in preparing for adaptation
- Promote and cooperate in R&D
- Report on emissions and other actions (so-called 'national communications', annually for Annex I countries and less frequently for others)
- Assist developing countries financially in their actions (rich industrialized countries, so-called Annex II countries)

Compliance: Review of reports by the secretariat and by visiting expert review teams

Institutions:

- Conference of the Parties (COP), the supreme decision making body; voting rules for decisions never agreed so de facto decisions only by consensus
- Bureau (officials, elected by the COP, responsible for overall management of the process)
- Two Subsidiary Bodies (for Implementation and for Scientific and Technological Advice) to prepare decisions by the COP
- Financial mechanism, operated by the Global Environment Facility of Worldbank, UNDP and UNEP, filled by Annex II countries on voluntary basis; two special funds: a Least Developed Country Fund and Special Climate Change Fund, mainly to finance adaptation plans and capacity building, but also technology transfer and economic diversification
- Expert groups on Technology Transfer, Developing Country National Communications, Least Developed Country National Adaptation Plans
- Secretariat (located in Bonn, Germany)

Other elements:

- Requirement to regularly review the need for further action

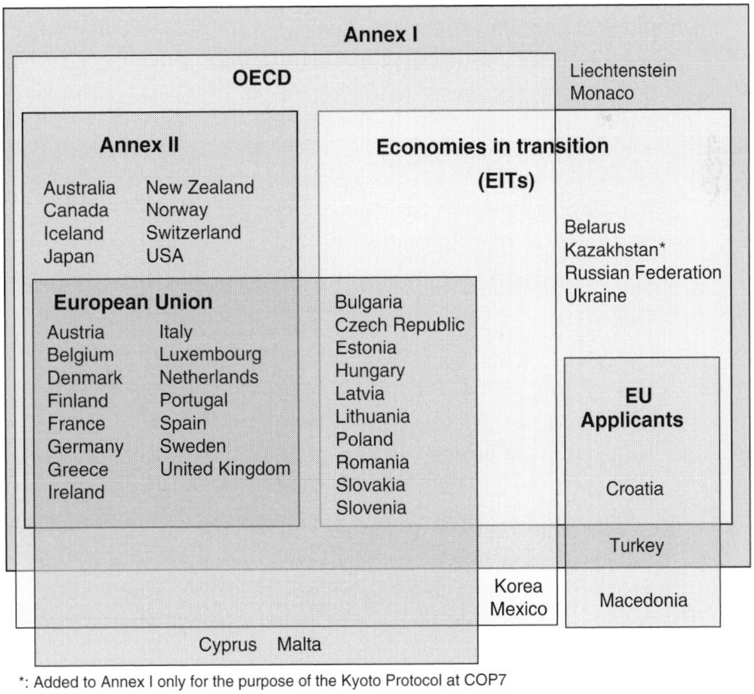

Annex I
OECD
Liechtenstein
Monaco
Annex II
Australia New Zealand
Canada Norway
Iceland Switzerland
Japan USA
Economies in transition (EITs)
Belarus
Kazakhstan*
Russian Federation
Ukraine
European Union
Austria Italy
Belgium Luxembourg
Denmark Netherlands
Finland Portugal
France Spain
Germany Sweden
Greece United Kingdom
Ireland
Bulgaria
Czech Republic
Estonia
Hungary
Latvia
Lithuania
Poland
Romania
Slovakia
Slovenia
EU Applicants
Croatia
Turkey
Macedonia
Korea
Mexico
Cyprus Malta

*: Added to Annex I only for the purpose of the Kyoto Protocol at COP7

Figure 12.1 Country groupings under the UNFCCC, OECD, and EU.

Source: IPCC Fourth Assessment Report, Working Group III, figure 13.2.

The Annex I and Annex II countries are listed in Figure 12.1, together with other relevant groupings. Former Eastern European and former Soviet Union countries have a special status under the Convention as so-called 'countries with economies in transition'.

Kyoto Protocol

At the first Conference of the Parties to the Convention in 1995 a decision was taken that further action was needed to address climate change. It was agreed to start negotiations towards a protocol (an annex to the Convention) that would commit industrialized countries (the so-called Annex I countries) to further reduce their greenhouse gas emissions. The industrialized countries had not yet done much in terms of emission reductions at the time. Therefore developing countries were deliberately exempted from further action in light of the 'common and differentiated responsibility' principle of the Convention. The USA explicitly agreed with that as one of the Parties to the Convention.

These negotiations led in 1997 to the agreement of the COP on the so-called Kyoto Protocol. It reaffirms in fact the basic agreement of the Convention and adds a number of elements: quantified emission caps for Annex I countries, so-called flexible mechanisms to allow for cost-effective implementation (emission trading between Annex I countries, a clean development mechanism on projects done in developing countries, and joint implementation on projects in Annex I countries), a compliance mechanism, and a new Adaptation Fund, which gets its funding from a levy on CDM projects (see Box 12.2).

Box 12.2 The Kyoto Protocol

Principles: same as the Convention
Goals: same as the Convention
Participation: 180 countries and the European Union (United States is not a Party)
Actions:

- Annex I countries together reduce emissions to 5% below 1990 level, on average over the period 2008–2012; specific emission caps for individual countries (see footnote 1)
- Option to use flexible mechanisms, i.e international trading of emission allowances (see footnote 2), or using the emissions reductions from projects in developing countries (through the Clean development Mechanism) or other Annex I countries (Joint Implementation)
- Option to develop coordinated policies and measures
- Strengthened monitoring and reporting requirements for countries with reduction obligations

Compliance: Shortage in emission reduction to be compensated in period after 2012, with 30% penalty

Institutions:

- COP of the Convention, acting as the Meeting of the Parties of the Protocol (CMP) as decision making body
- Use all other Convention institutions
- Compliance Committee, with consultative and enforcement branch
- Executive Board for the Clean Development Mechanism
- Joint Implementation Supervisory Committee
- Adaptation Fund, managed by the Adaptation Fund Board; administration by GEF and Worldbank; fund gets its money from a 2% levy on CDM projects

Other elements:

- Requirement to review the need for strengthening the actions

Footnote 1: see http://unfccc.int/kyoto_protocol/items/3145.php

Footnote 2: not to be confused with domestic emission trading systems as discussed in Chapter 11

The negotiating process

You may wonder how a negotiating process with 180 countries could ever produce a result. The secret is that countries operate in blocs. As in every UN negotiation developing countries coordinate positions in the so-called 'Group of 77 and China'. Now being a group that consists of 130 countries, including China, their joint position on any issue has enormous clout. The country holding the Chair of the G-77 in New York (which rotates every year) often speaks for the G-77 and China as a whole. Because the group is so large and covers countries with widely varying interests, it is not able to formulate common positions on all issues. Then the so-called 'regional groups' become important. Under the Climate Convention these are Africa, (developing) Asia, Latin America (including Central America and the Caribbean), Eastern Europe, and the Western Europe and Other Countries Group (USA, Japan, Australia, New Zealand, and Western Europe). These groups are proposing candidates for official functions in the Convention (on a rotational basis) and coordinate positions (except for the Eastern Europe and WEOG groups because they coordinate in different subgroups). Cutting across these regional groups, and often more important when it comes to coordinating positions, there is OPEC (Oil Producing and Exporting Countries), the European Union (the 27 Member States), the Association of Small Island States (Pacific and Caribbean small islands), the Umbrella Group (a loose grouping of Australia, Canada, Iceland, Japan, New Zealand, Norway, the Russian Federation, Ukraine, and the USA), and the Environmental Integrity Group (Mexico, the Republic of Korea, and Switzerland).

In practice, these latter groups and the G77/China and small island states groups are the ones negotiating, reducing the number of players to a manageable number. In addition there are normally 'friends of the chair's groups' on any important piece of negotiation, which allows bringing together the most important players to work out a compromise. By using 'lead countries' in the various groups for specific issues that work in parallel, coordination within the groups and efficiency of the process is managed. And on top of that the practice is now very much to organize much of the actual negotiations through so-called 'contact groups', open meetings where the lead countries and other interested countries try to find compromises. Sometimes these are supplemented with 'informal meetings' that are only open to negotiators if it is important to discuss matters behind closed doors. Bilateral discussions and other ad hoc off line discussions, sometimes at the level of Ministers or Heads of State, complement the range of tools available.

This does not mean the negotiating process is simple. There are many issues to deal with and with a lot of discussions moving in parallel, crosslinks need to be looked after. Things come up that have not been properly coordinated, countries may suddenly go against an earlier coordinated position, and non-governmental organizations try to influence the negotiations by lobbying and by publishing rumours or positions taken.

For the negotiations on a new agreement for the period after 2012, additional arrangements have been made. They will be discussed below.

Why the USA pulled out of Kyoto

The USA agreed with the agreement reached in Kyoto in December 1997 after Vice-president Al Gore came personally to Kyoto to instruct the US negotiators to be more accommodating on the reduction targets for GHG emissions. As a result the USA agreed to reduce its emissions by an average of 7% below 1990 by 2008–2012. The European Union accepted −8% and Japan and Canada −6%. The USA got a lot of what it had asked for: a so-called basket of gases, allowing countries the flexibility to decide what kind of reductions they would prefer to meet their target; inclusion of afforestation and reforestation as a 'sink' for CO_2; and the so-called flexibility mechanisms that allow countries to trade emission allowances between them and to use investments in projects in developing countries to compensate for reductions they would not realize at home (through the so-called Clean Development Mechanism).

But this happened against the background of a strong anti Kyoto sentiment in the US Congress. In July 1997 the US Senate had adopted the so-called Byrd-Hagel resolution with a 95–0 vote, saying that the USA should only be part of a new Protocol if the US economy would not be harmed and, more importantly, also developing countries would take on emissions reduction commitments[4]. This was in direct conflict with the negotiating mandate that was agreed in Berlin in 1995 with US support. As a result there was no chance to get the Kyoto Protocol agreed in the US Congress, because the US Constitution requires international treaties to be approved by a 2/3 majority in the Senate. Nevertheless the Clinton Administration agreed with the outcome in Kyoto.

International negotiations under the UNFCCC continued after the Kyoto meeting on the 'nuts and bolts' of the Kyoto Protocol. The exact way the flexible mechanisms should operate, the amount of credits that countries could take from new forest plantings, the arrangements for technology transfer, and the detailed financial provisions were determined, so that countries would have a clear view of what the Protocol exactly meant for them[5]. Most countries waited during that period for so-called 'ratification' (the formal approval through Parliament or otherwise). President Clinton and vice-president Gore basically kept silent during that period and did not attempt to convince the people or the Congress about the need for and the value of the Kyoto Protocol. They never formally asked the Senate for ratification.

When the Bush administration took office in January 2001, international negotiations on the details of the Kyoto Protocol were still going on and the newly appointed officials of the USA were participating in them. In March 2001 President George Bush, materializing his campaign stance about climate change, announced the USA would not ratify the Protocol. The reasons given were: it would seriously harm the US economy and developing countries were exempt from emission reductions. The economic argument was surprising, because the IPCC's Third Assessment Report that was about to be published clearly showed the economic costs of implementing the Kyoto Protocol to be very modest[6]. Special interests, i.e. the coal and oil industry, apparently had a lot of influence. The other argument, i.e. that developing countries were exempt from emissions reductions, was a direct consequence of the negotiating mandate of 1995. Australia followed suit in not ratifying the Kyoto Protocol, although it in fact was implementing climate policy to meet its agreed target (of +8% compared to 1990 by 2008–2012).

These withdrawals shocked the international community and disrupted the ongoing negotiations It is possible that as a result of this shock, agreement was reached in June 2001 among all other countries on the outstanding details of the Kyoto Protocol implementation. Speeches at that meeting frequently mentioned the victory of multilateral approaches to solving global problems. The USA answered by saying it would follow its own policies to tackle the problem, but everybody knew there was no credible US federal policy to reduce greenhouse gas emissions. This situation would continue during the Bush presidency.

How the Kyoto Protocol eventually became a reality

The Kyoto Protocol text says it would become effective after formal approval by 55 countries ('ratification' in Convention speak), provided that these countries also represent 55% of the 1990 emissions of CO_2 from industrialized (Annex I) countries. With the USA and Australia out (accounting for 36 and 2% of the Annex I emissions, respectively), it meant Russia (with 17%) was the crucial factor to make the Protocol a reality[7]. But Russia was in no hurry to ratify it.

There were serious voices in Russia claiming that climate change would be beneficial: fewer cold days, longer growing season, higher grain yields. Those voices were strongly embedded in the Russian scientific community. As a remnant from the communist period there was also a tradition for scientists to first come to an agreement on the facts before making recommendations to the government. Three consecutive reports from the UN Intergovernmental Panel on Climate Change (IPCC), the latest from 2001, were not enough to convince the Russian scientific community. Worse, the Russian vice-chairman of the IPCC and Member of the Russian Academy of Ecology, Professor Yuri Izrael, actively lobbied against the Kyoto Protocol on the grounds that it was not justified by science. At the World Climate Conference in Moscow in October 2003, top scientists and policy makers from all over the world strongly pleaded for Russian ratification. The conference itself however was dominated by Russian sceptics and chaired by the same Professor Izrael. Presidential economic advisor Illarionov was sent to undermine the conclusions of the IPCC and to argue that Russia could not afford to reduce its emissions. He made use of an error in the Russian translation of the IPCC's latest report where the costs of controlling climate change were shown to be 100 times as high as in the original IPCC report. The conference culminated in a speech by president Putin as part of a high level panel session. It was a short non-committal speech, where he gave no indication whatsoever about Russian ratification. He was about to leave when the then Head of the UNFCCC Secretariat, Ms Joke Waller, made a strong appeal to him to ratify the Protocol in the interest of humanity. Then, very unusually, he sat down again and gave a long personal response, showing his concern about the problem of climate change and outlining that Russia would carefully consider the ratification of the Kyoto Protocol. Something had touched him. It took high level talks with European leaders, amongst others, about Russia's interest to join the World Trade Organization and the EU support for it, to get Russia's ratification in November 2004[8].

Are countries meeting their emission reduction obligations?

The countries that ratified Kyoto are collectively on track to meet the agreed emissions reduction of 5% below 1990 by 2008–2012. In 2005 their emissions were 15% below the 1990 level[9] (see Figure 12.2). There are large differences however: the countries with economies in transition were about 35% below and the non-EIT countries 3% above. Individual countries show even greater differences: Latvia was 59% below its 1990 level (with a 2008–2012 target of −8%), while Spain was 53% above (with a target of +15%[10]). Including land use changes does not change this picture radically, except for Latvia, which had negative overall emissions, i.e. fixation of CO_2 in forests was bigger than the emissions of GHGs to the atmosphere from all other sources. It is striking that Canada's emissions were 54% above 1990 in 2005 (with a target of −6%), while the USA, which is not part of Kyoto, only saw a 16% increase above 1990 in 2005[11]. It confirms the complete lack of implementation of the Kyoto Protocol obligations in Canada. For comparison: China and India roughly doubled their emissions over that same period[12]. Projections for the period 2008–2012 show that the overall picture will roughly remain the same[13]. Figure 12.3 shows the performance of individual countries.

These numbers do not necessarily mean countries will not meet their obligations. The Kyoto Protocol has provisions to trade emission allowances, or in other words, to buy emission allowances on the carbon market in case of a domestic shortfall. That can be done through 'country-to-country' deals (say Russia selling part of its surplus to Japan or Canada), or through Clean Development Mechanism projects in developing countries and Joint Implementation projects in other Annex I countries. So in theory any country could

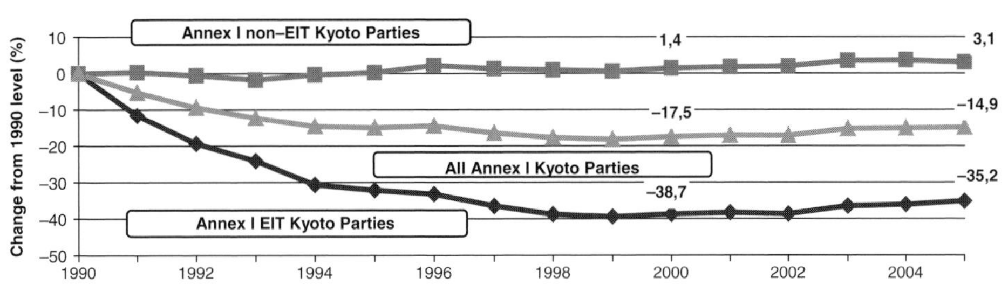

Figure 12.2 **Greenhouse gas emissions from Annex I Kyoto Parties 1990–2005; excluding sinks and sources from land use and land use change.**
Source: UNFCCC greenhouse gas emission trends 1990–2005.

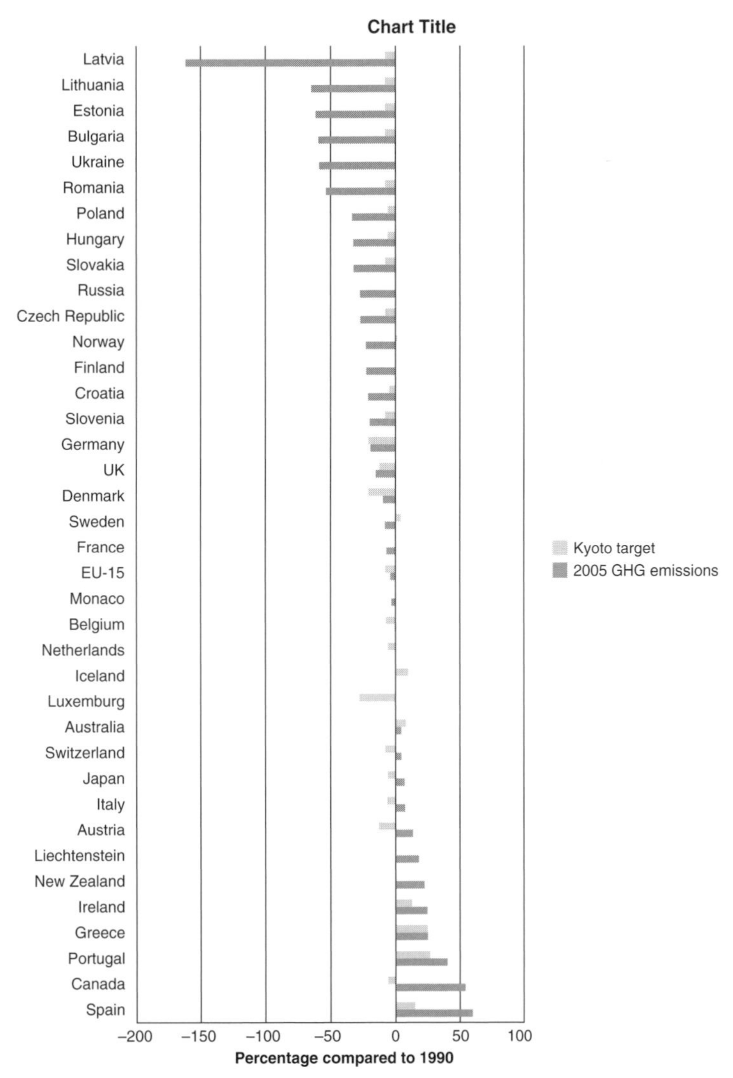

Figure 12.3 **Performance of the 36 individual Kyoto Annex I parties until 2005 and their Kyoto Protocol obligations; total net GHG emissions, including from land use and land use change.**
Source: UNFCCC greenhouse gas emission trends 1990–2005.

still meet its obligations by making the required purchases in time for the 2008–2012 targets. For most countries that seems a realistic prospect, but for some, such as Canada, it would require a very big political change.

Global emissions have continued to rise: they grew about 25% between 1990 and 2005. While the Kyoto countries are now 15% below 1990, the USA (good for about 25% of global emissions) is 16% above and all non-Annex I countries together increased their emissions by about 75% over the period 1990–2005.

Clean Development Mechanism

One of the successes of the Kyoto Protocol is the Clean Development Mechanism. It creates the possibility for Annex I countries to meet part of their commitments through emissions reductions from projects in developing countries. These projects would at the same time contribute to sustainable development in developing countries (the so-called 'host countries'). The principle is simple: any emission reduction project in a developing country that otherwise would not have happened is lowering global emissions and could therefore replace a comparable action in an industrialized country. It is a market mechanism. If it is cheaper to realize reductions in developing countries, it lowers the costs for industrialized countries. The clause 'that otherwise would not have happened' is of course crucial. If projects would have happened anyway, trading the resulting emission reductions no longer is a net global reduction. So the effectiveness of the CDM depends strongly on this so-called 'additionality' issue.

The CDM process

The CDM process is fairly complex. It consists of a project development phase and a project implementation phase (see Figure 12.4). The project development phase has several steps, including approval of the so-called Designated National Authority (DNA) of the country where the project is taking place and validation by an independent organization. It results in the registration of a project with the UNFCCC CDM Executive Board, meaning that also the method of calculating the emission reduction from the project is approved. Issuance of so-called Certified Emission Reduction units (CER, the 'currency' of the CDM, equivalent to 1 tonne of CO_2-eq avoided) only happens when an independent organization has indeed certified the actual reductions[14].

How has the CDM developed?

As of January 1 2009 there were 4474 CDM projects in the pipeline (i.e. either submitted to or registered by the CDM Executive Board). Out of these, 1370 were registered and for

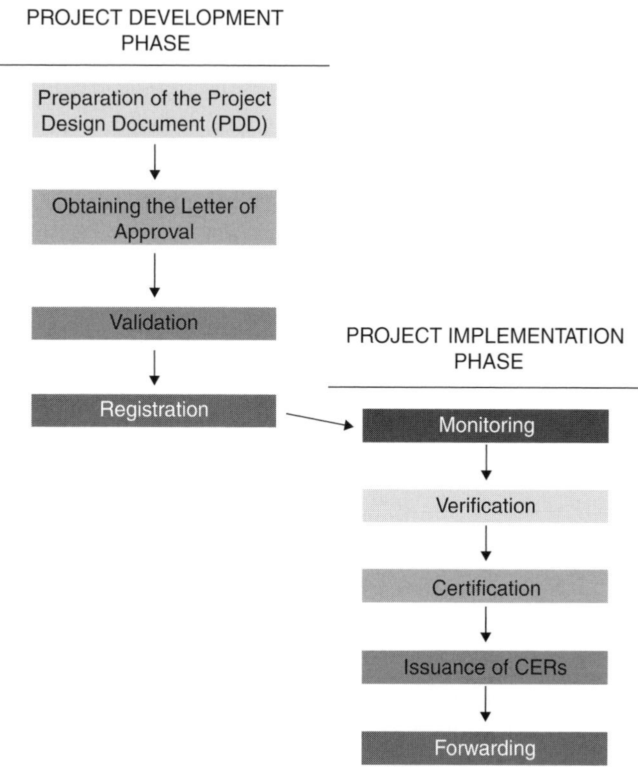

Figure 12.4 **Simplified diagram of CDM process.**
Source: BakerMcKenzie CDM Rulebook, http://cdmrulebook.org/PageId/305.

465 CERs were issued[15]. Together they are good for a reduction of about 0.3 $GtCO_2$-eq per year in the period 2008–2012 and about 0.7 $GtCO_2$-eq per year from 2013 to 2020. Given their relatively low price, the CDM CERs are very likely to be bought by Annex I countries to meet their obligations. To put things in perspective: the 0.3 $GtCO_2$-eq/year is about 50% of the total reduction that Kyoto Annex I countries are supposed to realize[16]. In other words, domestic emission reductions in these countries will be only half of what they would have been without the CDM, if indeed all available CERs are bought. The rest of the required reductions is offset by CDM credits.

CDM projects are covering a wide range of mitigation activities. The number of projects on renewable energy is the highest, with much smaller numbers for landfill gas (methane) recovery and destruction of HFC-23 at HCFC plants and N_2O at chemical plants (see Chapter 8 for a more detailed discussion). In terms of tonnes of CO_2-eq reduction expected before the end of 2012, renewable energy projects represent 36% and HFC-23 and N_2O projects 26%, reflecting the high Global Warming Potential of HFC-23. Figures 12.5 and 12.6 give an impression of the strong growth of the CDM (number of projects registered) and the relative contributions of various types of projects.

Projects are concentrated in a limited number of countries. Figure 12.7 shows that China, India, Brazil, and Mexico together host about 70% of all CDM projects. These countries

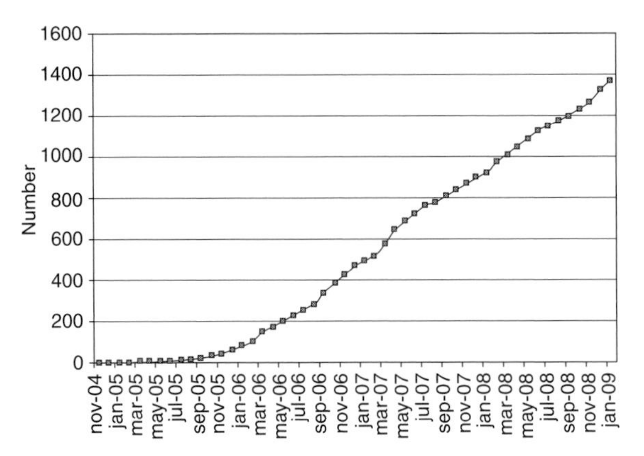

Figure 12.5 **Number of registered CDM projects over time.**
Source: UNEP Risoe CDM/JI Pipeline Overview, http://www.cdmpipeline.org/overview.htm#4.

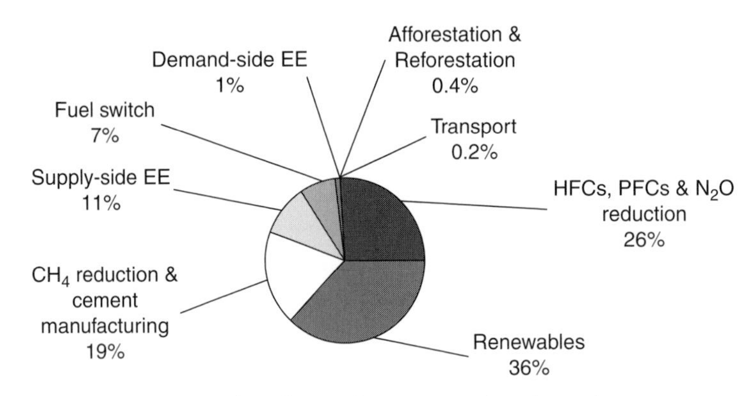

Figure 12.6 **CERs expected until 2012 from CDM projects in each sector.**
Source: UNEP Risoe CDM/JI Pipeline Overview, http://www.cdmpipeline.org/overview.htm#4.

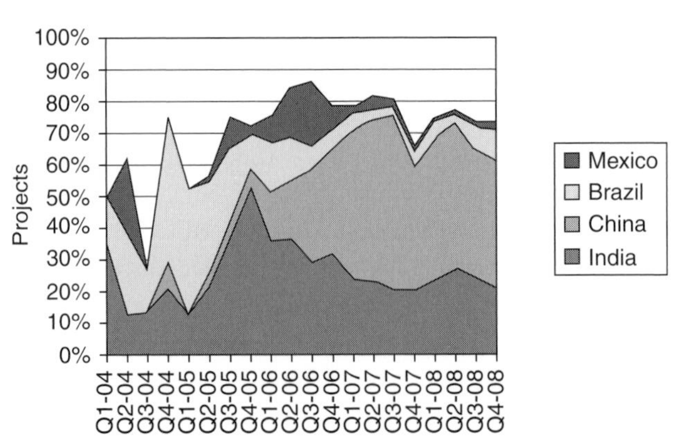

Figure 12.7 **CDM Projects in the pipeline in Brazil + Mexico + India + China as a fraction of all projects.**
Source: UNEP Risoe CDM/JI Pipeline Overview, http://www.cdmpipeline.org/overview.htm#4.

have organized their CDM activities well, making it easier for foreign buyers to get substantial tonnage without excessive administrative efforts. This also means that many countries hardly benefit from the CDM. All African countries together for instance only host 2% of the projects.

If we look at who the buyers of CDM CERs are we see that in 2007 about 80% of all CER acquisitions were made by the private sector; that means companies that fall under the EU ETS and brokers that are in the business of selling these CERs again to companies or countries that need them to comply with their Kyoto obligations. Direct government purchases of CERs for compliance purposes only covered 4% of all sales and the rest was bought by carbon funds operated by international development banks, such as the Worldbank. The total volume of the trade in CERs in 2007 had a value of about US$ 12 billion, including both the primary (sale of CERs from project) and secondary (resale of CERs to other entities) market. Prices of CERs varied between US$ 15 and 25 per tonne of CO_2-eq avoided, with secondary CERs and CERs from 'gold standard'[17] CDM projects fetching the highest prices[18].

How much of the projected CDM emission reductions are additional to what otherwise would have occurred?

According to the CDM rules emission reductions from CDM should be 100% additional. There is even a specific requirement to demonstrate that additionality in applying for an approval of a CDM project. But what is the real situation? This depends on what is considered to be the business as usual (or baseline) development. A number of hydropower projects have been approved under the CDM, many of which were already under development before the CDM came into being. Hydropower has been commercially attractive in many places for a long time. So why would certain hydropower projects be considered additional? A possible justification would be that the economic profitability (in terms of the time it takes to recoup the investment) might be less than what investors find acceptable. The CDM revenues can then make the difference between an unattractive and an attractive investment. But this is unlikely to be the case for most of the hydropower projects registered under CDM, given prior approval and comparable projects that were realized without CDM money[19].

Another interesting case is the destruction of HFC-23 from HCFC-22 production facilities. It is technically feasible to destroy HFC-23 in off-gas by using incinerators. The cost of this destruction, including investment and operating costs, is less than US$0.20 per tonne of CO_2-eq destroyed[20]. A number of HCFC-22 plants in the world have installed these devices. It is thus very hard to argue that this is something that cannot be seen as 'state of the art'. Nevertheless, HFC-23 destruction at 10 existing plants in China, India, and Korea was approved as a CDM project. Worse still is that the CERs from these projects were sold at market prices of up to US$15–20 per tonne of CO_2-eq avoided, meaning a substantial profit was made. And even worse is that attempts are being made to get CDM approval also for HFC destruction at newly built HCFC-22 plants. The counterargument from proponents of these CDM projects is that new HCFC-22 plants in developing countries are simply not being equipped with HFC destructors, because there is no economic or regulatory reason to do so.

Really worrisome is the CDM situation in China. Basically all new investments in hydropower, wind energy, and natural gas fired power plants are co-funded through the sale of CERs. Also the building of more efficient (so-called 'supercritical') coal fired power plants has now been accepted as eligible for CDM[21]. This means that almost anything China is doing to reduce its dependency on coal (which it is now also importing), reduce air pollution, and to improve efficiency of power plants is now done through CDM[22]. In other words, the assumption is that nothing of this would have been done in the absence of the CDM. That is hard to believe, since many of these installations have been built before without CDM funding and the self interest of China makes most of these projects completely viable. Given that such projects in other developing countries also will be eligible and the huge role of China and India in the CDM, this is a serious blow to the additionality of the CDM.

This issue is therefore on the table at the ongoing negotiations for a new international agreement for the period after 2012 (see below). There are strong voices calling for a serious reform of the CDM to repair these weaknesses.

Institutional infrastructure

Another key achievement of the Climate Change Convention and the Kyoto Protocol is that an elaborate institutional infrastructure has been built to deal with climate change. Apart from a carbon market with a wide range of players and institutions, there is a whole machinery of reporting on emissions, vulnerabilities to climate change, and planning and implementation of adaptation and mitigation activities (mandated by the Convention and the Protocol). Countries have implemented registries of greenhouse gas emissions and policies to control emissions. Because of the CDM many developing countries have done that as well.

The infrastructure however goes much further. A series of international public–private partnerships has been established to promote the development and diffusion of low carbon technologies and practices (see Box 12.3). Pure private initiatives have sprung up around the world where private sector companies work together to promote actions or NGOs collaborate with private companies (see Box 12.4).

Box 12.3 Public private partnerships

International Partnership for a Hydrogen Economy: Announced in April 2003, the partnership consists of 15 countries and the EU, working together to advance the global transition to the hydrogen economy, with the goal of making fuel cell vehicles commercially available by 2020. The Partnership will work to advance the research, development, and deployment of hydrogen and fuel cell technologies and to develop common codes and standards for hydrogen use. See: http://www.iphe.net

Carbon Sequestration Leadership Forum: This international partnership was initiated in 2003 and has the aim of advancing technologies for pollution free and GHG free coal fired power plants that can also produce hydrogen for transportation and electricity generation. See: http://www.cslforum.org

Generation IV International Forum: This is a multilateral partnership fostering international cooperation in research and development for the next generation of safer, more affordable, and more proliferation resistant nuclear energy systems. This new generation of nuclear power plants could produce electricity and hydrogen with substantially less waste and without emitting any air pollutants or GHG emissions. See: http://nuclear.energy.gov/genIV/neGenIV1.html

Renewable Energy and Energy Efficiency Partnership: Formed at the World Summit on Sustainable Development in Johannesburg, South Africa, in August 2002, the partnership seeks to accelerate and expand the global market for renewable energy and energy-efficiency technologies. See http://www.reeep.org

Asia-Pacific Partnership on Clean Development and Climate: Inaugurated in January 2006, the aim of this partnership between Australia, China, India, Japan, Republic of Korea, and the USA is to focus on technology development related to climate change, energy security, and air pollution. Eight public/private task forces are to consider: (1) fossil energy, (2) renewable energy and distributed generation, (3) power generation and transmission, (4) steel, (5) aluminium, (6) cement, (7) coal mining, and (8) buildings and appliances. See: http://www.asiapacificpartnership.org

IEA Implementing Agreements: Since its creation in 1974, the IEA has provided a structure for international cooperation in energy technology research and development and deployment. Its purpose is to bring together experts in specific technologies who wish to address common challenges jointly and share the fruit of their efforts. Within this structure, there are currently some 40 active programmes, known as the IEA Implementing Agreements. Almost three decades of experience have shown that these Agreements are contributing significantly to achieving faster technological progress and innovation at lower cost. They help to eliminate technological risks and duplication of effort, while facilitating processes like harmonization of standards. Special provisions are applied to protect intellectual property rights. The focus is on technologies for fossil fuels, renewable energies, efficient energy end-use, and fusion power. Effective dissemination of results and findings is an essential part of the mandate of each Implementing Agreement. See http://www.iea.org/textbase/techno/framework_text.pdf

(Source: IPCC Fourth Assessment Report, Working Group III, ch 13)

Box 12.4 Private initiatives

Business Leader Initiative on Climate Change (BLICC): Under this initiative, five European companies monitor and report their GHG emissions and set a reduction target. See http://www.respecteurope.com/rt2/BLICC/

Carbon Disclosure Project: Under this project, 940 companies report their GHG emissions. The project is supported by institutional investors controlling about 25% of the global stock markets. See http://www.cdproject.net

Carbon Trust: The Carbon Trust is a not-for-profit company set up by the UK government to reduce carbon emissions. The Trust provides technical assistance, investment funds, and other services to companies on emission reduction strategies and for the development of new technologies. See http://www.thecarbontrust.co.uk/default.ct

Cement Sustainability Initiative: Ten companies have developed 'The Cement Sustainability Initiative' for 2002–2007 under the umbrella of the World Business Council for Sustainable Development. This initiative outlines individual or joint actions to set emissions targets and monitor and report emissions

Chicago Climate Exchange: The Chicago Climate Exchange is a GHG emission reduction and trading pilot programme for emission sources and offset projects in the USA, Canada, and Mexico. It is a self-regulatory, rules-based exchange designed and governed by the members who have made a voluntary commitment to reduce their GHG emissions by 4% below the average of their 1998–2001 baseline by 2006. See http://www.chicagoclimatex.com

Offset programmes: There are many organizations that offer services to offset the emissions of companies, communities, and private individuals. These organizations first calculate the emissions of their participants and then undertake emission reduction or carbon sequestration projects or acquire and retire emission reduction units or emission allowances. See http://ecosystemmarketplace.com/pages/article.news.php?component_id=5794&component_version_id=8505&language_id=12

Pew Center on Climate Change Business Environmental Leadership Council: Under this initiative, 41 companies establish emissions reduction objectives, invest in new, more efficient products, practices and technologies, and support actions to achieve cost-effective emission reductions. See: http://www.pewclimate.org/companies_leading_the_way_belc/

Top 10 consumer information system: This NGO-sponsored programme provides consumers with information on the most efficient consumer products and services available in local markets. The service is available in 10 EU countries, with plans to expand to China and Latin America. See http://www.topten.info

WWF Climate Savers: The NGO World Wide Fund of Nature (WWF) has built partnerships with individual leading corporations that pledge to reduce their global warming emissions worldwide by 7% below 1990 levels by the year 2010. Six companies have entered this programme. See http://www.panda.org/about_wwf/what_we_do/climate_change/our_solutions/business_industry/climate_savers/index.cfm

(Source: IPCC Fourth Assessment Report, Working Group III, ch 13)

New agreements beyond 2012

It is obvious that further steps are needed to curb global emissions after the Kyoto Protocol's commitments for 2008–2012 expire. The Kyoto protocol has a provision that says such an agreement should be ready before the start of the first commitment period, i.e. not later than 2007. We did not make that deadline. The reason was the unwillingness of the USA to start negotiating. The best possible outcome of the Conference of the Parties in December 2005 in Montreal was a decision that the Kyoto Annex I countries (i.e. without the USA) would start negotiations on further reductions for Annex I countries after 2012 under the Kyoto Protocol, while a general dialogue would start amongst all countries about possible next steps under the Convention – quite a complex structure that was invented to circumvent the USA resistance to real negotiations. Surprisingly, the dialogue went well during the years 2006 and 2007 and by COP13 in Indonesia in December 2007 the pressure to start real negotiations on a new agreement had increased dramatically. Climate change concerns were at the top of the political agenda, not least because of the new report of the IPCC that was published in 2007 and the Nobel Peace Prize given to the IPCC and Al Gore.

It led to decisions to establish a new negotiation group, in addition to the already existing Ad-Hoc Group on the Kyoto Protocol. This became the Ad-Hoc Group on Long Term Cooperation. The mandate of this group was a hard fought result that is known as the Bali Action Plan (Box 12.5). It is a carefully balanced text that sets the stage for negotiations on a new agreement with new commitments by developed and developing countries, dropping the rigid distinction between Annex I and non-Annex I countries from the current Kyoto Protocol. It covers mitigation, adaptation, technology, and financial support to developing countries. Of course this negotiating mandate does not specify the outcome. That has to emerge from the actual negotiations. What are the most contentious issues?

Box 12.5 **Summary of the Bali Action Plan**

Main aim: to launch a comprehensive process to enable the full, effective, and sustained implementation of the Convention through long term cooperative action

Timeframe: agreement to be reached at COP15 in Copenhagen, December 2009

Main elements of what should be part of the eventual agreement:

(a) A shared vision for long term cooperative action, including a long term global goal for emission reductions, to achieve the ultimate objective of the Convention

(b) Enhanced national/international action on mitigation of climate change, through:

- (i) Measurable, reportable, and verifiable nationally appropriate mitigation commitments, or actions, including quantified emission limitation and reduction objectives, by all developed country Parties, while ensuring the comparability of efforts among them
- (ii) Nationally appropriate mitigation actions by developing country Parties in the context of sustainable development, supported and enabled by technology, financing, and capacity-building, in a measurable, reportable, and verifiable manner
- (iii) Reducing emissions from deforestation and forest degradation; and conservation, sustainable management of forests and enhancement of forest carbon stocks in developing countries
- (iv) Cooperative sectoral approaches
- (v) Opportunities for using markets, to enhance the cost-effectiveness of mitigation actions
- (vi) Economic and social consequences of response measures
- (vii) Strengthening the catalytic role of the Convention towards multilateral bodies, the public and private sectors and civil society

(c) Enhanced action on adaptation through:
   (i) International cooperation to support urgent implementation of adaptation actions, including through technical and financial support, specific projects, integration into national planning, and other ways to enable climate-resilient development and reduce vulnerability of all Parties
   (ii) Risk management and risk reduction strategies, including insurance
   (iii) Disaster reduction strategies in developing countries that are particularly vulnerable to the adverse effects of climate change
   (iv) Economic diversification to build resilience
   (v) Strengthening the catalytic role of the Convention
(d) Enhanced action on technology development and transfer to support action on mitigation and adaptation through:
   (i) Removal of obstacles to, and provision of financial and other incentives for, scaling up of the development, diffusion and transfer of technology to developing country Parties
   (ii) Cooperation on research and development of current, new, and innovative technology
   (iii) Mechanisms and tools for technology cooperation in specific sectors
(e) Enhanced action on the provision of financial resources and investment through:
   (i) Improved access to adequate, predictable, and sustainable financial resources and financial and technical support, and the provision of new and additional resources
   (ii) Positive incentives for developing country Parties for the enhanced implementation of national mitigation strategies and adaptation action
   (iii) Implementation of adaptation actions on the basis of sustainable development policies
   (iv) Mobilization of public- and private-sector funding and investment
   (v) Financial and technical support for capacity-building in the assessment of the costs of adaptation in developing countries

(Source: http://unfccc.int/resource/docs/2007/cop13/eng/06a01.pdf#page=3)

How much should emissions be reduced?

The long term goal of the Climate Convention is to stabilize greenhouse gas concentrations in the atmosphere at 'safe' levels (see Chapter 3). There is growing support amongst the G8 countries to translate that into a reduction of global emissions by 50% by 2050. But there is still a difference of opinion about the base-year. Is it compared to 1990, which would be just about consistent with a road to stabilization at about 450 ppm CO_2-eq, or an average global temperature increase in the long term of 2–2.4°C, or is it compared to 2005, which would make it more like a 3°C scenario?

The other important point is the interim target for 2020. To keep the possibility open of staying on a 2 degree course, global emissions should start declining not later than about 2015 (see Chapter 3). That is a serious additional constraint on the longer term emission reduction goal. And it is only for global emissions. What would it mean for emission reductions of developed and developing countries?

Who does what?

Table 12.1 shows the summary from IPCC[23] of the various studies that looked into that question, with different assumptions about what is an equitable distribution of the effort. For the 450 ppm CO_2-eq scenario the resulting numbers for allowable emissions fall in a fairly narrow band: 25–40% below 1990 level for developed (Annex I) countries by 2020 and 80–95% by 2050. For developing countries in Latin America, the Middle-East, and East Asia a deviation from the baseline emissions is needed. Since the deviation for all developing countries together is about 15–30%[24], for the more advanced regions mentioned it will be about one third higher: 20–40%. This is a deviation from the baseline, so growth of emissions would still be possible. China for instance could under this regime still increase its emissions between 1990 and 2020 by 2–3 times instead of 3–4 times if it did not take any action. For higher stabilization levels of 650 ppm CO_2-eq action is less urgent and applying fairness criteria then means developing countries can continue on a business as usual trajectory for some time. Under no circumstance can developed countries alone reduce emissions sufficiently to achieve stabilization at any level, since emissions eventually have to go down to almost zero.

This immediately raises concerns about the ability of China to do this without harming its social and economic development. Is it fair to ask such an effort from China and other developing countries? The numbers mentioned above do come from studies where equity was an explicit requirement, so the answer in principle should be 'yes'. However, this ignores practical problems of access to the latest technology, capacity in the country to organize drastic change, and financial resources to do the necessary investments. International assistance from developed countries would therefore be needed.

Another big issue is how the efforts should be distributed amongst developed and amongst developing countries. The Kyoto Protocol has differentiated emission reduction targets for individual countries (see above). There was no particular system behind these numbers. They came about in pretty much an ad hoc manner. For a new agreement it would be better to have an agreed formula that could be applied over time as countries' situations change. Many proposals have been made for such formulas. Emissions per capita (responsibility for the problem), income per capita (ability to pay for the solution), and relative costs or easiness of taking action (opportunity to contribute to the solution) are the most frequently used principles. Generally speaking, formulas that use a combination of those principles have the best chance of being acceptable, because feelings about fairness principles are often very strong.

Within the European Union some experience has been gained. The distribution of efforts under the Kyoto Protocol was based on a combination of ability to pay (lower income Member States were allowed to increase their emissions) and opportunities for action (Member States with the worst energy efficiency were asked to do more)[25].

Table 12.1. Range of allowed emissions,[a] compared to 1990 for stabilization at different levels for Annex I and non-Annex I countries, as reported by studies with different assumptions on fair sharing efforts

Scenario category	Region	2020	2050
A 450ppm CO_2-eq[b]	Annex 1	–25% to –40%	–80% to –95%
	Non-Annex 1	Substantial deviation from baseline in Latin America, the Middle East, East Asia, and Centrally-Planned Asia[c]	Substantial deviation from baseline in all regions
B 550ppm CO_2-eq	Annex 1	–10% to –30%	–40% to –90%
	Non-Annex 1	Deviation from baseline in Latin America and the Middle East, East Asia	Deviation from baseline in most regions, especially in Latin America and the Middle East
C 650ppm CO_2-eq	Annex 1	0% to –25%	–30% to –80%
	Non-Annex 1	Baseline	Deviation from baseline in Latin America and the Middle East, East Asia

[a] The aggregate range is based on multiple approaches to apportion emissions between regions (contraction and convergence, multistage, Triptych and intensity targets, among others). Each approach makes different assumptions about the pathway, specific national efforts, and other variables. Additional extreme cases – in which Annex I undertakes all reductions, or non-Annex I undertakes all reductions – are not included. The ranges presented here do not imply political feasibility, nor do the results reflect cost variances.

[b] Only the studies aiming at stabilization at 450ppm CO_2-eq assume a (temporary) overshoot of about 50ppm (See Den Elzen and Meinshausen, 2006).

[c] Later calculations put this deviation at 15–30% below baseline on average for developing countries (see note 24).

Source: IPCC Fourth Assessment Report, Working Group III, ch 13.

Recently EU Member States together have agreed to reduce GHG emissions by 20% below 1990 levels by 2020 unilaterally. It included a table for sharing out the reductions for the small emitters (the big emitters being under the EU Emission Trading System). For an average reduction of about 10% between 2005 and 2020, contributions from Member States vary between a growth of 20% to a reduction of 20%. Ability to pay has played a strong role as can be seen from Figure 12.8.

Amongst developing countries a differentiation is also needed. Incomes per capita and emissions per capita are so different that treating all countries as equal would be unfair[26].

Many proposals have been tabled for a fair distribution across developed and developing countries[27]. Four approaches have received a lot of attention:

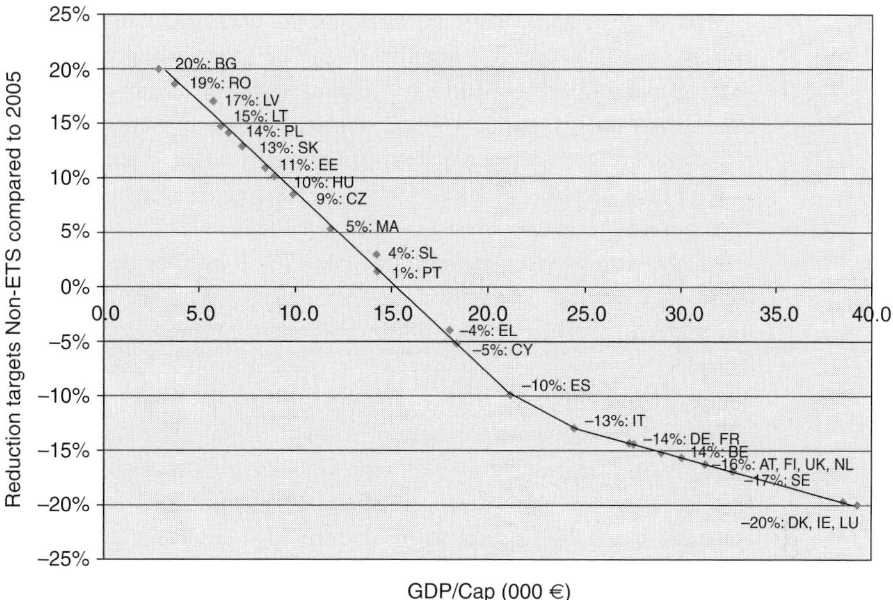

Figure 12.8 **Effort sharing between EU27 Member States as agreed in the Energy and Climate Change Package, showing income per capita to be the dominant criterion. BG=Bularia, RO=Rumania, LV=Latvia, LT=Lithuania, PL=Poland, SK=Slovakia, EE=Estonia, HU=Hungary, CZ=Czech Republic, MA=Malta, SL=Slovenia, PT=Portugal, EL=Greece, CY=Cyprus, ES=Spain, IT=Italy, DE=Germany, FR=France, BE=Belgium, AT=Austria, FI=Finland, UK=United Kingdom, NL=Netherlands, SE=Sweden, DK=Denmark, IE=Ireland, LU=Luxemburg.**

Source: European Commission, Staff working document, Impact assessment, document accompanying the Package of Implementation measures for the EU's objectives on climate change and renewable energy for 2020, Brussels, January 2008.

- *Equal emissions per capita*: A convergence of per capita emissions at a low level, consistent with the desired stabilization level of atmospheric concentrations; countries with high per capita emissions go down to the common low level over a period of say 50 years and low per capita emissions countries can increase their emissions to the common level
- *Multi-stage*: A system of 'graduation' to stronger contributions to emission reductions over time as incomes and emissions of countries increase; this is usually called a 'multi-stage approach'
- *Triptych*: applying different principles to the three different parts of countries' economies: for industry a convergence of carbon efficiency; for energy supply making use of low carbon opportunities; and for the rest (transport, buildings, agriculture, and small business) per capita emission conversion. This approach played a constructive role in the process of distributing Kyoto targets amongst EU member states as indicated above
- *Greenhouse development rights*: Contribution of countries according to the 'luxury emissions' of the high income part of the population; this approach takes the income and emission distribution in a country as the basis; the more high emission people, the bigger the effort a country is required to make[28]

Most of these approaches are based on the idea of sharing 'pain'; that is why these exercises are often called 'burden sharing'. The Triptych approach has elements that look at things more from an opportunity point of view (improving energy efficiency, applying low carbon energy options). True opportunity sharing approaches are not available, reflecting that discussions are still predominantly about sharing 'burden'.

With the exception of the Greenhouse Development Rights approach, most formulas give similar outcomes, when averaged over regions[29]. That is the very reason for the relatively narrow ranges reported in Table 12.1. It must be pointed out though that in all approaches specific assumptions have to be made. Changing those assumptions can alter the results. In specific versions of the GDR approach emission reductions shift strongly to developed countries, leading to reduction percentages for Annex I countries of about 55% compared to 1990 in 2020[30].

Figure 12.9 shows an example of a 'multi-stage' calculation for a 450ppm CO_2-eq stabilization scenario that is fairly representative of outcomes for that stabilization level. In this version of multi-stage countries are grouped in three categories: Category I: countries with a high per capita income and high per capita emissions; they reduce their emissions in absolute terms; Category II: middle income/middle emission countries: they slow down the growth of their GHG emissions; Category III: low income/low emission countries: they are exempt from taking action. Over time countries move to a higher category, when their income and/or emission level reaches the 'trigger' for graduation. In this example the trigger is set at a combination of income per capita and emissions per capita. As is shown in the figure, by 2020 South Asia (mainly India and Pakistan) are still exempt from taking action, Africa is supposed to take very limited action, while Latin America, the Middle East, and East Asia need to deviate from their baseline. All those regions however are still able to increase their emissions considerably compared to 1990. For 2050 emissions reductions become much bigger of course. But even then countries in Africa and South Asia (e.g. India) would still have room for much higher emissions than in 1990.

At present (mid 2009) the negotiations are difficult. The new US president Obama has announced he wants to reduce US GHG emissions to their 1990 level by 2020 and legislation is being discussed in Congress, which is a drastic change from the US positions thus far. However, it is still far from what the EU has indicated it is willing to do (20–30% below 1990 by 2020) and far from what is needed to stay on course for limiting warming to less than 2°C. At the same time there is still cautiousness of other developed countries to announce an ambitions reduction target ahead of advanced developing countries stating their position[31]. There is also strong resistance amongst developing countries to adopt a new classification system that would replace the Annex I versus non-Annex I system under the Convention. This makes it unlikely that a 'formula' can be agreed in Copenhagen that would govern the graduation of countries to more ambitious commitments as they develop. The debate seems to be shifting to a system of 'differentiation through actions', meaning that specific developing country action plans could be tailored to their development stage, without creating a classification system. The ambition level of these country action plans would then have to be consistent with the overall reduction compared to baseline as indicated in Table 12.1.

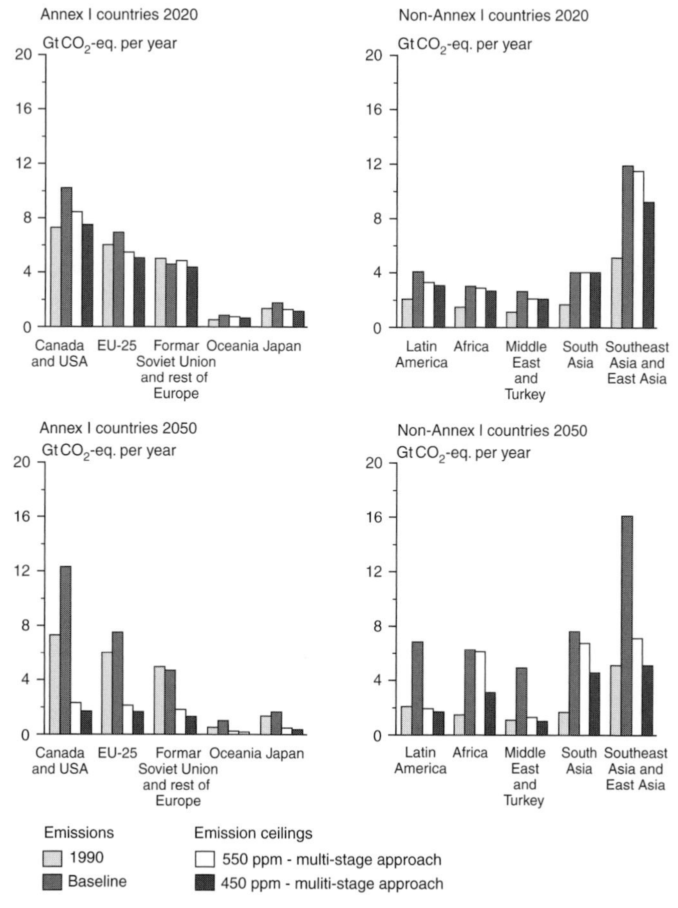

Figure 12.9 **Emission ceilings for developed and developing countries under a 'multi-stage' equity distribution for stabilization levels of 450 and 550 ppm CO_2-eq.**

Source: from climate objectives to emission reduction; overview of the opportunities for mitigating climate change, Netherlands Environmental Assessment Agency, 2006.

What kind of actions can countries commit to?

Emission ceilings are the simplest form of commitment to action. That is the way things were done under the Kyoto Protocol (see above). Absolute ceilings can however be problematic for countries with strongly varying economic growth rates and the costs of meeting them cannot be predicted accurately. Alternatives exist in the form of emissions per unit of production, for instance per unit of GDP for a country as a whole or per tonne of steel or per kWh electricity produced for a sector. Moving to such targets of course means the resulting emissions are no longer certain. With strong economic growth or growth of production, emissions come out higher. Nevertheless such relative or dynamic targets are being considered for developing countries, where uncertainties of growth are high and increasing costs can be problematic. Emission trading systems can handle such dynamic targets, although the system becomes more complex.

As already mentioned, sectoral targets are also being considered for developing countries. The rationale is that for sectors that are large emitters and for which information on emissions from individual plants are available, commitments on limiting their emissions might be feasible, while such a commitment for a country as a whole might be seen as too risky. If for instance the electricity sector, the steel, cement and fertilizer industry, and the transport sector were covered under sector targets, more than 40% of the emissions on average would be controlled.

Another way is to commit to the use of the cleanest technology, for instance by requiring minimum fuel efficiency standards for cars, and energy efficiency standards for washing machines, refrigerators, TVs, computers, and other products. This can also be applied to steel, glass, and cement making processes and other manufacturing processes. These so-called 'best available technology' approaches have been used widely in controlling other environmental problems (see also Chapter 11). Technology commitments could also be in the form of information sharing or joint demonstration programmes, although the impact of those on emission reductions would be hard to measure.

Yet another approach is to commit to implementing policies and measures. When doing that the result in terms of GHG emissions would not be exactly known, but could be estimated roughly. In fact, the Kyoto Protocol has an almost forgotten article 2 that creates the possibility for coordinated action on policies and measures (Box 12.6). This would make sense to make standards for traded products more effective. One interesting option that is being discussed for a new post-Kyoto agreement is a so-called 'Sustainable Development Polices and Measures' (SDPAM) approach. In this approach policies to enhance the sustainability of development are the starting point and reduction of greenhouse gas emissions is a co-benefit[32]. In Chapter 4 many examples of this thinking were presented, such as improving energy security by improving the efficiency of energy use and replacing imported fossil fuels with domestic renewable energy sources.

Box 12.6

Coordinated policies and measures: the revival of article 2 of the Kyoto protocol?

The history of Kyoto Protocol article 2.1.b and 2.4 is interesting. At the time of the negotiation of the Kyoto Protocol the USA was only interested in an emission ceilings approach with maximum flexibility of how to realize such a target. They did not want to specify any common action on policies and measures. The European Union however felt that both targets and common measures (such as on taxes or product standards) would be useful instruments and insisted on a provision in the Protocol (often derided by the other players at the negotiating table). As a result the Kyoto Protocol now has articles 2.1.b and 2.4 that explicitly create the opportunity for common policies and measures. It has remained a completely dead letter so far. More recently the thinking in the USA has changed. Proposals for a joint carbon tax have been made and coordinated policies on product standards are no longer taboo. They mostly come from economists who argue that a joint tax is the most economic, efficient way of reducing emissions. At the same time taxes remain unpopular in most countries and in particular in the USA. In light of the need to come up with ways for developing countries to contribute to global emission reduction and the realization that R&D on low carbon energy and products needs to be strengthened, a renewed interest in policies and measures is visible and a new life of the policies and measures article might be upon us.

Finance

To get a meaningful agreement for the period after 2012, substantial financing will be needed to assist developing countries with mitigation and adaptation measures and to promote technology development and diffusion. There is however no good idea of how much is needed. What is available is rough estimates of the additional investments needed.

For mitigation in a scenario where global emissions are back at 2005 levels by 2030 (more or less equivalent to stabilization at 450–500 ppm CO_2-eq) additional investments of about 200 billion US$/year in 2030 will be needed, according to estimates of the UNFCCC secretariat. For adaptation the amount needed will be about 50–180 billion US$/year, in this scenario where the most serious damages from climate change are avoided. Together we talk about 250–400 billion US$/year by 2030. More recent numbers from the IEA suggest this amount might be twice as high[33]. This looks like a huge number. Compared to the total annual investments in the world it is however only 1–2% (2–4% if IEA is right). As a percentage of world GDP it is even lower: less than 1%, even for the higher numbers.

Developing countries will need about 40–50% of the additional investment. And that is where financing problems exist. So what are the prospects of finding 100–200 billion US$/year (or maybe twice as much) in additional investments for these countries? To put things in perspective it is good to look at current investment flows in developing countries. Investment for energy and transportation through Official Development Assistance (ODA) was about 13 billion US$ in the year 2005. Foreign Direct Investment (FDI) in developing countries was about 380 billion US$ on average in 2006, while domestic investment in developing countries was about 1300 billion[34]. In 2030 these numbers are estimated to be three times as high. So the additional investment in developing countries by 2030 would be in the order of 5–10% of total investment by 2030. Not a very big number, but nevertheless additional international funding is needed if we do not want badly needed development investments to suffer. If we assume that all of the additional investment will have to come from international sources, then there are several possibilities: more ODA, loans from Development Banks, and FDI, supplemented with Climate Change Funds.

Direct carbon market finance (through CDM or International Emissions trading) is left out of the equation. The reason is that the carbon market is an offsetting mechanism. In other words: what is paid for (by the developed countries) for reductions in developing

Table 12.2. Financial flows relevant to the need for assisting developing countries to address climate change

Source	Approximate current and potential financial flows (billion US$/yr)
Domestic investment developing countries	1300
FDI developing countries	380
ODA for energy and transportation	13
Oil exporting country investments	500
Auctioning air/marine allowances	40
Auctioning all industrialised country allowances	hundreds
Tax on air travel	15
Tax on currency transactions	20

Source: UNFCCC, http://unfccc.int/cooperation_and_support/financial_mechanism/items/4053.php; Tirpak, D., Adams, H., Climate Policy, vol 8 (2008), pp 135–151; Miller A., Climate Policy vol 8 (2008), pp. 152–169.

countries is directly subtracted from the reductions in the developed countries themselves. It is a zero sum game. To stay on track towards a 450 ppm CO_2-eq stabilization, developing countries will have to deviate from their baseline emissions by 15–30% (see above). The realization of that contribution can therefore not be financed through the carbon market. Unless of course developed countries take a deeper reduction than the 25–40% below 1990 by 2020 that is consistent with their share of the global effort. But that seems unrealistic.

Investment is not the same as costs. A considerable part of the investments in emission reductions has benefits that make these investments profitable (through saved energy, improved air quality and lower health care costs, reduced oil imports, or otherwise). It means the net incremental costs of these investments are much lower. For the 2010–2020 period on average an estimate is 100–140 billion US$/year in incremental costs for emission reductions, adaptation and technology research, development, and demonstration together[35].

Climate change funds will therefore have to provide substantial amounts of money. Current flows from Climate Change Funds under the Climate Change Convention and the Kyoto Protocol are very modest: Adaptation Fund: US$80–300 million/year, Least Developed Country Fund and Special Climate Change Fund together something like US$15 million/year. Huge increases in these and other funds will be needed. One potentially interesting new source is the revenue from auctioning emission allowances in developed countries. As discussed in Chapter 11, the tendency in domestic emission trading systems is to move towards full auctioning of allowances to companies. That will generate hundreds of billions of dollars each year if generally applied, depending of course on the stringency of climate policy. It does not mean of course that countries will transfer all those revenues to international funds[36]. Domestic needs may get priority. That has triggered ideas to auction a certain percentage of allowances directly under the UNFCCC, so that the money does not go through national treasuries of individual countries. Other sources of funding may have to be found. Various proposals have been made, for instance a tax on international currency transactions, on air travel, or on fuels for international shipping (see Table 12.2).

Technology

Modern low carbon technology is essential for controlling climate change. We know that a large part of the opportunities for emissions reduction can be found in developing countries. We also know that there are many barriers to the use of these modern low carbon technologies in these countries (Chapter 10). Removing these barriers is therefore critical. This is commonly called the problem of 'technology transfer'.

What can international agreements do to remove these barriers? It is helpful to make a distinction between diffusion of existing technologies and the development of new ones. Existing technologies, such as energy efficient cars, appliances and industrial equipment, are readily available in developed countries (although they may not be universally applied there). But these technologies are much rarer in developing countries. Exceptions are recently built large scale manufacturing plants for steel, cement, or fertilizer, where often the most modern and efficient technology is being used[37]. Lack of knowledge of investors, high initial investments, insufficient maintenance expertise, banks that shy away from investments they are not familiar with, and absence of government regulations are some of the most important reasons for this[38].

International agreements can do something about creating the need for investments (in the form of countries committing to action), making it easier to access international financing (see the discussion above) and sharing the experience of countries by creating databases and best practice examples[39]. Much of the international action to assist countries in implementing modern low carbon technology is happening outside the Climate Change Convention. The IEA operates a series of so-called Implementing Agreements that allow IEA member and non-member countries and other organizations to engage in sharing information about implementing specific low carbon technologies[40]. There are currently 42 of these cooperative arrangements. There are also many public-private partnerships active in this field, such as the Renewable Energy and Energy Efficiency Partnership (REEEP), funded by national governments, businesses, development banks and NGOs[41], and the Renewable Energy Network for the 21st Century (REN21), connecting governments, international institutions, non-governmental organizations and industry associations[42]. See also Box 12.3.

Development of low carbon technology is different from diffusion. Development means scaling up of promising results from research and demonstrating it at semi-commercial scale. It also means technology improvements based on R&D that can significantly reduce costs.

Traditionally, new technology was developed in industrialized countries and then diffused to developing countries. Although that is still happening, it is no longer the only mechanism. Technology is now also developed in more advanced developing countries. Japan moved from a country good at copying and cheaply producing electronic products in the 1960s to the place where much of the innovation in these products is taking place today[43]. China is following that pattern and has already become the producer of the best and lowest cost supercritical coal fired power plants, the main manufacturer of electric bikes, solar water heaters, and solar panels (see also Chapters 5 and 6). It is set to take the number 1 position in wind turbine manufacturing in 2009[44]. Innovation capacity is rising fast, reflected by the tripling of R&D expenditures from 0.5% to 1.5% of GDP since 1990[45]. In India, Suzlon, one of the world's biggest wind turbine manufacturers, acquired a German firm, strengthening its market power and its innovative capacity.

What does this mean for the role of international agreements in promoting the development of new low carbon technologies? This role is probably limited. Arrangements such as the IEA Implementing Agreements can help share information. In the pre-competitive research stage, higher government R&D budgets can help. Doubling or tripling global energy related R&D budgets (aiming at low carbon technologies and energy efficiency) could be made part of the financial arrangements of a new agreement. Providing support to developing countries to build up their innovation capacity should be part of that effort. The model of the CGIAR, the Consultative Group on International Agricultural Research, might be useful. It is a network of 16 international research centres, spread over all regions of the world, aiming at providing food security to all people. It is funded by bilateral and multilateral donors and private foundations[46].

When technologies enter the stage of development, scaling up, and market introduction, commercial interest will dominate and the role of governments changes. International cooperation (not necessarily within the UNFCCC) could speed up the market introduction of new low carbon technologies by setting up larger demonstration programmes, with supporting government funding (this could also be one of the purposes of a new funding system).

Measuring, reporting, and verifying

The accountability for commitments in a new agreement has received a lot of attention during the negotiations on the Bali Action Plan. It is captured in the so-called 'MRV clause': actions committed to should be *measurable, reportable, and verifiable* (see Box 12.5). Since this applies to both developed and developing countries, this is certainly a step up from the current arrangements in the Convention and the Kyoto Protocol. There we do have a system of so-called national communications and review, but the requirements for developing countries are not very stringent. What is also important is that the MRV requirement for a new agreement applies to financial and technical support by developed countries as well. It is likely that a new system of reporting of those actions will be set up. This was seen by developing countries as a major step forward.

How these MRV clauses are going to be implemented is as yet unclear. It is likely reporting systems will build on the existing system of national communications, by making them more frequent and provide more direct guidance on what they should contain. An obvious improvement would be to have frequent inventories of greenhouse gas emissions from developing countries (currently not required and most developing countries have only submitted one inventory that is completely outdated by now). Frequent reporting on actions already taken in developing countries (see Chapter 4) would ensure that such actions can be taken into account when discussing appropriate actions by developing countries.

The 'measurable' clause will have an impact on the form of agreed actions, because they should indeed be measurable, ruling out vague, non-committal formulations.

The 'verification' part gets us to the discussion on review and compliance. The current review requirements for developed countries require an administrative and a so-called 'in-depth review', involving a team of experts visiting the country. This is still a relatively soft review process. It is also increasingly difficult to find qualified experts for such country visits. For developing country national communications there is only a limited administrative review. Upgrading the system of review, by a more rigorous administrative review, as well as by a professionalized country visit programme, would make a lot of sense.

Contours of a Copenhagen Protocol

As discussed above, a new global agreement for the period after 2012 (when the emission reduction commitments under the Kyoto Protocol expire) is very important. Between 2012 and 2020 global emissions have to start going down if we want to avoid the most serious climate change impacts. In line with the Bali Plan of Action and the discussion above, the following elements are good candidates to end up in an agreement in Copenhagen, at the end of 2009, or later:

- A *long term orientation* in the form of a non-binding agreement to have global greenhouse gas emissions in 2050 50% below the 1990 level and developed countries (the current ones listed under Annex I of the Convention plus a few new ones) indicating a reduction of 80–95% over that timeframe.
- *Mitigation arrangements* for developed and developing countries, including on forest preservation, that are composed of the following elements:
    - Developed countries have absolute caps on their emissions for 2020 , consistent with at least 25–30% reduction of their 1990 levels on average
    - Developing countries (except the least developed) have committed to low carbon development plans and specific (sectoral) action plans, in conjunction with technical and financial assistance from newly established funds (replacing CDM for those countries)
    - Least developed countries participate in mitigation activities through a (revised) CDM
    - Developing countries that protect their forests (as part of their low carbon development plan) receive funding for implementing appropriate policies and creating incentives for forest owners
    - Strengthened reporting and review

- Provisions on financing and implementing *adaptation actions*, particularly in vulnerable countries, through integration in national development planning and policy. Countries would prepare a climate resilient development plan that is integrated with the low carbon development plan.
- *Technology development* arrangements that strengthen R&D efforts in developed countries and set up cooperative R&D mechanisms on low carbon technologies in developing countries. Internationally coordinated demonstrations of emerging low carbon technologies would be an important part of this[47]. Technology diffusion and transfer would be part of the mitigation arrangements and focus on information exchange and financing investments, including intellectual property rights payments, where needed.
- A solid *system of financing* mitigation, adaptation, and technology actions in developing countries. The backbone of this system will have to be formed by substantial new international funds, but the system as a whole will consist of a multitude of bilateral, multilateral, and private funding. The carbon market will play an important role in moving capital to low cost mitigation options, but it will not replace the basic funding for developing country actions.
- A *legal form* that links it to the Climate Convention or the Kyoto Protocol. Given the resistance of the USA against the Kyoto Protocol it is likely to become a new Protocol under the Convention or even a more complex set of separate decisions by the Parties to the Climate Convention and the Kyoto Protocol (if for instance the USA is unable to get a two thirds majority in the US Senate to approve a new Protocol). Useful elements of the Kyoto Protocol will then be copied into this new legal instrument.

Hopefully the notion of new opportunities created by a low carbon economy will have spread sufficiently to make such an agreement attractive for all countries.

Notes

1. http://www.un.org/documents/ga/res/45/a45r212.htm.
2. Bolin B. A History of the Science and Politics of Climate Change: The Role of the Intergovernmental Panel on Climate Change, Cambridge University Press, 2007.
3. http://unfccc.int/resource/docs/convkp/conveng.pdf.
4. http://www.nationalcenter.org/KyotoSenate.html.
5. This was eventually laid down in the Marrakesh accords in 2001, unfccc.int/cop7/documents/accords_draft.pdf.
6. The IPCC Working Group III Report of 2001 estimated the economic cost of implementing the Kyoto Protocol for the USA at a GDP reduction of less than 0.5% by 2010 in a system of global emissions trading, compared to what it otherwise would have been; in other words the economy would not grow by say 25% over the period 2000–2010, but by something like 24.5%
7. http://www.climnet.org/EUenergy/ratification/1990sharestable.htm.
8. http://english.pravda.ru/main/18/88/354/14495_kyoto.html.
9. Excluding land use change emissions (relatively small); see http://unfccc.meta-fusion.com/kongresse/071120_pressconference07/downl/201107_pressconf_sergey_konokov.pdf.

10. The 15 EU Member States redistributed their collective −8% target amongst individual Member States to allow for specific national circumstances; see http://reports.eea.europa.eu/eea_report_2007_5/en/Greenhouse_gas_Emission_trends_and_projections_in_Europe_2007.pdf.
11. http://unfccc.int/ghg_data/ghg_data_unfccc/time_series_annex_i/items/3814.php.
12. http://cait.wri.org/.
13. http://unfccc.meta-fusion.com/kongresse/071120_pressconference07/downl/201107_pressconf_katia_simeonova_part1.pdf.
14. http://cdm.unfccc.int/index.html.
15. http://www.cdmpipeline.org/overview.htm#4.
16. http://www.feem.it/NR/rdonlyres/2C130D3B-124F-427E-9FAE-E958F0E83263/838/0800.pdf.
17. The 'gold standard', developed by the World Wildlife Fund and other NGOs sets strict requirements for CDM projects in terms of their additionality and contribution to sustainable development in the host countries; see http://www.cdmgoldstandard.org/index.php.
18. See Point Carbon 2008 report at http://www.pointcarbon.com/polopoly_fs/1.912721!Carbon_2008_dfgrt.pdf.
19. Haya B. Failed mechanism: how the CDM is subsidizing hydro developers and harming the Kyoto Protocol, International Rivers, 2007, see http://internationalrivers.org/en/climate-change/carbon-trading-cdm/failed-mechanism-hundreds-hydros-expose-serious-flaws-cdm; see also http://www.indiatogether.org/2008/jul/env-cdm.htm.
20. IPCC Special Report on Safeguarding the Ozone Layer and the Global Climate System, 2000.
21. The acceptance of the methodology for supercritical coal fired power plants was limited to a maximum of 15% of a country's power supply, see http://cdm.unfccc.int/index.html.
22. Wara M, Victor D. A realistic policy on international carbon offsets, Stanford University Programme on Energy and Sustainable Development, Working paper no. 74, April 2008, http://iis-db.stanford.edu/pubs/22157/WP74_final_final.pdf.
23. IPCC AR4, WG III, chapter 13.
24. den Elzen M., Hoehne N. Reductions of greenhouse gas emissions in Annex I and non Annex I countries for meeting concentration stabilisation targets, Climatic Change, vol 91 (2008), pp 249–274.
25. The allocations were based on the so-called Trytich approach, see Phylipssen et al, Energy Policy, vol 26(12) (1998), pp 929–943.
26. There is strong political resistance amongst the G77 and China Group to differentiate amongst them.
27. IPCC Fourth Assessment Report, WG III, Chapter 13, table 13.2.
28. http://www.ecoequity.org/docs/TheGDRsFramework.pdf.
29. The GDR approach was not covered in the ranges reported by IPCC.
30. Hoehne N., Moltmann S. Distribution of emission allowances under the Greenhouse Development Rights and other effort sharing approaches, Heinrich Boell Foundation, Berlin, 2008.

31. Some announcements for 2020: Australia 15–25% below 2000, Canada 20% below 2006, Japan 25% below 1990, Russia 10–15% below 1990.
32. Winkler et al. Climate Policy, vol 8 (2008), pp 119–134.
33. IEA, WEO 2008.
34. http://unfccc.int/cooperation_and_support/financial_mechanism/items/4053.php; Tirpak D, Adams H. Bilateral and multilateral financial assistance for the energy sector of developing countries, Climate Policy, vol 8 (2008), pp 135–151.
35. ClimateWorks Foundation, project Catalyst results, http://www.project-catalyst.info.
36. http://www.euractiv.com/en/climate-change/experts-warn-eu-climate-change-trade-war/article-175426.
37. See chapter 8.
38. IPCC, Special Report on Methodological and Technological Aspects of Technology Transfer, 2000.
39. http://unfccc.int/ttclear/jsp/index.jsp.
40. http://www.iea.org/textbase/techno/index.asp.
41. http://www.reeep.org/31/home.htm.
42. http://www.ren21.net/.
43. Patent applications from Canon multiplied 10-fold between the 1960s and the 1980s/1990s, see Suzuki J, Kodama F. Research Policy 33 (2004), pp 531–549.
44. http://www.climatechangecorp.com/content.asp?contentid=5344.
45. http://www.oecd.org/dataoecd/54/20/39177453.pdf.
46. http://www.cgiar.org/.
47. De Conink H, Stephens JC, Metz B. Global learning on carbon capture and storage: a call for strong international cooperation on CCS demonstration, Energy Policy (2009), doi:10.1016/j.enpol.2009.01.020.

Index

Note: Page numbers in *italic* denote figures. Page numbers in **bold** denote tables.